I. Todhunter

A History of the Mathematical Theories of Attraction and the Figure of the Earth

From the Time of Newton to that of Laplace Vol. 2

I. Todhunter

A History of the Mathematical Theories of Attraction and the Figure of the Earth
From the Time of Newton to that of Laplace Vol. 2

ISBN/EAN: 9783741170317

Manufactured in Europe, USA, Canada, Australia, Japa

Cover: Foto ©Thomas Meinert / pixelio.de

Manufactured and distributed by brebook publishing software
(www.brebook.com)

I. Todhunter

A History of the Mathematical Theories of Attraction and the Figure of the Earth

Si Poisson a été d'une fécondité extraordinaire, c'est qu'il était au courant de ce qui avait été fait avant lui, au courant, par exemple, des immenses travaux des Euler et des d'Alembert; c'est qu'il ne s'est jamais sottement obstiné à perdre son temps et ses forces à la recherche de ce qui était déjà trouvé.

Que l'exemple de Poisson serve de leçon à ces esprits irréfléchis qui, sous le prétexte de conserver leur originalité, dédaignent de prendre connaissance des découvertes de leurs devanciers, et restent sur les premiers dégrés de l'échelle, tandis que, avec moins d'orgueil, ils se seraient élevés au sommet.

Arago, *Œuvres Complètes*, Tome II. page 656.

A HISTORY

OF THE

MATHEMATICAL THEORIES OF ATTRACTION

AND

THE FIGURE OF THE EARTH,

FROM THE TIME OF NEWTON TO THAT OF LAPLACE.

BY

I. TODHUNTER, M.A., F.R.S.

IN TWO VOLUMES.

VOLUME II.

London:
MACMILLAN AND CO.
1873.

CHAPTER XIX.

LAPLACE'S FIRST THREE MEMOIRS.

741. The investigations of Laplace on Attractions and the Figure of the Earth fall naturally into five divisions. The first division consists of three memoirs, which treat the subjects without the use of what we now call the *Potential Function*, or of that branch of analysis which we now call *Laplace's Functions*. The second division consists of a separate volume which uses the Potential Function. The third division consists of various memoirs which use both the Potential Function and Laplace's Functions. The fourth division is formed by the republication of the preceding researches in the first and second volumes of the *Mécanique Céleste*. The fifth division consists of researches subsequent to the publication of the second volume of the *Mécanique Céleste*; they are reproduced in the fifth volume of the *Mécanique Céleste*.

We shall consider in the present Chapter Laplace's first three memoirs.

742. We begin with the seventh volume of the *Mémoires de Mathématique...par divers Savans*...1773: the date of publication is 1776. This volume contains two memoirs by Laplace, which among other subjects treat largely of Probability: see pages 473...475 of my *History...of Probability*. The part of the volume with which we are now concerned is entitled *Sur la figure de la Terre*; it occupies pages 524...534. It is not stated when these investigations were sent to the Academy; but from the title of the volume in which they appear we see that Laplace was not a member of the Academy when they were sent.

743. Laplace begins thus on page 524:

Lorsque Newton voulut déterminer la figure de la Terre, il considéra cette Planète comme une masse fluide homogène, et il supposa que la figure qu'elle a prise en vertu de son mouvement de rotation est celle d'un sphéroïde elliptique. Cette supposition étoit fort précaire; les Géomètres en ont ensuite démontré la possibilité; mais si la figure nécessaire pour l'équilibre, au lieu d'être elliptique, eût été d'un autre genre, on auroit été fort embarrassé pour la déterminer, parce qu'il est beaucoup plus facile de s'assurer si une figure donnée convient à l'équilibre, que de chercher immédiatement celles qui peuvent y convenir. Ce dernier Problème est sans contredit un des points les plus intéressans du Système du Monde; voici quelques recherches qui y sont relatives.

744. Thus the following is the problem to be discussed: a mass of homogeneous fluid in the form of a figure of revolution nearly spherical rotates with uniform angular velocity round its axis of figure and remains in relative equilibrium; determine the form. I call this problem Legendre's, because he was the first to solve it with tolerable success.

745. Let there be a circle of radius unity; let ψ be the angle which the radius to any point makes with a fixed radius: so that the ordinate of this point is $\sin\psi$. Produce this ordinate until it becomes $\sin\psi + \frac{\alpha y}{\sin\psi}$, where α is very small, and y is some function of ψ. Put x for $\cos\psi$. Then Laplace arrives at a differential equation between y and x of an infinite order, to determine the required generating curve; that is a differential equation involving $\frac{d^iy}{dx^i}$, $\frac{d^ry}{dx^r}$, ... and so on *ad infinitum*.

746. The preceding notation does not look very promising; in fact Laplace does not explicitly start with it, but arrives at it as he proceeds. Unless y is very small when ψ is very small the process is not satisfactory. Moreover Laplace in order to form his differential equation expands a function into a series without discussing whether the series is convergent.

747. The main result at which he arrives deserves notice. He wishes to know whether equilibrium would subsist for any other form besides an exact sphere when there is no rotation. He cannot completely solve this problem; but he shews that y cannot consist of a series of the form $ax^\lambda + bx^\mu + cx^\nu + \ldots$, where $\lambda, \mu, \nu, \ldots$ are numbers in *descending* order of magnitude. That is y cannot consist of a *finite* number of terms each involving a power of x; nor can y be an infinite series of *descending* powers of x: but he does not shew that y cannot be an infinite series of *ascending* powers of x.

When the fluid is supposed to rotate Laplace's demonstration amounts to shewing that among all series, finite or infinite, which can be arranged in descending powers of x, the only admissible form of y is $ax^2 + bx + c$; where a, b, and c are constants.

748. Laplace's demonstration is difficult, but satisfactory; that is to say after the points to which we have drawn attention in Art. 746, no very serious objection will occur to a reader.

After finishing his demonstration, Laplace says on his page 534:

...Je dois observer ici que M. d'Alembert a déjà fait une remarque semblable pour le cas où les exposans de x sont des nombres entiers et positifs (*voyez le tome* V *des Opuscules* de ce grand Géomètre).

These words are quite consistent with the supposition, that Laplace had found the error which we have pointed out in D'Alembert's process; because *to make a remark* is far less than *to demonstrate*. See Art. 576.

749. Laplace concludes with these words:

Il seroit utile d'étendre ces recherches au cas où les couches de la masse fluide sont inégalement denses; c'est ce que je me propose de faire dans un autre Mémoire.

The intention here expressed was not carried into effect until the publication of Laplace's seventh memoir in the Paris *Mémoires* for 1789.

750. All that Laplace's first memoir contains on our subject is reproduced with better notation in his second memoir to which

we shall proceed in our next Article: it is therefore unnecessary to treat the first memoir with much detail. In the second memoir we shall find that the radius vector of the generating curve is denoted by $1 + af(\cos\psi)$, where a is very small; so that what was called $\sin\psi + \frac{ay}{\sin\psi}$ in Art. 745 is equal to the $\sin\psi\,[1 + af(\cos\psi)]$ of the second memoir: that is $\frac{y}{\sin\psi}$ of the first memoir is $\sin\psi f(\cos\psi)$ of the second memoir, or y of the first memoir is $\sin^2\psi f(\cos\psi)$ of the second memoir. In the second memoir y is put for $f(\cos\psi)$.

751. In the Paris *Mémoires* for 1772, *Seconde Partie*, published in 1776, we have a memoir by Laplace entitled *Recherches sur le Calcul Intégral et sur le Système du Monde;* at a later part of the volume there are some *Additions* to this memoir: among these Additions we have a section entitled *De l'Équilibre des Sphéroïdes homogènes*, which occupies pages 536...554 of the volume.

752. The problem proposed to be discussed is the same as that of the preceding memoir: see Art. 744. Laplace was not able to solve the problem completely; but he reproduced his former demonstration, somewhat improved, that for a large number of figures the relative equilibrium was impossible.

753. But although he did not in this memoir arrive at the necessary form for equilibrium, yet he obtained a very remarkable result: namely, that the law of the variation of gravity, whatever be the form of equilibrium, is the same as for an oblatum. We will give in substance the method by which Laplace obtains this result.

We may remark that Laplace investigates the polar expression for an element of mass, namely in the usual modern notation $r^2 dr \sin\theta\, d\theta\, d\phi$: see his page 539. The investigation is in fact the same as we now have in our elementary books: see *Integral Calculus*, third edition, Art. 207.

In his first memoir Laplace used this polar expression but did not investigate it; he merely says, "on trouvera facilement...": see his page 525. See also Art. 710.

734. Let there be a curve differing very little from a circle, and symmetrical with respect to a diameter. Let half the length of this diameter be unity, and let the length of a radius vector inclined at an angle ψ to the diameter be $1 + \alpha f(\cos\psi)$, where f denotes any function, and α is a very small quantity the square of which we shall neglect. Suppose a solid formed by the revolution of this curve round the diameter which divides it symmetrically; take this diameter for the direction of the axis of x: then the equation to the surface will be

$$\sqrt{(x^2+y^2+z^2)} = 1 + \alpha f\left\{\frac{x}{\sqrt{(x^2+y^2+z^2)}}\right\} \text{ .... } (1).$$

We propose to find the attraction of the solid at a point situated on its surface; this point without loss of generality we may take in the plane of (x, y): let ψ be the angle between the radius vector of this point and the axis of revolution.

Put $x = \xi\cos\psi + \eta\sin\psi$, $y = \xi\sin\psi - \eta\cos\psi$; thus (1) becomes

$$\sqrt{(\xi^2+\eta^2+z^2)} = 1 + \alpha f\left\{\frac{\xi\cos\psi + \eta\sin\psi}{\sqrt{(\xi^2+\eta^2+z^2)}}\right\} \text{ .... } (2).$$

We can now pass easily to polar coordinates which have their origin at the attracted point: put

$$\xi = h - r\sin\theta\cos\phi,\quad \eta = r\sin\theta\sin\phi,\quad z = r\cos\theta,$$

where $$h = 1 + \alpha f(\cos\psi).$$

[I use θ for Laplace's p, and $\frac{\pi}{2} - \phi$ for his q.]

Thus (2) becomes

$$\sqrt{(h^2 - 2hr\sin\theta\cos\phi + r^2)}$$
$$= 1 + \alpha f\left\{\frac{h\cos\psi - r\sin\theta\cos\phi\cos\psi + r\sin\theta\sin\phi\sin\psi}{\sqrt{(h^2 - 2hr\sin\theta\cos\phi + r^2)}}\right\} \ldots (3).$$

We proceed to find from (3) the value of r to the order of approximation which we require.

If $\alpha = 0$, we should get $r = 2\sin\theta\cos\phi$; assume then

$$r = 2\sin\theta\cos\phi + \rho,$$

where ρ will be very small.

Substitute in (3) and we obtain

$$af(\cos\psi) - 2a\sin^2\theta\cos^2\phi f(\cos\psi) + \rho\sin\theta\cos\phi = af(u),$$

where u stands for $\cos\psi - 2\sin^2\theta\cos\phi\cos(\phi+\psi)$; so that

$$\rho = \frac{a(2\sin^2\theta\cos^2\phi - 1)}{\sin\theta\cos\phi} f(\cos\psi) + \frac{a}{\sin\theta\cos\phi} f(u).$$

We may also arrange the value of ρ thus,

$$\rho = 2a\sin\theta\cos\phi f(\cos\psi) + a\frac{f(u) - f(\cos\psi)}{\sin\theta\cos\phi};$$

and this shews that ρ remains small even when $\sin\theta\cos\phi$ is very small: for then we have $f(u)$ very nearly equal to $f(\cos\psi)$.

Now the attraction at the point resolved along the radius vector

$$=\iint r\sin^2\theta\cos\phi\, d\theta\, d\phi = \iint (2\sin\theta\cos\phi + \rho)\sin^2\theta\cos\phi\, d\theta\, d\phi;$$

the limits for θ are 0 and π; the limits for ϕ are $-\left(\frac{\pi}{2}+\beta\right)$ and $\frac{\pi}{2}-\beta$, where β is some function of ψ, which is of the order of a. It is easy to see that for our approximation we may proceed as if β were zero. Denote this resolved attraction by A: thus

$$A = 2\iint \sin^3\theta\cos^2\phi\, d\theta d\phi + af(\cos\psi)\iint \sin\theta(2\sin^2\theta\cos^2\phi - 1)\, d\theta d\phi$$
$$+ a\iint \sin\theta f(u)\, d\theta\, d\phi.$$

The first and second integrations may be easily effected; with respect to these it is *exactly* true that we may proceed as if β were zero: and we obtain

$$A = \frac{4\pi}{3} - \frac{2a\pi}{3} f(\cos\psi) + a\iint \sin\theta f(u)\, d\theta\, d\phi.$$

Let B denote the attraction resolved in the meridian plane at right angles to the radius vector; then

$$B = \iint (2\sin\theta\cos\phi + \rho)\sin^2\theta\sin\phi\, d\theta\, d\phi$$

$$= 2\iint \sin^2\theta\cos\phi\sin\phi\,d\theta\,d\phi + 2af(\cos\psi)\iint \sin^2\theta\sin\phi\cos\phi\,d\theta\,d\phi$$
$$+ a\iint \frac{f(u)-f(\cos\psi)}{\cos\phi}\sin\theta\sin\phi\,d\theta\,d\phi.$$

It is easy to see that the first and second integrals vanish; so that

$$B = a\iint \frac{f(u)-f(\cos\psi)}{\cos\phi}\sin\theta\sin\phi\,d\theta\,d\phi:$$

this integral is finite, for $f(u)-f(\cos\psi)$ vanishes when $\cos\phi$ vanishes.

The last integral may be transformed. By integration by parts we have

$$\int [f(u)-f(\cos\psi)]\sin\theta\,d\theta$$
$$= -\cos\theta\,[f(u)-f(\cos\psi)] - 4\int \cos^2\theta\sin\theta f'(u)\cos\phi\cos(\phi+\psi)\,d\theta;$$

when this is taken between the limits 0 and π the first term vanishes; so that we have

$$B = -4a\iint \cos^2\theta\sin\theta f'(u)\sin\phi\cos(\phi+\psi)\,d\theta\,d\phi$$
$$= -2a\iint \cos^2\theta\sin\theta f'(u)\,[\sin(2\phi+\psi)-\sin\psi]\,d\theta\,d\phi.$$

Now $\qquad u = \cos\psi - \sin^2\theta\,[\cos(2\phi+\psi)+\cos\psi];$

so that $\qquad \int \sin(2\phi+\psi)f'(u)\,d\phi = \dfrac{f(u)}{2\sin^2\theta},$

and this vanishes when taken between the limits $-\frac{\pi}{2}$ and $\frac{\pi}{2}$. Thus finally

$$B = 2a\sin\psi\iint \cos^2\theta\sin\theta f'(u)\,d\theta\,d\phi \;\ldots\ldots\ldots\ldots(4).$$

755. We shall now shew that

$$\frac{dA}{d\psi} = \frac{2a\pi}{3}f(\cos\psi)\sin\psi - \frac{B}{2}\;\ldots\ldots\ldots\ldots\ldots(5).$$

We have $\frac{dA}{d\psi} = \frac{2\pi}{3} f'(\cos\psi)\sin\psi + a\iint \sin\theta f'(u)\frac{du}{d\psi}\,d\theta\,d\phi,$

and $$u = \cos^2\theta\cos\psi - \sin^2\theta\cos(2\phi+\psi),$$

so that $$\frac{du}{d\psi} = -\cos^2\theta\sin\psi + \sin^2\theta\sin(2\phi+\psi).$$

Hence

$$\frac{dA}{d\psi} = \frac{2\pi}{3} f'(\cos\psi)\sin\psi - a\sin\psi\iint \cos^2\theta\sin\theta f''(u)\,d\theta\,d\phi$$

$$= \frac{2\pi}{3} f'(\cos\psi)\sin\psi - \frac{B}{2}.$$

This is the first appearance in Laplace's writings of a theorem which he seems to have valued highly: see Art. 652. We shall meet the theorem again several times: it appears in a different form in the *Mécanique Céleste*, Livre III. § 10.

756. Now let us suppose that the attracting body is a fluid, or at least that there is a superficial stratum of fluid. Then for relative equilibrium the resolved part of the force along the tangent to the meridian must vanish. This part consists of the resolved parts of A and B, together with the centrifugal force.

The direction of A is nearly at right angles to the tangent; the cosine of the angle between the directions is $-af'(\cos\psi)\sin\psi$. The direction of B makes only an indefinitely small angle with the tangent. Hence, denoting the angular velocity by ω, we have

$$B - Aaf'(\cos\psi)\sin\psi - \omega^2\sin\psi\cos\psi = 0;$$

that is, neglecting the square of a,

$$B = \frac{4\pi a}{3} f'(\cos\psi)\sin\psi + \omega^2\sin\psi\cos\psi \text{ ....(6).}$$

Let P denote the gravity at the point considered;

then approximately $P = A - \omega^2\sin^2\psi,$

therefore $\dfrac{dP}{d\psi} = \dfrac{dA}{d\psi} - 2\omega^2 \sin\psi \cos\psi$

$$= \frac{2a\pi}{3} f'(\cos\psi)\sin\psi - \frac{B}{2} - 2\omega^2 \sin\psi\cos\psi \text{ by (5)}$$

$$= -\frac{5}{2}\omega^2 \sin\psi \cos\psi \text{ by (6).}$$

Therefore $P = \text{constant} - \frac{5}{4}\omega^2 \sin^2\psi = P_0 - \frac{5}{4}\omega^2 \sin^2\psi$, where P_0 denotes the force of gravity at the pole. This is the result which, as we stated in Art. 753, Laplace established.

757. We shall now form the differential equation of an infinite order at which Laplace arrives.

Put $f(\cos\psi) = y$, and $\cos\psi = x$, so that $f'(\cos\psi) = \dfrac{dy}{dx}$.

Then $f'(u) = f'\{\cos\psi - 2\sin^2\theta\cos\phi\cos(\phi+\psi)\}$
$= f'(\cos\psi - s)$ say;

hence expanding by Taylor's Theorem this becomes

$$\frac{dy}{dx} - s\frac{d^2y}{dx^2} + \frac{s^2}{\underline{|2}}\frac{d^3y}{dx^3} - \frac{s^3}{\underline{|3}}\frac{d^4y}{dx^4} + \dots$$

Then equating the values of B given by (4) and (6), and dividing by $\sin\psi$, we obtain

$$2a\iint \cos^2\theta \sin\theta \left\{\frac{dy}{dx} - s\frac{d^2y}{dx^2} + \frac{s^2}{\underline{|2}}\frac{d^3y}{dx^3} - \dots\right\} d\theta\, d\phi$$
$$= \frac{4\pi a}{3}\frac{dy}{dx} + \omega^2 x.$$

The term $\dfrac{dy}{dx}$ will disappear from this equation, because $\iint \cos^2\theta\sin\theta\, d\theta\, d\phi$ between the proper limits $= \dfrac{2\pi}{3}$. Thus we have

$$\iint \cos^2\theta \sin\theta \left\{s\frac{d^2y}{dx^2} - \frac{s^2}{\underline{|2}}\frac{d^3y}{dx^3} + \frac{s^3}{\underline{|3}}\frac{d^4y}{dx^4} - \dots\right\} d\theta\, d\phi = -\frac{\omega^2 x}{2a} \dots (7).$$

758. The integrations with respect to θ and ϕ in (7) can be easily effected. Consider first the integration with respect to ϕ. We have

$$\epsilon^n = [2\sin^2\theta\cos\phi\cos(\phi+\psi)]^n = \sin^{2n}\theta\,[\cos\psi+\cos(2\phi+\psi)]^n.$$

Thus we require $\int_{-\frac{\pi}{2}}^{\frac{\pi}{2}} [\cos\psi+\cos(2\phi+\psi)]^n\,d\phi$.

Now $[\cos\psi+\cos(2\phi+\psi)]^n = \cos^n\psi + n\cos^{n-1}\psi\cos(2\phi+\psi)$

$$+\frac{n(n-1)}{\underline{|2}}\cos^{n-2}\psi\cos^2(2\phi+\psi)+\ldots$$

When we integrate between the limits the terms which involve *odd* powers of $\cos(2\phi+\psi)$ disappear, and those which involve even powers are easily obtained. For example, take

$$\int_{-\frac{\pi}{2}}^{\frac{\pi}{2}} \cos^4(2\phi+\psi)\,d\phi\,;$$

put t for $2\phi+\psi$, then we get

$$\frac{1}{2}\int_{-\pi+\psi}^{\pi+\psi}\cos^4 t\,dt,$$

and this obviously $=\frac{1}{2}\int_0^{2\pi}\cos^4 t\,dt$

$$=\frac{4}{2}\int_0^{\frac{\pi}{2}}\cos^4 t\,dt = 2\,.\,\frac{3\,.\,1}{4\,.\,2}\,.\,\frac{\pi}{2} = \frac{3\,.\,1}{4\,.\,2}\pi.$$

In this way we easily see that $\int_{-\frac{\pi}{2}}^{\frac{\pi}{2}}[\cos\psi+\cos(2\phi+\psi)]^n\,d\phi$

$$=\pi\left\{x^n+\frac{n(n-1)}{2^2}x^{n-2}+\frac{n(n-1)(n-2)(n-3)}{2^2\,.\,4^2}x^{n-4}\right.$$

$$\left.+\frac{n(n-1)(n-2)(n-3)(n-4)(n-5)}{2^2\,.\,4^2\,.\,6^2}x^{n-6}+\ldots\right\}.$$

[This is not Laplace's method; but seems to me rather simpler.]

It is obvious that the integration with respect to θ can be easily effected.

759. We shall now shew that except in the particular case of $y = ax^2 + bx + c$, the equation (7) cannot be satisfied by a value of y of the form $ax^\lambda + bx^\mu + cx^\nu + \ldots$ where λ, μ, ν, ... are in *descending* order of magnitude.

For substitute this assumed form of y in (7); then the term which involves the highest power of x on the left-hand side will be found to be

$$\pi a\lambda x^{\lambda-1}\int d\theta \sin\theta\cos^2\theta\left\{(\lambda-1)\sin^2\theta - \frac{(\lambda-1)(\lambda-2)}{\lfloor 2}\sin^4\theta \right.$$
$$\left. + \frac{(\lambda-1)(\lambda-2)(\lambda-3)}{\lfloor 3}\sin^6\theta - \ldots\right\}.$$

If $\lambda = 2$ the coefficient of $x^{\lambda-1}$, that is of x, must be equated to $-\frac{\omega^2}{2\pi}$. But if λ is not $= 2$, the coefficient of $x^{\lambda-1}$ must be equated to zero.

Thus $\lambda\int_0^\pi \sin\theta\cos^2\theta\,[1-(1-\sin^2\theta)^{\lambda-1}]\,d\theta = 0,$

that is $\lambda\int_0^\pi \sin\theta\cos^2\theta\,[1-\cos^{2\lambda-2}\theta]\,d\theta = 0,$

that is $\lambda\left\{\frac{2}{3} + \frac{(-1)^{2\lambda+1}}{2\lambda+1} - \frac{1}{2\lambda+1}\right\} = 0.$

If we suppose $(-1)^{2\lambda+1} = -1$, this reduces to $\lambda\left(\frac{2}{3} - \frac{2}{2\lambda+1}\right) = 0$; so that $\lambda = 0$, or $\lambda = 1$.

If we suppose $(-1)^{2\lambda+1} = 1$, it reduces to $\frac{2\lambda}{3} = 0$; so that $\lambda = 0$. Laplace strangely multiplies up by $2\lambda + 1$, so that he introduces the solution $2\lambda + 1 = 0$, which gives $\lambda = -\frac{1}{2}$: this he rejects because it would make y impossible when x is negative, and infinite when $x = 0$. But the apparent solution $2\lambda + 1 = 0$, really does not occur; but is introduced by Laplace without reason.

Hence it follows that if we put $y = ax^2 + bx + c + \eta$, then η cannot take the form of a series, finite or infinite, arranged according to descending powers of x.

760. It will be observed that the principle of Laplace's demonstration resembles that which D'Alembert adopted in the fifth volume of his *Opuscules Mathématiques*. But, as we have seen, D'Alembert went astray in the details of his process: see Art. 570. Laplace repeats the remark relative to D'Alembert which we quoted in Art. 748.

761. Laplace then extends his result, and shews that y cannot be a fraction, the numerator and denominator of which are series arranged in *descending powers* of x.

762. Laplace's own statements of his theorems are liable to objection. He does not say explicitly that his series are arranged in *descending powers* of x; but this limitation is obvious in his demonstration. He proves that y cannot be a series which has a highest power of x; but he does not prove that y cannot be an infinite series of ascending powers of x. See Art. 747.

763. All that this memoir contains is comprised in the theorem which Legendre first demonstrated that the generating curve *must* in fact be an ellipse: for thus the form of the surface and the law of gravity are definitely settled.

Laplace refers on his page 543 to *l'excellent Ouvrage de M. Clairaut sur la figure de la Terre.*

764. In the Paris *Mémoires* for 1775, published in 1778, we have a memoir by Laplace entitled *Recherches sur plusieurs points du Système du Monde.* One section of this memoir is *Sur la loi de la Pesanteur à la surface des sphéroïdes homogènes en équilibre;* this occupies pages 75...89 of the volume.

765. The investigations here given are an extension of those in the volume for 1772. There Laplace had found the law of the variation of gravity on the assumption that the spheroid is a figure of revolution; here he considers the case in which the spheroid is not assumed to be a figure of revolution.

This result is really involved in the former; for Laplace subsequently shewed that the spheroid must be a figure of revolution, in fact an oblatum.

766. Let O be the centre of a sphere which nearly coincides with the spheroid, M any point at the surface; we propose to obtain expressions for the attraction at M.

Let R be any other point on the surface. Let $OR = 1 + a\mu'$, where a is a very small quantity, and μ' is a function of the elements which determine the position of R; these elements may be conveniently the colatitude ψ', and the longitude λ'. Then $OM = 1 + a\mu$, where μ is what μ' becomes, when ψ' and λ' become ψ and λ respectively.

Now make M the origin of the usual polar coordinates r, θ, and ϕ; where $MR = r$.

Let A denote the resolved attraction along MO; let B denote the resolved attraction at right angles to MO, and in the meridian plane of M, towards the pole; and let C denote the resolved attraction at right angles to the directions of A and B. Then

$$A = \iint r \sin^2\theta \cos\phi \, d\theta \, d\phi,$$

$$B = \iint r \sin^2\theta \sin\phi \, d\theta \, d\phi,$$

$$C = \iint r \sin\theta \cos\theta \, d\theta \, d\phi.$$

[I use θ for Laplace's p, and $\frac{\pi}{2} - \phi$ for his q.]

767. We have to find an expression for r. Laplace uses a diagram for this purpose.

In the diagram OP is the diameter from which the colatitudes are measured.

RZ is perpendicular to the plane MOP; RL and ZL are perpendiculars to OM.

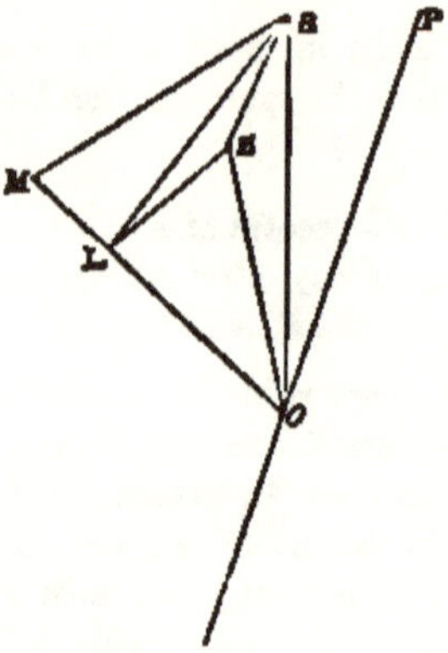

$MR = r$, $RZ = r\cos\theta$, $ML = r\sin\theta\cos\phi$, $ZL = r\sin\theta\sin\phi$.

Now $OR^2 = OL^2 + RL^2 = (OM - r\sin\theta\cos\phi)^2 + RL^2$;

that is $$OR^2 = OM^2 - 2OMr\sin\theta\cos\phi + r^2;$$

therefore

$$(1+\alpha\mu')^2 = (1+\alpha\mu)^2 - 2(1+\alpha\mu)\,r\sin\theta\cos\phi + r^2.$$

Solve this quadratic in r, and neglect powers of α above the first: thus

$$r = (1+\alpha\mu)\sin\theta\cos\phi \pm \left\{(1+\alpha\mu)\sin\theta\cos\phi + \frac{\alpha(\mu'-\mu)}{\sin\theta\cos\phi}\right\}.$$

The upper sign must be taken; for the lower sign would lead to a value of r of the order α: therefore

$$r = 2(1+\alpha\mu)\sin\theta\cos\phi + \frac{\alpha(\mu'-\mu)}{\sin\theta\cos\phi}.$$

768. The limits of the integrations, without introducing any error of the order we are retaining, may be taken to be 0 and π for θ, and $-\frac{\pi}{2}$ and $\frac{\pi}{2}$ for ϕ. Hence using the value of r found in the preceding Article we have

$$A = \frac{4\pi}{3} - \frac{2\pi a\mu}{3} + a\iint \mu' \sin\theta \, d\theta \, d\phi,$$

$$B = a\iint (\mu' - \mu) \frac{\sin\theta \sin\phi}{\cos\phi} d\theta \, d\phi,$$

$$C = a\iint (\mu' - \mu) \frac{\cos\theta}{\cos\phi} d\theta \, d\phi.$$

Here B is estimated *towards* the pole; if we estimate B *from* the pole, as Laplace does, we must change the sign of the expression.

769. Laplace then incautiously says that

$$B = -a\int \frac{\mu' \sin\theta \sin\phi}{\cos\phi} d\theta \, d\phi, \quad C = a\int \frac{\mu' \cos\theta}{\cos\phi} d\theta \, d\phi;$$

but he never uses these erroneous forms, and probably the introduction of them is only a misprint. These erroneous forms would make B and C infinite.

770. In order to effect the integrations in the values of A, B, and C, supposing μ' a known function of ψ' and λ', it would be necessary to connect ψ' and λ' with θ and ϕ: to this we proceed.

It will be seen that μ' enters with the coefficient a; and thus we may use approximations in our equations.

In the diagram of Art. 767, by projecting on OP the straight line OR, and also the broken line made up of OL and LZ, we have exactly

$$OR \cos\psi' = OL \cos\psi + LZ \sin\psi$$
$$= (OM - r \sin\theta \cos\phi) \cos\psi + r \sin\theta \sin\phi \sin\psi;$$

that is $\quad OR \cos\psi' = OM \cos\psi - r \sin\theta \cos(\phi + \psi).$

Hence to the order which we have to retain

$$\cos\psi' = \cos\psi - 2 \sin^2\theta \cos\phi \cos(\phi + \psi) \quad \text{..........(1)}.$$

Again, $\quad \sin(\lambda' - \lambda) = \frac{RZ}{OR \sin\psi'} = \frac{r \cos\theta}{OR \sin\psi'}.$

Hence to the order which we have to retain

$$\sin(\lambda' - \lambda) = \frac{2 \sin\theta \cos\theta \cos\phi}{\sin\psi'} \quad \text{.................. (2)}.$$

Equations (1) and (2) theoretically give us the required connection of ψ' and λ' with θ and ϕ.

771. Laplace proceeds to establish the following result:

$$\frac{dA}{d\psi} = -\frac{2\pi a}{3}\frac{d\mu}{d\psi} + \frac{B}{2}.$$

This corresponds to equation (5) of Art. 755. There is a difference in sign however, because Laplace here estimates B *from* the pole, whereas in his second memoir he estimated it *towards* the pole: we shall for convenience keep here to the method of the third memoir.

Laplace arrives at his result by four pages of integration and differentiation: we shall not reproduce this investigation, as Laplace himself afterwards gave a simpler process, and we shall have to return to the subject.

This is the second appearance in Laplace's writings of the theorem to which we have referred in Art. 755: it is here extended to the case of a body not of revolution.

772. In the same manner as the theorem in Art. 771 is established, we may shew that

$$\frac{dA}{\sin\psi\, d\lambda} = -\frac{2\pi a}{3}\frac{d\mu}{\sin\psi\, d\lambda} + \frac{C}{2}.$$

773. If we now consider the condition of relative equilibrium of a fluid mass, or of a mass covered with a superficial stratum of fluid, we arrive as in Art. 756 at the following results:

$$B + \omega^2 \sin\psi \cos\psi = \frac{4\pi a}{3}\frac{d\mu}{d\psi},$$

$$P = P_1 - \frac{5}{4}\omega^2 \sin^2\psi,$$

where B is now estimated *from* the pole.

And we have now besides

$$C = \frac{4\pi a}{3\sin\psi}\frac{d\mu}{d\lambda}.$$

These general results must have been very interesting at the time they were given: but they have since lost much of their value, because it is known that under the supposed conditions the mass can only take the form of an oblatum: this was established by Laplace in his fourth memoir, and the demonstration is reproduced in the *Mécanique Céleste*: see Livre III. § 26.

774. Laplace extends his process; for he supposes that besides the attraction of the mass itself there may be external forces: and in particular he considers the case of an external body which is in relative equilibrium with respect to the attracted body. In this case we have as usual to find, not the absolute attraction at a point, but the excess of this attraction above that at the centre of gravity of the mass. Laplace's method is peculiar: see his pages 87 and 88. The result is correct, but a reader will probably verify it, as he easily may, before he accepts it.

775. On pages 261...267 of the Paris *Mémoires* for 1776, we have an *Addition* to the memoir we are now considering. This Addition gives us a simple proof of the theorems of Arts. 771 and 772 in the more general form which they take when the force of attraction is supposed to vary as the nth power of the distance. The investigation depends on the same principles as that in the *Mécanique Céleste*, Livre III. § 10; and the results may be said to be summed up in equation (1) of that section, inasmuch as they will follow from that equation by differentiation. But the Potential Function is not used in the pages now under notice. These pages are in substance reproduced in the fourteenth section of Laplace's *Théorie...de la Figure des Planetes* of which we shall hereafter give an account.

776. I will notice some remarks on page 262 by which Laplace tries to shew that if there is one figure of equilibrium for a rotating spheroid there will be an infinite number.

Let $1 + \alpha y$ be the radius vector to any point of the surface, where α is indefinitely small (*infiniment petit*), and y is any function of θ the colatitude and ϖ the longitude. Let αB denote the tangential attraction; then for equilibrium we must have $B = 0$. Let y be such a function of θ and ϖ as to satisfy this: then he says the condition will also be satisfied when we put $\theta + a$ for θ, and $\varpi + b$ for ϖ; where a and b are constants. Let y' be what y becomes by this change. Then he says that there will be equilibrium if the radius vector be $1 + \alpha y'$; and therefore also if the radius vector be $1 + \alpha y + n\alpha y'$ where n is any constant.

I presume that Laplace does not suppose any rotation here, or any action but that of the fluid mass itself. Thus in fact we have a case something like that of the coexistence of small motions in Dynamics.

If we suppose the mass to rotate then the argument will have to be slightly modified. We cannot then change θ into $\theta + a$, but we can change ϖ into $\varpi + b$; and thus as before we shall still have an infinite number of solutions. Compare the *Mécanique Céleste*, Livre III. § 26.

777. The first three memoirs by Laplace on our subjects are superseded by his investigations of the third division: see Art. 741. The first three memoirs may be considered to attach themselves to the researches of D'Alembert, and to continue those researches. In his writings of the third division Laplace may be said to derive great assistance from Legendre. It will be necessary as we proceed to be particular with the chronology of the memoirs by Laplace and Legendre; for we shall find indications that Legendre was not quite satisfied with Laplace's silence as to the matter of priority: see also Pontécoulant's *Système du Monde*, Vol. III. page x.

778. The order then which we shall have to adopt after thus considering Laplace's first three memoirs is the following:

Legendre's first memoir; this is in the *Mémoires...par divers Savans...* Vol. X.

Laplace's treatise *De la Figure des Planetes;* this is contained in a work published in 1784.

Legendre's second memoir; this is in the Paris *Mémoires* for 1784.

Laplace's fourth memoir; this is in the Paris *Mémoires* for 1782.

Laplace's fifth memoir; this is in the Paris *Mémoires* for 1783.

Laplace's sixth memoir; this is in the Paris *Mémoires* for 1787.

Legendre's third memoir; this is in the Paris *Mémoires* for 1788.

Legendre's fourth memoir; this is in the Paris *Mémoires* for 1789.

Laplace's seventh memoir; this is in the Paris *Mémoires* for 1789.

When a volume of the Paris *Mémoires* is said to be for a specified year, it by no means follows that all the memoirs which it contains were written during or before the specified year. The order in which we have placed these writings of Laplace and Legendre is the order of their production, as will appear by our extracts from them as we proceed.

CHAPTER XX.

LEGENDRE'S FIRST MEMOIR.

779. A VERY important memoir by Legendre is contained in the tenth volume of the *Mémoires...présentés par divers Savans...* The date of publication of the volume is 1785. The memoir, however, must have been communicated to the Academy at an earlier period; for in the treatise *De la Figure des Planetes*, which was published in 1784, Laplace refers to the researches of Legendre which constitute the present memoir: see page 96 of Laplace's treatise.

Legendre's memoir is entitled *Recherches sur l'attraction des sphéroïdes homogènes;* it occupies pages 411...434 of the volume.

780. Legendre begins thus:

M. Maclaurin est le premier qui ait déterminé l'attraction d'un Sphéroïde elliptique pour les points situés dans son intérieur ou à sa surface. Les propositions qu'il a établies à ce sujet, et d'où résulte une solution si simple du problême de la figure de la Terre, servent de base à son excellente Pièce sur le Flux et le Reflux de la Mer, et sont connues de tous les Géomètres. Le même Auteur a considéré aussi l'attraction des Sphéroïdes elliptiques sur les points situés au dehors; mais il s'est borné aux points situés sur l'axe ou sur l'équateur pour les Sphéroïdes de révolution, et seulement aux points placés dans la direction d'un des trois axes, lorsque le Sphéroïde a toutes ses coupes elliptiques. Ces deux objets se trouvent compris dans un théorême remarquable, dont M. Maclaurin donne l'énoncé, art. 653 de son Traité des Fluxions; théorême dont MM. d'Alembert et de la Grange ont donné depuis la démonstration; le premier, dans les Mémoires de Berlin, année 1774, et dans le tome VII. de ses Opuscules; le second, dans les Mémoires de Berlin, année 1775.

Il ne paroît pas que les Géomètres aient poussé plus loin leurs recherches sur cette matière intéressante; car, quoique M. de la Grange ait considéré le problême dans toute sa généralité (Mém. de Berlin, année 1773), l'intégration n'a réussi à ce grand Géomètre que dans les cas déjà résolus par M. Maclaurin. C'est dans la vûe de concourir à la perfection de cette théorie, que j'ai entrepris les Recherches dont je vais rendre compte.

We see from this extract that, as we have stated in Art. 260, Legendre underrates what Maclaurin really did demonstrate.

781. We will now give in substance Legendre's treatment of Maclaurin's theorem in attractions.

Let the equation to an ellipsoid be $\frac{x^2}{a^2}+\frac{y^2}{b^2}+\frac{z^2}{c^2}=1$: required the attraction at a point on the prolongation of the axis of x at the distance h from the origin.

Let r be the distance of any point of the ellipsoid from the attracted particle, ϕ the angle between r and its projection on the plane of (x, z), and ψ the angle between this projection and the axis of x. The element of mass is $r^2\cos\phi\, d\phi\, d\psi\, dr$; and thus the attraction resolved along the axis on which the attracted particle is situated is

$$\frac{r^2\cos\phi\, d\phi\, d\psi\, dr}{r^2}\times\cos\phi\cos\psi,$$

that is $$\cos^2\phi\cos\psi\, d\phi\, d\psi\, dr.$$

We first integrate for r; the limits are r_1 and r_2, where r_1 and r_2 are the limiting radii vectores drawn from the attracted particle to the ellipsoid in the direction assigned by the angles ϕ and ψ. Thus r_1 and r_2 are the roots of the quadratic equation

$$\frac{(h-r\cos\phi\cos\psi)^2}{a^2}+\frac{r^2\sin^2\phi}{b^2}+\frac{r^2\cos^2\phi\sin^2\psi}{c^2}=1;$$

hence we find that

$$r_2-r_1=\frac{2abc\sqrt{\{c^2(a^2-h^2)\sin^2\phi+b^2\cos^2\phi\,(a^2\sin^2\psi+c^2\cos^2\psi-h^2\sin^2\psi)\}}}{b^2c^2\cos^2\phi\cos^2\psi+a^2c^2\sin^2\phi+a^2b^2\cos^2\phi\sin^2\psi}.$$

Now $(r_2 - r_1)\cos^2\phi\cos\psi$ can be integrated with respect to ψ; for we may put the expression to be integrated in the form

$$\frac{Kd\psi\cos\psi\sqrt{(A^2 - B^2\sin^2\psi)}}{1 + L\sin^2\psi}$$

where A, B, K, and L do not involve ψ.

The integral with respect to ψ is to be taken between such limits as make $r_2 - r_1$ vanish, that is between the limits given by

$$\sin\psi = \pm\frac{A}{B}.$$

Assume $\sin\psi = \frac{A}{B}\sin\zeta$; thus the integral becomes

$$\frac{KA^2}{B}\frac{\cos^2\zeta d\zeta}{1 + \frac{LA^2}{B^2}\sin^2\zeta}.$$

The limits for ζ are $-\frac{\pi}{2}$ and $\frac{\pi}{2}$. This gives for the value of the integral

$$\frac{\pi KB}{L}\left\{\sqrt{\left(1 + \frac{LA^2}{B^2}\right)} - 1\right\}.$$

Substitute for A, B, K, and L; and then the required attraction becomes

$$\frac{2\pi ach}{a^2 - c^2}\int\left\{\sqrt{\left(\frac{c^2\sin^2\phi + b^2\cos^2\phi}{a^2\sin^2\phi + b^2\cos^2\phi}\right)} - \frac{\sqrt{(h^2 - a^2 + c^2)}}{h}\right\}\cos\phi d\phi.$$

The limits of ϕ are such as make $r_2 - r_1$ vanish when $\psi = 0$; hence they are obtained by putting $A = 0$: we shall find that this gives

$$\sin\phi = \pm\frac{b}{\sqrt{(h^2 - a^2 + b^2)}}.$$

Assume $\sin\phi = \frac{b\sin\theta}{\sqrt{(h^2 - a^2 + b^2)}}$; then the limits of θ will be $-\frac{\pi}{2}$ and $\frac{\pi}{2}$. Put M for $\frac{4\pi abc}{3}$, that is the volume of the ellipsoid. Hence finally the attraction is equal to

$$\frac{3Mh}{(a^2-c^2)\sqrt{(h^2-a^2+b^2)}}\int_0^{\frac{\pi}{2}}\left\{\sqrt{\left(\frac{h^2-a^2+b^2+(c^2-b^2)\sin^2\theta}{h^2-a^2+b^2+(a^2-b^2)\sin^2\theta}\right)}\right.$$
$$\left.-\frac{\sqrt{(h^2-a^2+c^2)}}{h}\right\}\cos\theta d\theta.$$

Now for confocal ellipsoids a^2-b^2, a^2-c^2, and b^2-c^2 are constant; hence for such ellipsoids the attraction at the assigned point varies as the mass of the ellipsoid.

782. Legendre expresses his belief that the theorem of Maclaurin respecting the attraction of confocal ellipsoids holds whatever be the position of the attracted particle; see his page 413. This we now know is true, for it was demonstrated by Laplace. But Legendre was at this time unable to give a complete demonstration; and so confined himself to the case of ellipsoids of revolution.

In order to prepare the way for this demonstration he first establishes a very remarkable theorem, namely: *if the attraction of a solid of revolution is known for every external point which is on the prolongation of the axis it is known for every external point.*

783. Legendre's demonstration of the important result just stated is conducted by the aid of series. We here for the first time meet those famous coefficients which it is usual to call *Laplace's coefficients;* and we see that to Legendre really belongs the honour of introducing them.

These functions might with propriety be called Legendre's functions when they involve only one variable, and Laplace's functions when they involve two: Legendre himself seems to acquiesce in this in a passage at the beginning of his fourth memoir. But in consideration of the great use which Laplace has made of these coefficients, and of the important extension which he has given to the theory of them, I shall continue to use the common English title of Laplace's coefficients for them, after having formally recognised the rights of Legendre.

We may observe that Legendre's researches with respect to Laplace's coefficients are reproduced with extended generality in his *Exercices de Calcul Intégral*, Vol. II. 1817, pages 247...273.

784. An important work on the branch of analysis of which Laplace's coefficients is the origin is Heine's *Handbuch der Kugelfunctionen*, published at Berlin in 1861: to this I shall occasionally refer. The preface to Heine's work gives the evidence for the priority of Legendre to Laplace in the introduction of these coefficients; and also for the recognition of this fact by Jacobi and Dirichlet.

785. The following is the definition of Laplace's coefficients: Let $(1 - 2\alpha\cos\psi + \alpha^2)^{-\frac{1}{2}}$ be expanded in ascending powers of α; and let $P_n\alpha^n$ denote the general term: then P_n is a function of $\cos\psi$, and is called Laplace's coefficient of the nth order.

786. In the present memoir Legendre has occasion to use only the coefficients of an *even* order, because he supposes that his attracting body is symmetrical with respect to the equator. He writes down the values of P_2, P_4, P_6, and P_8; and from these the general form of the coefficient for an even order may be easily perceived. We have, as shewn by Heine in his page 6:

$$P_n = \frac{1 \cdot 3 \cdot 5 \ldots (2n-1)}{\lfloor n} \left\{ x^n - \frac{n(n-1)}{2(2n-1)} x^{n-2} + \frac{n(n-1)(n-2)(n-3)}{2 \cdot 4 \cdot (2n-1)(2n-3)} x^{n-4} - \ldots \right\},$$

where $x = \cos\psi$. Legendre's values are particular cases of this general expression.

This general expression may be easily obtained by first expanding $(1 - 2\alpha x + \alpha^2)^{-\frac{1}{2}}$ in the form

$$1 + \frac{1}{2}\alpha(2x - \alpha) + \frac{1 \cdot 3}{2 \cdot 4}\alpha^2(2x - \alpha)^2 + \frac{1 \cdot 3 \cdot 5}{2 \cdot 4 \cdot 6}\alpha^3(2x - \alpha)^3 + \ldots,$$

and then selecting the term in α^n from each of these.

787. Legendre arrives at an important property of Laplace's coefficients. Suppose that for x or $\cos\psi$ we write

$$\cos\theta\cos\theta' + \sin\theta\sin\theta'\cos\omega,$$

then P_n becomes a function of θ, θ', and ω; integrate with respect to ω from 0 to 2π; then the result is a function of θ and θ'. This result will be of the form $2\pi f(\cos\theta) f(\cos\theta')$; that is it will be the product of a certain function of $\cos\theta$ into the same function of $\cos\theta'$; and $f(\cos\theta)$ is in fact P_n, when x is put for $\cos\theta$. Legendre demonstrates this for the case which he requires, that is for the case of a coefficient of an *even* order; and he obtains the correct form for the function which we denote by f. Legendre's demonstration shews his energy and perseverance, but to a modern student it will appear laborious and uninviting.

The property here considered follows immediately from the general expressions which have been since given for Laplace's coefficients when treated as functions of θ, θ', and ω, in the manner above indicated. See the *Mécanique Céleste*, Livre III. § 15.

788. Another important theorem respecting the coefficients is briefly indicated by Legendre on his page 420: it is demonstrated in Legendre's second memoir, being the last of the seven theorems which are there investigated: see Art. 830.

789. In this memoir we meet for the first time the function V which we now call the *Potential*, and which denotes the sum of the elements of a body divided by their distances from a fixed point. The introduction of this function Legendre expressly assigns to Laplace. The following are the circumstances.

A point is situated outside a solid of revolution. Legendre has to determine the attractions of the solid at the point, along the radius vector which joins the point to the centre of the solid, and at right angles to this direction. He has found a series for the former; and he says the latter might be determined by similar investigations; then he adds:

...mais on y parvient bien plus facilement à l'aide d'un Théorème que M. de la Place a bien voulu me communiquer: voici en quoi il consiste.

Then follows the theorem which is enunciated and immediately demonstrated. The theorem is that the attraction along the radius vector is $-\frac{dV}{dr}$, and the attraction at right angles to the radius vector is $-\frac{dV}{rd\theta}$; where r is the radius vector, and θ

the angle which it makes with the axis of the solid: these attractions being estimated towards the centre, and the pole respectively.

These statements relative to the function V are now well known and given in elementary books.

790. We may observe that the name *Potential* was first used by the late George Green, in his *Essay on the Application of Mathematical Analysis to the Theories of Electricity and Magnetism*, published in 1828: see his page 9, or page 22 of the volume in which Green's works were collected and reprinted in 1871. Gauss used the word in his memoir entitled *Allgemeine Lehrsätze in Beziehung auf die...Anziehungs-und-Abstossungs-Kräfte*, published in 1840. As Gauss does not refer to any previous authority we are, I presume, to infer that he had independently selected the name.

791. Let us briefly indicate the demonstration of Legendre's remarkable theorem enunciated in Art. 782. We shall use the *potential* throughout, whereas Legendre himself only used it partially. But substantially the demonstration we shall give is Legendre's.

Let r and θ, as in Art. 789, be the polar coordinates of the attracted particle; let r' and θ' be the corresponding two polar coordinates of an element of the attracting body; let ω be the difference of longitude, as we may call it, of the attracted point and the attracting element. Then, taking the density as unity, the element of the attracting mass is $r'^2 \sin\theta' \, d\theta' \, d\omega \, dr'$. Thus

$$V = \iiint \frac{r'^2 \sin\theta' \, d\theta' \, d\omega \, dr'}{\sqrt{(r^2 - 2rr' \cos\psi + r'^2)}},$$

where $$\cos\psi = \cos\theta \cos\theta' + \sin\theta \sin\theta' \sin\omega.$$

Expanding the denominator in ascending powers of $\frac{r'}{r}$ we have

$$V = \iiint \frac{r'^2}{r} \left\{1 + P_1 \frac{r'}{r} + P_2 \frac{r'^2}{r^2} + P_3 \frac{r'^3}{r^3} + \ldots\right\} \sin\theta' \, d\theta' \, d\omega \, dr'.$$

We integrate first with respect to r'; and, since the attracting body is assumed to be symmetrical with respect to its equator, the limits will be $-s$ and s, where s is the radius vector of the solid corresponding to a colatitude θ'. Thus

$$V = 2\iint \frac{s^3}{r}\left\{\frac{1}{3} + \frac{P_2}{5}\frac{s^2}{r^2} + \frac{P_4}{7}\frac{s^4}{r^4} + \dots\right\} \sin\theta'\, d\theta'\, d\omega.$$

Now we integrate with respect to ω between the limits 0 and 2π. Then by the theorem of Art. 787, we obtain for V a series of which the first term is $\frac{\mathrm{M}a\sigma\pi}{r}$; and for the following terms the general form is

$$\frac{4\pi f_{2n}(\cos\theta)}{(2n+3)\, r^{2n+1}} \int f_{2n}(\cos\theta')\, s^{2n+3} \sin\theta'\, d\theta'.$$

The limits for θ' are 0 and $\frac{\pi}{2}$. The integration could not be effected until the form of the attracting body is assigned so as to make s a known function of θ'. We shall denote the integral by L_{2n}; it will be some numerical quantity. Thus the general term of V is

$$\frac{4\pi f_{2n}(\cos\theta)\, L_{2n}}{(2n+3)\, r^{2n+1}}.$$

Here $f_{2n}(\cos\theta)$ is a certain known function of $\cos\theta$ which is independent of the form of the body; moreover this function does not vanish when $\theta = 0$. Now if the attraction is known at all points which are on the prolongation of the axis, it follows that V must also be known for all such points. Hence all the quantities of which L_{2n} is the type must be known. Therefore V is known for all external points; and therefore the attraction is also known for all external points.

The demonstration is in substance reproduced by Laplace in the *Mécanique Celeste*, Livre III. § 17.

792. It must be observed that the preceding demonstration is satisfactory only so long as r is greater than the greatest radius vector of the body, so that we may be sure of having convergent series throughout. The subject is discussed by Poisson in the *Connaissance des Tems* for 1829; he shews that the formulæ used by Legendre and Laplace are correct, to the third order of the standard small quantity inclusive. I have extended Poisson's investigation in a paper published in the *Proceedings of the Royal Society*, Vol. XX.

793. We can now demonstrate Legendre's extension of Maclaurin's theorem respecting the attraction of an ellipsoid of revolution.

Put $b = c$ in the result of Art. 781; then it becomes

$$\frac{3M}{a^2-b^2}\int\left\{\frac{h}{\sqrt{[h^2-a^2+b^2-(b^2-a^2)\sin^2\theta]}}-1\right\}\cos\theta\, d\theta;$$

and $[h^2-a^2+b^2-(b^2-a^2)\sin^2\theta]^{-\frac{1}{2}}$ can be expanded by the Binomial Theorem in a convergent series of powers of

$$\frac{(b^2-a^2)\sin^2\theta}{h^2-a^2+b^2},$$

so that the attraction can be expressed in a convergent series which is a function of a^2-b^2, and thus remains the same for confocal ellipsoids of revolution.

Thus for points on the prolongation of the axis of revolution the attractions of confocal ellipsoids at the same point are as the masses of the ellipsoids. Then by the aid of Art. 791 it follows that this will hold for *any* external point.

Legendre himself integrates before expansion; this does not affect the essence of his method. The integral takes different forms according as a is greater or less than b.

794. In conclusion we may affirm that no single memoir in the history of our subject can rival this in interest and importance. During forty years the resources of analysis, even in the hands of D'Alembert, Lagrange, and Laplace, had not carried the theory of the attraction of ellipsoids beyond the point which the geometry of Maclaurin had reached. Legendre now extended the chief result of that geometry, by shewing that it was true in the case of an ellipsoid of revolution for any external point. The introduction of the coefficients now called Laplace's, and their application to the remarkable theorem of Art. 782, commence a new era in mathematical physics. Moreover the existence and the value of the potential function were now first manifested.

It is not too much to say that this memoir is the foundation for all that Laplace added in the theories of Attraction and the Figure of the Earth to the works of Maclaurin and Clairaut.

CHAPTER XXI.

LAPLACE'S TREATISE.

795. We are now about to notice a work by Laplace entitled *Théorie du Mouvement et de la Figure Elliptique des Planetes*; this is a quarto volume which was published in 1784. The title-page and preface occupy xxiv pages; then the first part, which is on the theory of the elliptic motion of the planets, occupies pages 1...66; the second part, which is on the figure of the planets, extends from page 67 to page 150: an addition to the first part is given on pages 150...152, and a list of errata on page 153.

796. The volume is scarce, and seems but little known. The late Professor De Morgan in a note to his article *Table* in the *English Cyclopædia* remarked:

When a person is distinguished by one particular work, his other, and particularly his previous, writings, even on the same subject, go out of notice. How many persons, for instance, know that Laplace published (separately from the Memoirs of the Academy) a small work on the elliptic motion and on the figures of the planets, in 1784? (See Lalande, Bibl. Astron. ann. 1784.) And how many biographical accounts of Laplace mention it?

My copy of this work formerly belonged to Mechain, and subsequently to Arago. There are some irregularities in the paging: no pages occur numbered 105, 106, 145, 146; on the other hand, pages numbered 129, 130 occur twice.

According to a bookseller's catalogue, a German translation of the work by J. J. A. Ide was published at Berlin in 1800; but I have not seen it.

797. The circumstances which led to the publication of the work are thus stated by Laplace in the preface, pages xviii and xix:

Une des propriétés les plus remarquables de la loi de pesanteur qui a lieu dans la Nature, est de terminer les orbites des corps célestes par des lignes du second ordre, et leurs figures par des surfaces du second ordre, du moins lorsqu' on fait abstraction des petites inégalités qui troublent leurs mouvemens et leurs figures. Cette propriété m' a fait naître depuis long-tems l' idée d'exposer dans un ouvrage particulier, les principaux résultats du mouvement et de la figure elliptiques des Planètes; mais entraîné par d'autres occupations, j'aurois entiérement renoncé à ce travail, sans le désir qu'un Magistrat également distingué par son rang, par sa naissance, et par ses lumieres, m'a témoigné plusieurs fois, de voir les propriétés des mouvemens elliptiques et paraboliques, déduites de la seule considération des équations différentielles du second ordre qui déterminent à chaque instant, le mouvement des corps célestes autour du Soleil.

In a note at the foot of the page the name of the distinguished magistrate is given: M. de Saron, Président du Parlement, Honoraire de l'Académie Royale des Sciences.

The passage is very interesting as recording thus early a design which was afterwards carried out on a larger scale by the publication of the *Mécanique Céleste.*

798. In a manuscript note in my copy it is stated that M. Bochart printed the work at his own expense; I presume that this is another name for M. de Saron.

Poisson has recorded the fact that an ellipsoid of revolution, used by Coulomb in his experiments on electricity, was turned by M. de Saron. *Mémoires de l'Institut,* Vol. XIII. 1835, page 501.

799. The work which we are now about to examine may be said to form the transition between Laplace's first three memoirs which do not reappear in the *Mécanique Céleste,* and the subsequent memoirs which do. In the present work Laplace introduces what we call the *potential,* but not what we call *Laplace's functions;* although these functions had already been used by Legendre: see Art. 783.

800. The treatise *De la Figure des Planetes* is arranged in seventeen sections; the first seven sections relate to attractions, the next six to the relative equilibrium of a mass of rotating fluid, and the rest principally to the value of gravity at the surface of such a body.

801. The first section, on pages 67 and 68, is preliminary. The equation to an ellipsoid is given under the form

$$x^2 + my^2 + nz^2 = k^2.$$

This notation appears repulsive to modern readers, trained to study symmetry; but it has been adopted by very eminent mathematicians. Lagrange in his memoir of 1773, and Poisson in his memoir of 1835, also employ m, n, k in the sense here adopted.

802. The second section, on pages 69...73, defines the potential function V, and expresses by means of it the attraction of a body on a particle resolved parallel to three coordinate axes. As we have already seen in Art. 789, the function was introduced by Laplace into mathematical science.

803. In the third section, on pages 73...78, polar coordinates are employed. It is shewn that V may be expanded into an infinite series; and in particular some of the properties of this series, in the case of an ellipsoid, are noticed.

804. The fourth section, on pages 78...86, is very important. Laplace forms three equations involving V, and the differential coefficients of V taken with respect to m, n, k, and the coordinates a, b, c of the attracted particle. Then from these Laplace obtains a demonstration of the theorem which I call by his name, being the extension of Maclaurin's: see Art. 254. This is the first appearance of the demonstration in print; but we learn from page 97 of the treatise we are considering that the demonstration was communicated to the Academy in May, 1783: see Art. 806.

The demonstration is given in an improved form in Laplace's fourth memoir; and in this improved form it is reproduced in the *Mécanique Céleste*, Livre III. § 5 and § 6: we shall defer our remarks on it until we treat of the *Mécanique Céleste*.

The operations of the fourth section of Laplace's treatise are somewhat developed in a memoir by Plana in the *Memorie... Societa Italiana*, Vol. XV. Modena, 1811.

805. The fifth section, on pages 86...90, treats of the attraction of an ellipsoid on an internal particle. The attraction parallel to an axis of the ellipsoid is reduced to a single definite integral; thus Laplace values and appropriates the treasure which D'Alembert deliberately threw away: see Art. 631. This section is embodied in the *Mécanique Céleste*, Livre III. § 3.

We know that the integral can be expressed by means of elliptic functions; Laplace had convinced himself that it could not be expressed by the ordinary functions, but he did not publish his argument. After shewing that the integral, although *definite*, involved all the difficulty of the *indefinite* integral, he says on his page 90:

L'intégrale indéfinie des fonctions différentielles de la forme $\frac{x^2dx}{\sqrt{(1+\alpha x^2)(1+\beta x^2)}}$, est impossible, excepté dans les deux cas suivants, sçavoir lorsque l'une ou l'autre des quantités α, et β, est nulle, ou lorsqu'elles sont égales; je me suis assuré que dans tous les autres cas, l'intégrale ne peut pas être exprimée par une fonction finie de quantités algébriques, d'arcs de cercle et de logarithmes; ainsi l'expression intégrale que nous venons de trouver..., est la plus simple que l'on puisse donner à cette valeur, et il seroit inutile de chercher à la réduire en termes finis.

In his fifth section Laplace demonstrates what he calls a remarkable result, namely: that a particle placed within an elliptic shell of any thickness, and of which the outer and inner surfaces are perfectly similar, will be in equilibrium. This is apparently the first formal statement of the result; but, as we have seen in Art. 662, Frisi may be considered to have obtained it.

806. The sixth section, on pages 90...97, continues the subject of the attraction of an ellipsoid on an internal particle; and it effects the integrations in the particular case of an oblatum. This is embodied in the *Mécanique Céleste*, Livre III. § 7.

The section concludes with a sketch of the history of the problem of the attraction of an ellipsoid. The mistake is made with respect to Maclaurin which I have pointed out in Art. 200. Laplace next speaks of the very ingenious method by which Legendre had shewn that the theorem given by Maclaurin for the case of a point on the prolongation of the axis was true for any position of the point with respect to ellipsoids of revolution: see Art. 793. Then Laplace concludes thus:

...mais la méthode de M. le Gendre, fondée sur la considération des suites, n'est pas applicable aux ellipsoïdes qui ne sont point de révolution; il étoit cependant très-vraisemblable que relativement à ces sphéroïdes, le théorème de M. Maclaurin s'étendoit encore à un point situé d'une maniere quelconque au-dehors; mais l'impossibilité d'intégrer les attractions différentielles, du moins sous la forme que leur donnent les méthodes connues, rendoit assez difficile la démonstration de ce théorème: après quelques tentatives inutiles, j'y suis enfin parvenu par la méthode précédente dont j'ai fait part à l'Académie au mois de Mai 1783. En cherchant à transformer les attractions différentielles, on parviendroit, selon toute apparence, à les rendre intégrables, par un choix convenable des coordonnées; mais la méthode que j'ai suivie, m'ayant conduit assez simplement au résultat que je cherchois, je n'ai point tenté d'autres moyens, et j'ai pensé que le nouvel usage qu'elle présente, du calcul aux différences partielles, pourroit être utile dans d'autres circonstances, et par cette raison intéresser les Géomètres.

I do not understand what is meant by the confident expectation that the expression for the attraction could be integrated by a suitable transformation: this seems to contradict the statement made by Laplace in his fifth section: see Art. 805.

807. We now pass to the relative equilibrium of a rotating fluid mass. The seventh section, on pages 97...103, contains the general équations of fluid equilibrium. This section is embodied in the *Mécanique Céleste*, Livre I. § 17 and § 34.

The section is followed by a *Remarque* which criticises some of Newton's investigations. I do not understand this criticism; Laplace seems to assert that there is some fatal error of principle which attaches to Newton's investigations on the Figure of the Earth, the Tides, and Precession and Nutation: but in the fifth volume of the *Mécanique Céleste* it is stated on the other hand that Newton laid the true foundations of the theories of all these subjects.

Laplace's words in the present Treatise are:

Newton, dans sa théorie de la figure de la terre, suppose cette planete homogène et fluide à sa surface; il détermine dans cette hypothèse, l'applatissement qu'elle doit avoir pour être en équilibre en vertu de son mouvement de rotation, et de l'attraction de toutes ses parties.

Dans sa théorie du flux et du reflux de la mer, il cherche la figure que cette masse doit prendre pour être en équilibre en vertu de son mouvement de rotation, des attractions de toutes ses parties, et de celles du Soleil et de la Lune.

Ce grand Géomètre ne s'est pas apperçu que si les choses se passoient ainsi dans la nature, il ne pourroit y avoir, en vertu des attractions du Soleil et de la Lune, aucune tendance au mouvement dans l'axe de rotation de la terre, et qu'ainsi il n'y auroit ni précession des équinoxes, ni nutation dans l'axe terrestre.

808. The eighth section, on pages 103...113, treats of the relative equilibrium of a homogeneous mass of rotating fluid, acted on by distant bodies as well as by its own attraction. The problem is that which we have noticed in Art. 629. In the *Mécanique Céleste*, Livre III. § 23, the action of the distant bodies is treated in a simpler mode than in the present section. At the end of Art. 629, I have drawn attention to two circumstances which it seems to me that D'Alembert ought to have noticed; Laplace says nothing about the first, but he alludes to the second, though in scarcely an adequate manner. The present section contains an important remark, to which I have already referred in Art. 153.

809. The ninth section, on pages 113...116, applies the preceding section to the case of the Moon supposed fluid and homogeneous. Laplace arrives at the conclusion that the elongation of the Moon's diameter directed towards the Earth is four times as great as the elongation of the diameter which is at right angles to this and in the plane of the Moon's orbit. We will briefly indicate the process by which this is obtained. We have shewn in Art. 623, what Laplace assumes at the outset, namely that the axis of rotation will coincide with one of the principal axes of the Moon, and the radius vector to the Earth with another.

Take then $\cos\lambda = 0$, $\cos\mu = 0$, $n = 0$, $m = 0$; thus the axis of z is that of rotation, and the axis of x passes through the Earth, the centre of the Moon being the origin.

Then the last two equations of Art. 617 reduce to

$$a^2\left(A - \frac{2M}{R^3} - \omega^2\right) = b^2\left(B + \frac{M}{R^3} - \omega^2\right) = c^2\left(C + \frac{M}{R^3}\right).$$

In Art. 616, we have spoken of a part of the action of M which is not what we call a *disturbing* force; in the present problem this part is duly regarded, and is in fact balanced by what in common language is called the centrifugal force arising from the revolution of the Moon round the Earth. Moreover this revolution gives rise to the following relation connecting the quantities involved:

$$\frac{M}{R^2} = R\omega^2.$$

Thus the above equations become

$$a^2\left(A - \frac{3M}{R^3}\right) = b^2 B = c^2\left(C + \frac{M}{R^3}\right).$$

We now require the values of A, B, and C.

Put e^2 for $\frac{a^2 - b^2}{a^2}$ and e'^2 for $\frac{a^2 - c^2}{a^2}$; then e^2 and e'^2 being supposed small we have approximately by Art. 620,

$$A = \frac{V}{a^3}\left\{1 + \frac{3}{10}(e^2 + e'^2)\right\}.$$

In like manner we can express B and C; supposing that e^2 and e'^2 will not be sensibly changed if we take b^2 or c^2 for denominator instead of a^2, we have

$$B = \frac{V}{b^3}\left\{1 + \frac{3}{10}(e'^2 - 2e^2)\right\},$$

$$C = \frac{V}{c^3}\left\{1 + \frac{3}{10}(e^2 - 2e'^2)\right\}.$$

Hence we shall obtain finally

$$e^2 = \frac{15}{2}\varpi, \qquad e'^2 = 10\varpi,$$

where ϖ stands for $\frac{M}{V} \cdot \frac{a^3}{R^3}$, it being assumed that ϖ remains sensibly unchanged if we multiply it by $\frac{c^3}{a^3}$.

Thus $e''^2 - e'^2 = \frac{5}{2} \varpi$, so that $e''^2 = 4(e''^2 - e'^2)$; and this amounts to Laplace's statement that one elongation is four times the other.

Laplace supposes that $\frac{M}{V} = \frac{1}{70}$, and that $\frac{a}{R} = \sin 15' \, 45''$; and thus he finds that $a = \frac{29712}{29711} c$, and $b = \frac{118845}{118844} c$.

810. The tenth section, on pages 116...122, is devoted to the case of a homogeneous fluid mass rotating with uniform angular velocity, and acted on by its own attraction. Laplace says: "Il est visible qu'alors, le sphéroïde sera un ellipsoïde de révolution..." This is however more than he demonstrates, for he confines himself to demonstrating that the oblatum is a possible form of relative equilibrium: see Art. 168. Laplace obtains the equation which connects the angular velocity with the ellipticity of the generating ellipse: see Art. 202. Laplace's investigations are embodied in the *Mécanique Céleste*, Livre III., Chapitre III. On his page 121 Laplace says that we may presume the Earth to be homogeneous from the centre up to a few leagues from the surface; at a subsequent period he leaned to the opinion that the density increases as we approach the centre; see the *Mécanique Céleste*, Livre III., page 101, and Livre XI., page 12.

811. The eleventh section, on pages 122...125, resumes the equation of the preceding section which connects the angular velocity with the ellipticity. Laplace demonstrates certain results which he states thus:

Il suit delà que pour un mouvement de rotation donné, il y a toujours deux figures elliptiques applaties vers les pôles, qui satisfont à l'équilibre. Cette remarque intéressante sur la possibilité de plusieurs figures d'équilibre relatives à un même mouvement de rotation, est dûe à M. d'Alembert; mais il n'en avoit pas déterminé le nombre que j'ai trouvé se réduire à deux, par l'analyse précédente dont je fis part à cet illustre Géomètre dans le mois de Juillet de 1778.

The researches of D'Alembert on this subject are contained in the sixth and eighth volumes of his *Opuscules Mathématiques:* see Arts. 581, 585, and 637.

With respect to the first sentence of the preceding extract we may observe that the words *il y a toujours* require to be limited, for Laplace shews in his next section that if the angular velocity be too great, an oblatum is not a possible form of relative equilibrium.

Laplace demonstrates that corresponding to a given angular velocity there *cannot be more* than two oblata; but he does not explicitly shew that there will always be two oblata, provided the angular velocity be less than a certain limit. It would be very easy to supply this; but perhaps Laplace thought that it was unnecessary to repeat what had really been given by D'Alembert.

Laplace's demonstration is sound, but is less simple than that which he afterwards gave in the *Mécanique Céleste,* although depending on the same principles.

Laplace gives approximate investigations for determining the oblata in the extreme cases of a very small ellipticity and a very great ellipticity.

The section is reproduced in an improved form in the *Mécanique Céleste,* Livre III., Chapitre III.

812. The twelfth section, on pages 125...128, discusses the limiting value of the angular velocity for which an oblatum is possible, and gives numerical results for the case of the Earth. Also it is shewn that an oblongum is not a possible form of relative equilibrium. The section is substantially reproduced in the *Mécanique Céleste,* Livre III., Chapitre III.

813. The thirteenth section, on pages 128...131, applies the principle of conservation of areas, as we now call it, to the subject; the section is substantially reproduced in the *Mécanique Céleste,* Livre III., § 21.

The section is followed by a *Remarque* which deserves to be noticed. Laplace alludes to what had been shewn by Clairaut with respect to the Figure of the Earth considered as heterogeneous. Taking the excentricity of the strata as a small quantity of the first order, and neglecting small quantities of the second order, it was shewn that equilibrium might subsist with elliptical strata. Then Laplace proceeds thus:

Nous renvoyons sur cet objet à son excellent Ouvrage sur la figure de la Terre, et nous nous contenterons d'observer ici, que l'équilibre rigoureux est impossible dans l'hypothèse de l'ellipticité des couches; car il résulte des formules précédentes, que dans ce cas, l'attraction des couches intérieures du sphéroïde sur un point placé à la surface, a pour expression, une fonction transcendante des coordonnées de ce point; ainsi l'équation donnée par la condition de l'équilibre à la surface, seroit transcendante, et par conséquent ne pourroit coincider avec la supposition de l'ellipticité des couches;...

I am unable to understand the argument by which it is inferred that the equilibrium is strictly impossible in the case of elliptical strata: it seems to me that in the same way it might be asserted that the relative equilibrium of an ellipsoid of rotating fluid would be impossible, and this is contrary to Jacobi's theorem.

814. The fourteenth section, on pages 132...137, presents to us a matter to which Laplace seems to have attached great importance, and which has given rise to some controversy. It may be considered as consisting of a theorem which has already appeared three times in Laplace's writings: see Arts. 755, 771, and 775.

The section is in substance taken from the *Addition* to the third memoir, which we noticed in Art. 755; and it is embodied in the *Mécanique Céleste*, Livre III. § 10. The main result is an equation which is numbered (1) and is thus expressed in the *Mécanique Céleste*:

$$\frac{dV}{dr} = A' - \frac{n+1}{2a} A + \frac{n+1}{2a} V \text{ .... } (1).$$

This is obtained on the supposition that attraction varies as the nth power of the distance, and that V is the sum of the product of every element of mass into the $(n+1)$th power of the distance of the element from the attracted particle. The notation is different, but the mode of investigation in the present treatise is like that in the *Mécanique Céleste*.

If ds represents an infinitesimal length measured in any direction, then $-\frac{1}{n+1}\frac{dV}{ds}$ is the value of the attraction estimated in the direction of the element ds; Laplace makes this remark on his page 134, and it is now familiar to us from our elementary books.

In page 264 of the *Addition* to the third memoir, and in page 130 of the present treatise Laplace makes a remark with respect to the case in which $n = -1$, that does not seem quite safe. He says that in this case the vertical attraction is constant over the surface of the spheroid. But we cannot strictly apply our formulæ to this case; for instance, the expression for the attraction $-\frac{1}{n+1}\frac{dV}{ds}$ cannot be used when $n+1$ vanishes.

Let ds now represent an element of arc on the surface of the attracting body; then $-\frac{1}{n+1}\frac{dV}{ds}$ represents the attraction resolved along ds. And $-\frac{1}{n+1}\frac{dV}{dr}$ represents the attraction resolved along the radius r towards the origin, or resolved along the normal very approximately if the body is very nearly spherical; denote this by ϕ. Thus from (1) we have

$$\phi = A' - \frac{A}{2a} + \frac{V}{2a},$$

and as A and A' are constants on the surface we have

$$\frac{d\phi}{ds} = \frac{1}{2a}\frac{dV}{ds} \text{ .... (2).}$$

Hence from (2) we see that the attraction resolved along ds is $-\frac{2a}{n+1}\frac{d\phi}{ds}$. Thus we have a result which Laplace expresses in the following words:

A la surface de tout sphéroïde homogène infiniment peu différent d'une sphère dont le rayon est pris pour l'unité, l'attraction horizontale dirigée suivant un petit côté du sphéroïde, et multipliée par ce côté, est égale au produit de $-\frac{2}{n+1}$, par la différence des attractions verticales aux deux extrémités de ce côté.

This result substantially includes as particular cases the formulæ of Arts. 771 and 772.

815. The fifteenth section, on pages 137...140, offers some remarks on the general problem of determining all possible forms of equilibrium of a fluid mass which correspond to given forces: this is obviously much more difficult than the mere verification that a certain assigned form is admissible. Laplace says that the general solution is impossible, at least in the present state of analysis; and we may add that after the lapse of nearly a century the statement seems still applicable.

It will be convenient to give a general equation which Laplace forms.

Suppose that A, B, C are the attractions of the body itself parallel to the coordinate axes, on a particle whose coordinates are x, y, z. Let P, Q, R be the corresponding other forces which act. Then for equilibrium the following must be the differential equation to the free surface of the fluid:

$$(A+P)\,dx + (B+Q)\,dy + (C+R)\,dz = 0.$$

But $\quad A = -\frac{1}{n+1}\frac{dV}{dx}, \quad B = -\frac{1}{n+1}\frac{dV}{dy}, \quad C = -\frac{1}{n+1}\frac{dV}{dz}.$

Hence the equation becomes

$$0 = -\frac{1}{n+1}\,dV + P\,dx + Q\,dy + R\,dz;$$

therefore $\quad V = (n+1)\int (P\,dx + Q\,dy + R\,dz) + \text{a constant}.$

816. The sixteenth section, on pages 140...144, investigates the law of gravity at the surface of a nearly spherical mass of fluid in relative equilibrium.

Denote the gravity by p; then with the notation of the preceding Article

$$p^2 = (A+P)^2 + (B+Q)^2 + (C+R)^2.$$

If P, Q, and R are small compared with A, B, and C we have approximately

$$p = \sqrt{(A^2 + B^2 + C^2)} + \frac{AP + BQ + CR}{\sqrt{(A^2 + B^2 + C^2)}}.$$

Here $\sqrt{(A^2 + B^2 + C^2)}$ expresses the attraction exerted by the nearly spherical body itself; it will be approximately along the normal or the radius vector: we will denote it by ϕ.

Then in the second term of p it will be sufficient to put for A, B, and C approximate values which must hold inasmuch as the mass is nearly spherical. Taking a for the radius of the sphere which nearly coincides with the mass we have approximately

$$A = -\frac{x\phi}{a}, \qquad B = -\frac{y\phi}{a}, \qquad C = -\frac{z\phi}{a}.$$

Therefore $$p = \phi - \frac{1}{a}(Px + Qy + Rz).$$

By the aid of the value of ϕ which is given in equation (1) of Art. 814, and the value of V found in Art. 815, we obtain

$$p = \frac{n+1}{2a}\int(Pdx + Qdy + Rdz) - \frac{1}{a}(Px + Qy + Rz) + H,$$

where H is some constant, which will be known by a single observation of the value of gravity at the surface of the mass.

This result was obviously much valued by Laplace at the time; he says:

Nous voilà donc parvenus à déterminer directement la loi de la pesanteur, ce qui est d'autant plus remarquable, que la figure du sphéroïde dont cette loi paroît dépendre, nous est entiérement inconnue.

This is a generalisation of what he had before obtained for the ordinary law of attraction, that is for the case in which $n = -2$.

Laplace proceeds to consider the case in which P, Q, and R involve the action of other bodies as well as the so-called centrifugal force. If we confine ourselves to the latter, and take ω for the angular velocity, we have

$$P = 0, \qquad Q = \omega^2 y, \qquad R = \omega^2 z.$$

Hence $$p = \Pi + \frac{n-3}{4a} \omega^2 (y^2 + z^2).$$

If $n = 3$ we see that p is constant. If $n = -2$ we have the case of nature.

The main results of the section are reproduced in another form in the *Mécanique Céleste*, Livre III. § 36.

817. The seventeenth section, on pages 144...150, investigates what the law of attraction must be in order that a spherical shell may attract an external particle in the same manner as if the shell were condensed at its centre. The investigation, here given for the first time, was substantially reproduced in the *Mécanique Céleste*, Livre II. § 12. In the reproduction Laplace added an investigation of the law which makes the resultant attraction of the shell on an internal particle zero. Both have since passed into the elementary books.

818. On the whole we may say that the present treatise forms a valuable contribution to our subjects. In the theory of the attraction of ellipsoids we have for the first time the single definite integral by which the resolved attraction at any internal point is expressed; and also the important theorem of Laplace with respect to the attraction at an external point. The theory of the Figure of the Earth, considered as homogeneous, appears in the form which it has since retained; Laplace demonstrated the point left unsettled by D'Alembert as to the number of possible oblata corresponding to a given angular velocity, and shewed that an oblongum was not an admissible figure. To these results we must add the general expression for the force of gravity at the surface of a fluid spheroid, and the investigation as to the attraction of a spherical shell on an external particle. The treatise may be said still to survive in the pages of the *Mécanique Céleste*, where so much of it is reproduced; and it well deserves this abiding honour.

CHAPTER XXII.

LEGENDRE'S SECOND MEMOIR.

819. In the Paris *Mémoires* for 1784, published in 1787, there is a memoir by Legendre entitled *Recherches sur la Figure des Planètes;* it occupies pages 370...389 of the volume. The memoir was read to the Academy, on the 7th July, 1784.

820. The object of the memoir is to shew that under certain conditions the oblatum is the only form of relative equilibrium for a mass of rotating fluid: the conditions will appear in an extract given in the next Article. We will first reproduce a note bearing on the history of the subject which occurs at the beginning of the memoir. After referring to D'Alembert's *Opuscules Mathématiques*, Vols. v. and vii., and Laplace's memoir of 1772, Legendre says:

La proposition qui fait l'objet de ce Mémoire, étant démontrée d'une manière beaucoup plus savante et plus générale dans un Mémoire que M. de la Place a déjà publié dans le Volume de 1782, je dois faire observer que la date de mon Mémoire est antérieure, et que la proposition qui paroît ici, telle qu'elle a été lûe en juin et juillet 1784, a donné lieu à M. de la Place, d'approfondir cette matière, et d'en présenter aux Géomètres, une théorie complète.

821. Legendre states the conditions of his demonstration thus:

Je suppose, comme on paroît l'avoir fait jusqu'à présent, que la figure cherchée est celle d'un solide de révolution peu différent d'une sphère, et partagé en deux parties égales et semblables par son équateur. L'attraction de ce sphéroïde s'évalue facilement à l'aide des formules que j'ai données pour cet objet, (*Mémoires des Savans étrangers, tome* x.); et j'en tire l'équation du méridien exprimée par une suite infinie, équation d'une forme très-différente de celle qu'a trouvée M. de la Place, pour le cas où le sphéroïde ne diffère qu'infiniment peu de la sphère. Je fais voir ensuite que la série renfermée dans cette équation, est toujours convergente; que l'ellipse y est comprise suivant le théorème de Maclaurin, et qu'aucune autre courbe n'y peut satisfaire.

The equation obtained by Laplace to which Legendre here refers is I presume that we have given in Equation (7) of Art. 757.

822. Legendre begins with demonstrating seven theorems respecting the coefficients which he had introduced in his first memoir, and which we now call Laplace's coefficients.

Legendre says:

Pour démontrer ces diverses propositions, j'ai recours aux propriétés d'une espèce particulière de fonctions rationnelles, qui ne se sont point encore présentées aux Analystes, et qui paroissent mériter leur attention;...

823. Legendre, as in his first memoir, uses only coefficients of an *even* order. We will state the seven theorems he demonstrates, and give references for the demonstrations. See Art. 784. We assume that P_n has the meaning assigned in Art. 786, so that P_n is a known function of x.

824. When $x = 1$ then $P_m = 1$. This is obvious from the definition of P_m; for when $x = 1$ then P_m is the coefficient of a^m in the expansion of $\dfrac{1}{1-a}$.

825. If m be any positive integer less than n, then

$$\int_0^1 x^{2m} P_{2n}\, dx = 0.$$

Heine, page 37.

826. If m be any positive integer

$$\int_0^1 x^m P_{2n}\, dx = \frac{m(m-2)(m-4)\ldots(m-2n+2)}{(m+1)(m+3)\ldots(m+2n+1)}.$$

Heine, page 38.

827. If m and n are different positive integers

$$\int_0^1 P_m P_n \, dx = 0.$$

And $$\int_0^1 (P_n)^2 \, dx = \frac{1}{4n+1}.$$

Heine, page 34.

828. The function P_n can be decomposed into n factors of the form $x^2 - \alpha^2$, $x^2 - \beta^2$, $x^2 - \gamma^2$, ... where α, β, γ, ... are real unequal proper fractions.

Heine, page 23.

829. While x lies between 0 and 1 the function P_n is always less than unity.

Heine, page 8. The demonstration consists in developing P_n in a series of cosines of the multiples of θ, where $\cos\theta = x$; it is found that every term is positive, and so the greatest value is when $\theta = 0$: then, as we have seen in Art. 824, the greatest value is unity. Legendre gives the demonstration. I do not understand what Heine means on his page 8 by ascribing priority to Laplace in giving this form to P_n. Heine says: "Aus dieser Reihe, welche Laplace entwickelt, schliesst Legendre dass P^n seinen grössten Werth für $\theta = 0$ erhält." Heine refers in notes to Laplace's memoir of 1782, and to Legendre's of 1784; but the latter memoir was really the earlier in composition, as Heine shews in his Preface. Heine's P^n is what we call P_n; but it must be remembered that at present with Legendre only coefficients of an even order explicitly occur.

The passage of Laplace's memoir to which Heine refers is reproduced on page 41 of the *Mécanique Céleste*, Vol. II. Laplace does really no more than Legendre had done. Laplace formally writes down a general term, while Legendre writes down sufficient particular terms to render the general term obvious.

830. If k be any constant

$$\int_0^1 \frac{P_n \, dx}{(1+kx^2)^{n+\frac{3}{2}}} = \frac{(-k)^n}{(2n+1)(1+k)^{n+\frac{1}{2}}}.$$

Heine, page 43.

831. Having thus finished his preliminary analysis, Legendre proceeds to form the equation which determines the nature of the meridian curve, in order that relative equilibrium may subsist.

Let V be the potential of the mass for a point on its surface whose radius vector is r, and colatitude θ. Then by Art. 789, and Huygens's plumb-line principle, we have for the condition of relative equilibrium

$$V + \frac{\omega^2 r^2 \sin^2\theta}{2} = \text{constant} \ldots\ldots\ldots\ldots\ldots\ldots (1),$$

where ω is the angular velocity.

Legendre arrives at an equation precisely equivalent to this; and then he says on his page 379:

...Cette équation est la même qu'a donnée M. de la Place, dans le volume de l'Académie de 1772, et dans sa Théorie du mouvement et de la figure des planètes, *page* 137.

This is substantially true; but to prevent mistake, we must observe that in the volume for 1772, Laplace does not introduce the function V: the result he gives there coincides with our equation (6) of Art. 756, and is in fact what we should obtain by differentiating (1) with respect to θ.

Now put for V its value from Art. 791. Thus if M denote the mass, and u_{2n} stand for $\frac{4\pi L_{2n} f_{2n}(\cos\theta)}{(2n+3)\, r^{2n+1}}$, we have

$$\frac{\omega^2 r^2 \sin^2\theta}{2} + \left\{\frac{M}{r} + u_2 + u_4 + u_6 + \ldots\right\} = \text{constant} \ldots\ldots\ldots\ldots (2).$$

832. Next Legendre shews that equation (2) is satisfied when the meridian curve is an ellipse, at least if the angular velocity does not exceed a certain value; this occupies pages 382...387 of the original memoir. We will not reproduce this, for, as Legendre himself remarks, there can be no doubt of the truth of the proposition after Maclaurin's researches: and we shall be able to give the essence of Legendre's process without this subsidiary part.

833. We know then by Art. 262 that there is relative equilibrium for an oblatum if

$$X - Y\sqrt{(1 - e^2)} = jX = n\omega^2;$$

and with the values of X and Y given there this reduces to

$$\omega^2 = \frac{2\pi\sqrt{(1-e^2)}}{e^3}\left[(3-2e^2)\sin^{-1}e - 3e\sqrt{(1-e^2)}\right].$$

Hence substituting in (2) we know that if we take

$$\frac{a^2}{r^2} = 1 + \frac{e^2}{1-e^2}\cos^2\theta,$$

the following equation will be satisfied:

$$\frac{\pi\sqrt{(1-e^2)}}{e^3}\left[(3-2e^2)\sin^{-1}e - 3e\sqrt{(1-e^2)}\right] r^2\sin^2\theta$$

$$+\frac{M}{r} + u_2 + u_4 + u_6 + \ldots = \text{constant} \ldots\ldots\ldots\ldots\ldots (3).$$

Put $\sin\psi$ for e; then

$$\sqrt{(1-e^2)}\frac{(3-2e^2)\sin^{-1}e - 3e\sqrt{(1-e^2)}}{e^3} = \frac{(3-2\sin^2\psi)\psi - 3\sin\psi\cos\psi}{\sin^2\psi\tan\psi}$$

$$= \frac{(3+\tan^2\psi)\psi - 3\tan\psi}{\tan^3\psi}.$$

Now we have a known formula

$$\psi = \tan\psi - \frac{1}{3}\tan^3\psi + \frac{1}{5}\tan^5\psi - \ldots;$$

and thus the preceding expression becomes

$$4\left\{\frac{1}{3.5}\tan^2\psi - \frac{2}{5.7}\tan^4\psi + \frac{3}{7.9}\tan^6\psi - \ldots\right\}.$$

Put k for $\tan^2\psi$, and x for $\cos\theta$. Then equation (3) becomes

$$4\pi r^2(1-x^2)\left\{\frac{k}{3.5} - \frac{2k^2}{5.7} + \frac{3k^3}{7.9} - \ldots\right\}$$

$$+\frac{M}{r} + u_2 + u_4 + u_6 + \ldots = \text{constant} \ldots\ldots\ldots\ldots (4),$$

where
$$u_{2n} = \frac{4\pi P_{2n}}{(2n+3)r^{2n+1}}\int_0^1 P_{2n}r'^{2n+3}dx.$$

This we know is satisfied by $\frac{a^2}{r^2} = 1 + kx^2$.

It must be observed that a is a function of M and k, being determined by the equation

$$M = \frac{4\pi a^3}{3\sqrt{(1+k)}}, \ldots\ldots\ldots\ldots\ldots\ldots\ldots\ldots (5),$$

but for the sake of brevity we will retain a and not substitute for it in terms of M and k.

834. Thus equation (4) has been obtained subject only to two limitations: the series which we have used must be convergent: and the angular velocity of rotation must lie within those known limits for which an oblatum is a possible form of relative equilibrium.

Legendre obtains an equation which is substantially the same as (4). He has however divided the equation by M, and he has taken $a = 1$; these two steps however do not appear to me advantageous. Legendre himself gives so little explanation of his process that after this stage I am compelled to add much to his brief outline in order to render the whole matter intelligible.

835. Legendre himself does not distinctly state what he really demonstrates with more or less success; so that we must supply this omission. Suppose there to be a given mass of fluid; let this rotate with any angular velocity comprised between certain specified limits, then there is a corresponding oblatum, or rather two oblata; so that we have a *series* of oblata corresponding to a *series* of angular velocities. Now Legendre shews that there is no other *series* of figures possible except oblata. He does not shew that for isolated values of the angular velocity no solution exists except an oblatum. But it may seem very natural that if any solution besides an oblatum is possible, such solution will be possible *throughout some range* of angular velocity, and not merely for certain values of the angular velocity finite in number.

Moreover as to angular velocities beyond the specified limits, Legendre's process gives no information; so that at the utmost all that it proves is, if an oblatum is possible no other figure is possible; and as to the cases in which an oblatum is not possible it says nothing.

836. Equation (4) was obtained by considering one figure; but it will hold for all figures, provided we give to the word *constant* a proper interpretation. This word *constant* must be understood to mean constant with respect to x; so that it may

involve k. The essence of Legendre's process now is to consider that (4) must hold for all values of k within a certain range; so that if the left-hand side be supposed developed in powers of k, the coefficient of each power of k must be equal to a constant.

Moreover not only must (4) hold for all values of k within a certain range; but so also must the following

$$M = \frac{4\pi}{3}\int_0^1 r^3\,dx \text{ ....} (6).$$

837. We know then that (4) and (5) are satisfied by

$$\frac{a^2}{r^2} = 1 + kx^2.$$

Suppose if possible that they are also satisfied by

$$\frac{a^2}{r^2} = 1 + kx^2 + kp_1 + k^2p_2 + k^3p_3 + \dots$$

where $p_1, p_2, p_3, \dots$ are any functions of x. Substitute in (4), and pick out the term which involves k; we need not attend to such terms as depend only on the kx^2 which occurs in $\frac{a^2}{r^2}$, because we know that all such terms must have a constant value.

In u_m we have a term

$$-2\pi P_m a^2 k\int P_m p_1\,dx;$$

and in $\frac{M}{r}$ we have a term $\frac{M}{2a}kp_1$.

And from (5) we have

$$a = \left(\frac{3M}{4\pi}\right)^{\frac{1}{3}}(1+k)^{\frac{1}{6}} = c(1+k)^{\frac{1}{6}} \text{ say.}$$

Hence finally equating the coefficient of k to a constant, we have

$$\frac{Mp_1}{c} - 4\pi c^2\left\{P_2\int_0^1 P_2p_1\,dx + P_4\int_0^1 P_4p_1\,dx\right.$$

$$\left. + \dots + P_m\int_0^1 P_m p_1\,dx + \dots\right\} = \text{a constant};$$

therefore $$\frac{p_1}{3} - \left\{P_2 \int_0^1 P_2 p_1\, dx + P_4 \int_0^1 P_4 p_1\, dx + \dots \right.$$
$$\left. + P_{2n} \int_0^1 P_{2n} p_1 dx + \dots \right\} = \gamma \dots\dots\dots\dots\dots (7),$$

where γ is some constant.

We may express this result thus:

$$\frac{p_1}{3} = \gamma + \gamma_2 P_2 + \gamma_4 P_4 + \dots + \gamma_{2n} P_{2n} + \dots\dots\dots\dots\dots (8),$$

where γ_{2n} stands for $\int_0^1 P_{2n} p_1\, dx$.

Now multiplying (8) by P_{2n}, and integrating between the limits 0 and 1, we obtain by Art. 827,

$$\frac{1}{3}\int_0^1 P_{2n} p_1 dx = \frac{\gamma_{2n}}{4n+1},$$

therefore $$\frac{\gamma_{2n}}{3} = \frac{\gamma_{2n}}{4n+1},$$

therefore $\gamma_{2n} = 0$.

Hence we see that p_1 reduces to a constant which we denote by 3γ.

838. Now we shall shew that $\gamma = 0$.

For from (6) we have

$$M = \frac{4\pi a^3}{3}\int_0^1 \{1 + kx^2 + kp_1 + k^2 p_2 + \dots\}^{-\frac{3}{2}} dx,$$

that is by (5)

$$(1+k)^{-\frac{1}{2}} = \int_0^1 \{1 + kx^2 + kp_1 + k^2 p_2 + \dots\}^{-\frac{3}{2}} dx.$$

In this put for p_1 a constant 3γ; then equate the coefficients of k in the expansions of the two sides, and we have

$$-\frac{1}{2} = -\frac{3}{2}\int_0^1 (x^2 + 3\gamma)\, dx,$$

that is $$0 = -\frac{9}{2}\gamma,$$

so that γ is zero.

839. Having thus shewn that p_1 is zero, we may in precisely the same way shew that p_2 is zero; then that p_3 is zero; and so on.

Thus it is impossible that (4) can be satisfied by such a value of r as is determined by

$$\frac{a^2}{r^2} = 1 + kx^2 + kp_1 + k^2p_2 + k^3p_3 + \ldots$$

840. Such is Legendre's demonstration. I may remark that I do not see why he did not introduce in the assumed value of $\frac{a^2}{r^2}$, a term independent of k, say p_0; for his process would apply to p_0 and lead to the conclusion that p_0 must be zero.

841. In Art. 837 the method of shewing that γ_m is zero, deserves notice. This is the first appearance of a particular case of the general proposition now well known, that a given function can be expanded in only one series of Laplace's functions.

842. In estimating the force of the demonstration the main point to be considered is the convergence of the series employed.

The expansion of ψ in terms of $\tan\psi$ is convergent only so long as $\tan\psi$ does not exceed unity; so that e must be less than $\frac{1}{2}$. This range is much narrower than the range for which an oblatum is known to be a possible form of equilibrium; for in the extreme case in which an oblatum is possible the ratio of the axes is almost that of 2·72 to unity. See page 126 of Laplace's *Figure des Planetes*, and page 386 of this memoir by Legendre.

Legendre himself does not notice this instance of the subject of convergence.

Then there is the question whether the series which is used for V in Art. 831 is convergent; it is on this series that our fundamental equation (4) depends. Legendre does advert to this point and considers that he establishes the convergence of the series for V. We will now consider his arguments.

843. Since P_n changes sign n times between $x = 0$ and $x = 1$, he suggests that the positive and negative parts of $\int_0^1 P_n f(x)\,dx$ neutralise each other, especially when n is very large. This is unsatisfactory. For though we may thus be led to believe that $\int_0^1 P_n f(x)\,dx$ will be in general very small when n is very large, yet this does not shew that the series of which this is the nth term is convergent.

Legendre takes as a special case that in which $f(x) = x^m$. The value of $\int_0^1 P_n x^m\,dx$ is known from Art. 826. In this case we shall have, as Legendre remarks, for the value of $\int_0^1 P_n x^m\,dx$, when n is very great, approximately $\frac{A}{n^{m+\frac{3}{2}}}$, where A is a constant. This result can be obtained by the theorem called Stirling's Theorem. The series which has for its nth term the expression just given is certainly convergent. Thus in this special case Legendre establishes his statement.

Legendre says nothing as to the convergence of the expansions which are employed in the processes of Arts. 837 and 838.

Poisson in the *Connaissance des Tems* for 1829, page 360, alludes to Legendre's remark on the convergence of his series. Poisson holds justly that it is not sufficient to prove that the series which we finally obtain are convergent; it is necessary for the soundness of our demonstration that the series should be convergent throughout; hence the expression used by Legendre for V cannot be considered to be obtained with rigour: see Art. 792.

844. It remains to estimate the value of this investigation. Legendre himself seems to have fluctuated in his opinion.

In the extract we have given in Art. 821 Legendre apparently distinguishes between the hypothesis that a body is little different from a sphere (peu différent), and the hypothesis that a body differs infinitesimally from a sphere (infiniment peu). He considered that his own demonstration applied to the case of a body not restricted, like the body to which Laplace confined himself, to be indefinitely close to a sphere in form.

In the extract we have given in Art. 820 Legendre seems to allow that Laplace's demonstration in the volume for 1782 was more clever and more general than his own. I presume the greater generality means that Laplace did not restrict himself to the case of a solid of revolution. Laplace still retained the hypothesis of a body deviating extremely little from a sphere. In his memoir in the volume for 1789 Legendre seems to maintain the superiority of his own demonstration. The passage will be quoted hereafter. Legendre in fact claims as the merit of his own solution that he did not assume the body to be almost spherical.

This is the great merit of Legendre's process. He does not restrict himself to the case of a body very nearly spherical; and although the demonstration cannot be regarded as perfect, yet there is a good attempt at the problem in its most general form, so far as a surface of revolution is concerned.

845. Laplace says in the *Mécanique Céleste*, Vol. v. page 10:

M. Legendre a fait voir ensuite que si la figure est de révolution, elle doit, pour l'équilibre être elliptique; et j'ai reconnu que cela est exact, sans supposer une figure de révolution.

This passage seems to me unsatisfactory; it leaves out of sight the important fact that Laplace expressly limited himself to the case of a nearly spherical body, and that Legendre did not.

846. Ivory in the *Philosophical Transactions* for 1824, page 526, refers to Legendre's memoir. Ivory says: "To the mathematical processes employed by that eminent geometer, no objection can be made." Ivory however proceeds to object to the memoir for other reasons which depend on his own abnormal notions as to the conditions of fluid equilibrium. I hold on the contrary that Legendre is sound as to hydrostatical principles, but weak in the mathematical investigations, because his series are not necessarily convergent.

847. Jacobi in a paper in which he enunciated the theorem that a rotating ellipsoid of fluid might be in relative equilibrium spoke in the highest terms of Legendre's investigation; see Poggendorff's *Annalen*, Vol. XXXIII. 1834, pages 229...233. After observing that corresponding to a given angular velocity there might be two oblate as figures of relative equilibrium, one having a small ellipticity and the other a large ellipticity, Jacobi proceeds thus:

Die erste dieser Lösungen, die das wenig abgeplattete Umdrehungsellipsoid giebt, hat durch Legendre's bewundernswürdige Arbeiten über die Figur der Erde eine grössere Bedeutung erlangt. Dieser Mann, dessen Ruhm mit den Fortschritten der Mathematik zunimmt, hatte durch Einführung jener merkwürdigen Ausdrücke, durch welche wir heut in den Anwendungen die Functionen zweier Variabeln darstellen, die allgemeinsten Untersuchungen über diesen Gegenstand möglich gemacht. Er zeigte, dass unter allen Figuren, die nicht zu sehr von der sphärischen Gestalt abweichen, so dass es möglich ist, die Anziehung, welche auf einen Punkt der Oberfläche ausgeübt wird, nach den Potenzen dieser Abweichung zu entwickeln, das wenig abgeplattete Umdrehungsellipsoid, wie es Clairaut und Maclaurin bestimmt hatten, die einzig mögliche Figur des Gleichgewichts sey, und zwar nicht in irgend einer Annäherung, sondern in absoluter, geometrischer Strenge. Wenn man bedenkt, dass man hier aus Relationen zwischen dreifachen Integralen, deren Gränzen unbekannt sind und welche Constanten enthalten, zwischen denen eine unbekannte Relation statt findet, die Gleichung zwischen den drei Variabeln zu suchen hat, welche die Gränzen giebt und zugleich die unbekannte Relation zwischen den Constanten bestimmt, so staunt man über die Kühnheit und das Glück dieses Unternehmens. Es ist zu bedauern, dass der Autor der *Mécanique céleste* es nicht für zweckmässig fand, das merkwürdige Theorem in sein weitschichtiges Werk aufzunehmen.

Perhaps Jacobi was rendered partial towards Legendre by their common interest in the theory of elliptic functions, and by the kindness with which the veteran mathematician had received and appreciated the efforts of the rising genius: see the *Annales Scientifiques de l'Ecole Normale Supérieure*, Vol. VI. 1869. I have been swayed by Jacobi's opinion in endeavouring to render the essence of Legendre's investigation accessible to students.

CHAPTER XXIII.

LAPLACE'S FOURTH, FIFTH, AND SIXTH MEMOIRS.

848. LAPLACE'S fourth memoir on our subject is contained in the Paris *Mémoires* for 1782, published in 1785; it is entitled *Théorie des attractions des sphéroïdes et de la Figure des Planètes.* The memoir occupies pages 113...196 of the volume.

849. The memoir is divided into five sections; the last of these relates to the oscillations of a fluid of small depth surrounding a sphere; this belongs to the theory of the tides which we do not discuss in the present work. Thus we are concerned only with the other four sections.

Speaking generally, we may say that this memoir is reproduced in the *Mécanique Céleste*; most of it is verbally reprinted. We shall therefore confine ourselves to a brief account of it, reserving more detail for the analysis we shall give of the *Mécanique Céleste.*

850. The first section treats of the attraction of ellipsoids. With respect to this section Laplace says in his page 113:

> ...Je donne une théorie complète des attractions des sphéroïdes terminés par des surfaces du second ordre; cette théorie a déjà paru dans l'Ouvrage que j'ai publié sur le mouvement et sur la figure elliptique des Planètes; mais elle est ici présentée d'une manière plus directe et plus simple.

Laplace here has in view principally the demonstration of the theorem which I have called by his name; this demonstration was first published in the fourth section of his Treatise, but is given in the present memoir in a simpler form: see Art. 804.

Laplace rests his demonstration now on *one* partial differential equation instead of *three*, which he used in the Treatise.

The present section forms Chapter I. of the third Book of the *Mécanique Céleste*. I observe only two changes.

In the memoir Laplace merely states that his element of volume is a rectangular parallelopiped, of which the three dimensions are dr, rdp, and $r\sin p\,dq$: in the *Mécanique Céleste* he unnecessarily goes through the process of transforming the rectangular expression $dx\,dy\,dz$ into the above polar form.

The partial differential equation which occurs in the *Mécanique Céleste*, Livre III. § 5, towards the beginning of the section, is rather simpler than the corresponding form in the memoir; but the two are practically equivalent: the form at the end of this section of the *Mécanique Céleste* is identical with that which corresponds to it in the memoir.

851. The second section of the memoir treats on the development in a series of the attractions of any spheroids: this section is reproduced in the second Chapter of the third Book of the *Mécanique Céleste*.

In this section we have for the first time the partial differential equation with respect to the coordinates of the attracted particle which the potential V must satisfy: it is expressed by means of polar coordinates in the form

$$\frac{d}{d\mu}\left\{(1-\mu^2)\frac{dV}{d\mu}\right\}+\frac{1}{1-\mu^2}\frac{d^2V}{d\varpi^2}+r\frac{d^2Vr}{dr^2}=0.$$

In the memoir we are merely told that it is easy to convince oneself by differentiation that this equation holds. In the *Mécanique Céleste* we are also told that this equation is a transformation of the equation in rectangular coordinates

$$\frac{d^2V}{dx^2}+\frac{d^2V}{dy^2}+\frac{d^2V}{dz^2}=0.$$

And in the *Mécanique Céleste*, Livre II. § 11, there is a sketch of the process of transformation.

It is curious that the equation should first occur in the polar form, which is much less obvious and simple than the rectangular form.

On pages 138...143 of the memoir we have some investigations as to what we call Laplace's coefficients. A general expression is given for these coefficients in terms of two variables; but it involves an error: for Laplace assumes, on his page 141, that his $i+n$ is always an even number. The mistake was corrected by Legendre in his memoir of 1789, page 432. Laplace gives a correct form in the *Mécanique Céleste*, Livre III. § 15.

852. The third section of the memoir treats on the attraction of spheroids, which differ but little from spheres: this section is also reproduced in the second Chapter of the third Book of the *Mécanique Céleste*.

This section begins with a demonstration of the equation to which we have already drawn attention: see Art. 814. Here Laplace restricts himself to the ordinary law of attraction, and puts his equation exactly in the form of equation (2) of the *Mécanique Céleste*, Livre III. § 10, namely

$$-a\frac{dV}{dr}=\frac{2\pi a^2}{3}+\frac{1}{2}V:$$

the demonstration however here given is different, and we will reproduce it.

Let ρ denote the density; let r, θ, ϕ be the usual polar coordinates of a fixed point; and let r', θ', ϕ' be variable coordinates. Then the value of V at the point (r, θ, ϕ) is given by

$$V=\iiint\frac{\rho r'^2\sin\theta'\,dr'\,d\theta'\,d\phi'}{(r^2-2rr'\mu+r'^2)^{\frac{1}{2}}},$$

where μ stands for $\cos\theta\cos\theta'+\sin\theta\sin\theta'\cos(\phi-\phi')$.

Suppose that at the surface of the spheroid we have $r'=a(1+\alpha y')$, where α is very small, and y' is a function of θ' and ϕ'; and let y be the value of y' when for θ' and ϕ' we put θ and ϕ respectively. Then we may suppose the spheroid to consist of a sphere of the radius $a(1+\alpha y)$, and of an additional shell of which the variable thickness is $a\alpha(y'-y)$.

The potential of the sphere is easily found to be

$$\frac{4\pi a^3 (1+\alpha y)^3 \rho}{3r}.$$

The potential for the additional shell may be represented by an expression similar to the above for V; we may put $a\alpha(y'-y)$ instead of dr'.

Let V_1 denote this part of the potential, so that

$$V_1 = a\alpha \iint \frac{\rho r'^2 \sin\theta' \,(y'-y)\, d\theta'\, d\phi'}{(r^2 - 2rr'\mu + r'^2)^{\frac{1}{2}}}.$$

Hence, by differentiating with respect to r, we have

$$-\frac{dV_1}{dr} = a\alpha \iint \frac{\rho r'^2 (r - r'\mu) \sin\theta' \,(y'-y)\, d\theta'\, d\phi'}{(r^2 - 2rr'\mu + r'^2)^{\frac{3}{2}}}.$$

Now suppose the fixed point to be on the surface so that $r = a(1+\alpha y)$; then neglecting α^2 we get

$$-\frac{dV_1}{dr} = \frac{\alpha a}{2^{\frac{3}{2}}} \iint \frac{\rho \sin\theta' \,(y'-y)\, d\theta'\, d\phi'}{(1-\mu)^{\frac{1}{2}}}.$$

And to the same order of approximation we have

$$V_1 = \frac{a^2\alpha}{2^{\frac{1}{2}}} \iint \frac{\rho \sin\theta' \,(y'-y)\, d\theta'\, d\phi'}{(1-\mu)^{\frac{1}{2}}}.$$

Thus $$-a\frac{dV_1}{dr} = \frac{1}{2} V_1.$$

Then, denoting the whole potential by V, we have

$$V = \frac{4\pi a^3 (1+\alpha y)^3}{3r} + V_1$$

$$-\frac{dV}{dr} = \frac{4\pi a^3 (1+\alpha y)^3}{3r^2} - \frac{dV_1}{dr}.$$

Thus to our order of approximation

$$V = \frac{4\pi a^2}{3}(1 + 2\alpha y) + V_1,$$

$$-\frac{dV}{dr} = \frac{4\pi a}{3}(1 + \alpha y) - \frac{dV_1}{dr};$$

therefore

$$-a\frac{dV}{dr} = \frac{2\pi a^2}{3} + \frac{1}{2}V.$$

This method of investigating the equation coincides substantially with that recommended by D'Alembert: see Art. 652.

653. We may notice another process in the memoir which is not repeated in the *Mécanique Céleste*.

Suppose y to be a rational function of μ, $\sqrt{(1-\mu^2)}\cos\varpi$, and $\sqrt{(1-\mu^2)}\sin\varpi$; and it is required to transform y into a series of Laplace's Functions.

Suppose that y is of the ith degree; and assume

$$y = Y_0 + Y_1 + Y_2 + Y_3 + \ldots + Y_i \ldots\ldots\ldots\ldots\ldots(1),$$

where Y_r is a Laplace's function of the order r, so that

$$\frac{d}{d\mu}\left\{(1-\mu^2)\frac{dY_r}{d\mu}\right\} + \frac{1}{1-\mu^2}\frac{d^2Y_r}{d\varpi^2} + r(r+1)\,Y_r = 0.$$

Put y_1 for $\dfrac{d}{d\mu}\left\{(1-\mu^2)\dfrac{dy}{d\mu}\right\} + \dfrac{1}{1-\mu^2}\dfrac{d^2y}{d\varpi^2}$; then we see that

$$-y_1 = 1.2\,Y_1 + 2.3\,Y_2 + 3.4\,Y_3 + \ldots + i(i+1)\,Y_i \ldots\ldots(2).$$

In like manner let y_2 be derived from y_1 as y_1 was from y; and then y_3 in like manner from y_2; and so on. Thus we obtain equations of which the general type is

$$(-1)^r y_r = (1.2)^r\,Y_1 + (2.3)^r\,Y_2 + \ldots + \{i(i+1)\}^r\,Y_i.$$

The equations (1), and (2), and the other $i-1$ equations of the type just expressed, serve to determine $Y_0, Y_1, Y_2 \ldots Y_i$.

Another process is given instead of this in the *Mécanique Céleste*, Livre III. § 16.

834. The fourth section of the memoir treats on the Figure of the Planets; this section is reproduced in the fourth Chapter of the third Book of the *Mécanique Céleste*.

Suppose that X, Y, Z denote the accelerating forces parallel to the axes at the point (x, y, z) of a fluid in equilibrium. Let p denote the pressure, and ρ the density. Then

$$\frac{dp}{\rho} = Xdx + Ydy + Zdz.$$

Now Laplace proposes to consider that part of the right-hand member which arises from the action of a distant body.

Let S denote the mass of this distant body, s its distance from (x, y, z); then at first sight $-\frac{Sds}{s^2}$ might appear to be the term required. But Laplace makes the hypothesis that the centre of gravity of the fluid mass is at rest; and thus he wants not the *action* of the distant body, but what may be called the *disturbing* action. Hence we have to apply in the reversed direction the action of the distant body on the centre of gravity. Laplace does this in three ways in three different places. In the *Théorie... de la Figure des Planètes* he makes an approximate investigation which is correct though a little tedious: see page 108 of the work. This method is the same as D'Alembert used: see Art. 616. In the present memoir Laplace proceeds without approximation; but his method is wrong: see page 158 of the memoir. In the *Mécanique Céleste* he uses a brief and correct method without approximation; see Livre III. § 23.

Let σ denote the distance of S from the centre of gravity of the fluid; and let α, β, γ be the angles which σ makes with the axes. Then we have to subtract $\frac{S\cos\alpha}{\sigma^2}$, $\frac{S\cos\beta}{\sigma^2}$, and $\frac{S\cos\gamma}{\sigma^2}$ from the forces at (x, y, z) parallel to the axes of x, y, z respectively. Thus instead of $-\frac{Sds}{s^2}$ we now have

$$-\frac{Sds}{s^2} - \frac{S}{\sigma^2}(\cos\alpha dx + \cos\beta dy + \cos\gamma dz).$$

Hence the part of $\int(Xdx + Ydy + Zdz)$ which arises from the action of this distant body is

$$\frac{S}{s} - \frac{S}{\sigma^2}(x \cos \alpha + y \cos \beta + z \cos \gamma) + \text{constant}.$$

The error in the memoir is this: instead of subtracting $\frac{S \cos \alpha}{\sigma^2}$, Laplace subtracts $\frac{S(\sigma \cos \alpha - x)}{\sigma^2}$; and similarly for the other axes.

855. The memoir treats on the figure of a planet supposed homogeneous on pages 154...170. The theory here given is reproduced almost word for word in the *Mécanique Céleste*, Livre III. §§ 22...28. And with the exception of the correction noticed in the preceding Article the *Mécanique Céleste* adds nothing to the memoir.

On pages 179...186 of the memoir Laplace treats of the case in which the planet is not supposed homogeneous: but the memoir really gives very little on this head. In fact the memoir contains observations only on the value of gravity and of the length of a degree at different parts of the Earth's surface: these observations occur in the *Mécanique Céleste*, Livre III. § 33; but the application there made to Bouguer's hypothesis does not occur in the memoir.

856. On the whole we see that the theory of attraction and the theory of the homogeneous figure of the Earth are given in this memoir substantially as they were afterwards reproduced in the *Mécanique Céleste*.

857. The important property with respect to two Laplace's functions of different orders that

$$\int_{-1}^{1}\int_{0}^{2\pi} Z_n Z_m \, d\mu \, d\phi = 0$$

is first given in this memoir. The case in which $m = n$ is not explicitly considered; this was investigated by Legendre in his memoir of 1789: see Heine's *Handbuch der Kugelfunctionen*, page 265. But it ought to be noticed that the remarkable equation (1) of the *Mécanique Céleste*, Vol. II. page 44, which involves all that applies to the case in which $m = n$, is implicitly contained in the memoir of 1782; see page 152 of the memoir: but the equation is not explicitly brought into notice.

858. The memoir is a very valuable contribution to our subject. We may especially observe that here for the first time it is demonstrated, without assuming a figure of revolution, that the oblatum is the only form of relative equilibrium for a nearly spherical mass of rotating homogeneous fluid. To this matter we shall return hereafter.

859. Laplace's fifth memoir on our subject is contained in the Paris *Mémoires* for 1783, published in 1786; it is entitled *Mémoire sur la Figure de la Terre.* The memoir occupies pages 17...46 of the volume.

860. Laplace first considers some measures of lengths of degrees; see pages 18...23 of the memoir. He uses four, namely those in Peru, at the Cape of Good Hope, in France, and in Lapland: he quotes the lengths from Frisi's *Cosmographia.* In order to deduce from these measures the elements of the Earth's dimensions Laplace uses the method which is explained in the *Mécanique Céleste,* Livre III. § 39. Laplace obtains for the ratio of the axes of the oblatum that of 250 to 249; but he considers that the measured lengths do not agree very well with the elliptic figure.

Laplace finds that the observations of the lengths of the seconds pendulum agree reasonably well with theory; he uses a table which is given in Frisi's *Cosmographia,* Vol. II. page 139, and by the aid of Clairaut's theorem he deduces as the ratio of the axes of the oblatum that of 321 to 320.

These discussions as to the lengths of degrees and of the seconds pendulum appear in a much more elaborate form in the *Mécanique Céleste,* Livre III. §§ 39, 41, and 42. The method of treating discordant observations which is explained in § 39 does not appear to have found much favour.

861. Assume that the radius vector of the Earth is an expression of the form

$$1 + \alpha (Y_0 + Y_1 + Y_2 + \dots),$$

where a is small, and Y_r is a Laplace's function of the rth order. Laplace shews that if the centre of gravity is the origin Y_1 is zero; see pages 25...27 of the memoir. The process is reproduced substantially in the *Mécanique Céleste*, Livre III. § 31. We must observe that the density of the Earth is not assumed to be constant; the proposition had been already given in the fourth memoir assuming the density to be constant.

Laplace also investigates what consequences follow as to the form of Y_2 if we assume that the axis of rotation is a principal axis of the mass; see pages 28...30 of the memoir. The process is reproduced substantially in the *Mécanique Céleste*, Livre III. § 32.

862. Laplace in his pages 30...34 makes some remarks on the value of gravity and of the length of a degree of the meridian at different places on the Earth's surface; the remarks coincide in effect with those in the fourth memoir: see Art. 855. A numerical example is given in illustration; this we will reproduce: the formulæ which we shall use will be found in the *Mécanique Céleste*, Livre III. § 33.

The point to be illustrated is, that there may be deviations from the figure of an oblatum which will be sensible in the measures of the lengths of degrees of the meridian, though hardly sensible in the observations of the lengths of pendulums.

Suppose that the radius of the Earth is

$$1 + aY_2 + aY_6,$$

where a is very small, Y_6 is a Laplace's function of the sixth order, and $aY_2 = h\left(\mu^2 - \frac{1}{3}\right)$.

Then the formula for the length of a seconds pendulum is

$$L\left\{1 + aY_2 + j\left(\mu^2 - \frac{1}{3}\right) + 5a\,Y_6\right\};$$

that is

$$L\left\{1 + (j + h)\left(\mu^2 - \frac{1}{3}\right) + 5a\,Y_6\right\},$$

where $j = \frac{1}{289}$.

And from observation it follows that the value of h is such that $j + h = -\frac{7}{4} h$ nearly.

Let λ denote the ratio of αY_4 to αY_2. Then the corresponding ratio which occurs in the length of a seconds pendulum, namely the ratio of $5\alpha Y_4$ to $(j+h)\left(\mu^2 - \frac{1}{3}\right)$, is $-\frac{20}{7}\lambda$.

Again, the expression for the length of a degree of the meridian is

$$c\left\{1 - 5\alpha Y_2 - 41\alpha Y_4 + 2\alpha\mu \frac{d}{d\mu}(Y_2 + Y_4) - \frac{\alpha}{1-\mu^2} \frac{d^2}{d\varpi^2}(Y_2 + Y_4)\right\}.$$

With the above value of Y_2 this becomes

$$c\left\{1 + \frac{2h}{3} - 3h\left(\mu^2 - \frac{1}{3}\right) - 41\alpha Y_4 + \alpha\mu \frac{dY_4}{d\mu} - \frac{\alpha}{1-\mu^2} \frac{d^2 Y_4}{d\varpi^2}\right\}.$$

The ratio of $-41\alpha Y_4$ to $-3h\left(\mu^2 - \frac{1}{3}\right)$ is equal to $\frac{41\lambda}{3}$. This is numerically nearly five times $-\frac{20}{7}\lambda$.

Hence we may say that the disturbing effect of the term αY_4 is about five times as great on the length of a degree as it is on the length of the seconds pendulum.

803. The memoir after a few remarks on parallax proceeds on page 35 to the subject of precession and nutation, with which we are not concerned.

We may observe however that in his pages 38 and 39, Laplace investigates the form of a homogeneous solid which has every axis through the centre of gravity a *principal* axis. He comes to the conclusion that the fifth power of the radius vector measured from the centre of gravity, must be equal to a series of Laplace's functions in which the function of the *second* order does not appear. Laplace however ought to have excluded also the term of the *first* order. The oversight was corrected by Legendre in his fourth memoir, page 442.

On the whole the present memoir cannot be considered of very great importance. The new matter which it furnishes consists of the results noticed in Art. 861, and the method of treating discordant observations to which we alluded in Art. 860.

864. Laplace's sixth memoir on our subject is contained in the Paris *Mémoires* for 1787, published in 1789; it is entitled *Mémoire sur la Théorie de l'Anneau de Saturne*. The memoir occupies pages 249...267 of the volume; it consists of eight sections. The corresponding part of the *Mécanique Céleste* is the sixth Chapter of the third Book.

865. The memoir begins by stating some facts relative to the ring. Huygens is named as the person who first explained the appearances; and Cassini as the person who observed that the ring is divided into two nearly equal parts by a dark band. Then Short with a powerful telescope perceived several concentric bands. Laplace proceeds thus:

Ces observations ne permettent pas de douter que l'anneau de Saturne ne soit formé de plusieurs anneaux situés à peu-près dans le même plan; elles donnent lieu de croire que de plus forts télescopes y feront apercevoir un plus grand nombre d'anneaux.

La théorie de la pesanteur universelle qui s'accorde si bien avec les phénomènes que présentent les mouvemens et les figures des corps célestes, doit également satisfaire à ceux que nous offre l'anneau de Saturne; mais jusqu'ici personne n'a entrepris de déterminer sa figure d'après cette théorie; car l'explication que M. de Maupertuis a donnée de la formation des anneaux, dans son discours sur la figure des astres, n'étant pas fondée sur la loi de la gravitation mutuelle de toutes les parties de la matière, mais sur la supposition d'une tendance des molécules des anneaux vers plusieurs centres d'attraction; elle ne doit être regardée que comme une hypothèse ingénieuse, propre tout au plus à faire entrevoir la possibilité des anneaux dans le cas de la nature. En appliquant à cet objet, les recherches que j'ai données dans nos Mémoires de 1782, sur les attractions des sphéroïdes et sur la figure des planètes; je suis parvenu aux résultats suivans que je ne présente que comme un essai d'une théorie de l'anneau de Saturne, qui pourra être perfectionnée, lorsque de nouvelles observations faites avec de grands télescopes, auront fait connoître le nombre et les dimensions des anneaux dont il paroît formé.

Laplace then states the hypothesis he makes, namely that a film of fluid spread over the surface of the ring remains in equilibrium in virtue of the forces which act on it; and he gives a reason for the hypothesis, as in the *Mécanique Céleste*.

866. Laplace's second section is devoted to the function V, which we call the potential.

At a point (x, y, z) external to the attracting mass, V satisfies the equation

$$\frac{d^2V}{dx^2}+\frac{d^2V}{dy^2}+\frac{d^2V}{dz^2}=0 \qquad (1).$$

This is the first appearance of the equation in rectangular coordinates: see Art. 851. Laplace says:

Cette équation rapportée à d'autres coordonnées, est la base de la théorie que j'ai présentée dans mes Mémoires de 1782, sur les attractions des sphéroïdes et sur la figure des planètes.

For the case of a solid of revolution the equation (1) may be transformed into

$$\frac{1}{r}\frac{dV}{dr}+\frac{d^2V}{dr^2}+\frac{d^2V}{dz^2}=0 \qquad (2),$$

where $r^2=x^2+y^2$.

For the case of a sphere it may be transformed into

$$\frac{2}{r}\frac{dV}{dr}+\frac{d^2V}{dr^2}=0,$$

where $r^2=x^2+y^2+z^2$.

Laplace applies the last equation to determine the value of V for the case of a sphere. He uses it both for an external and internal particle; but his process is unsound with respect to the internal particle, as we now know that equation (1) is not true in that case: we shall return to this point. The correct process was given by Poisson in the *Connaissance des Tems* for 1829, page 362, and is now in elementary books. See *Statics*, Art. 240.

867. Laplace's third section contains an interesting process.

Let p_1 be the attraction of the ring at a point of the inner circumference, r_1 the distance of this point from the centre of

Saturn, S the mass of Saturn, ω the centrifugal force at a unit of distance, arising from the rotation of the ring. Then

$$p_1 \text{ must be greater than } \frac{S}{r_1^2} - \omega r_1.$$

In like manner if p_2 and r_2 refer to a point on the outer circumference of the ring,

$$p_2 \text{ must be greater than } \omega r_2 - \frac{S}{r_2^2}.$$

$$\text{Hence } p_1 + \frac{r_1}{r_2}\, p_2 \text{ must be greater than } \frac{S(r_2^3 - r_1^3)}{r_1^2 r_2^3}.$$

Let q be the attraction at the surface of Saturn, R the radius of Saturn. Then $q = \dfrac{S}{R^2}$.

$$\text{Hence } p_1 + \frac{r_1}{r_2}\, p_2 \text{ must be greater than } q\,\frac{R^2(r_2^3 - r_1^3)}{r_1^2 r_2^3}.$$

Now Laplace says that the mass of the ring is much less than that of Saturn, and a sphere must exert a greater attraction on a particle at its surface than a very flat body of the same mass. On both these accounts q must be much greater than p_1 or p_2. Hence it follows that $\dfrac{R^2(r_2^3 - r_1^3)}{r_1^2 r_2^3}$ must be a very small coefficient; and hence r_1 and r_2 must differ but slightly. But this would not be the case with Saturn if it formed a continuous ring; for observation shews that $r_1 = \dfrac{5}{3} R$ and $r_2 = \dfrac{7}{3} R$, and thus $\dfrac{R^2(r_2^3 - r_1^3)}{r_1^2 r_2^3} = \cdot 228805$. This is far too great to be admitted. Hence even if observation had not made known the division of Saturn's ring into several concentric rings theory would have been sufficient to convince us of it.

This investigation is not reproduced in the *Mécanique Céleste*. Plana doubted the validity of the inference; he published a paper on the subject in De Zach's *Correspondance Astronomique*, Vol. I. 1818, pages 346...350, in which he states his results: and he gives his process in detail in the Turin *Memorie*, Vol. XXIV. 1820. We will return presently to Plana's criticisms.

868. Laplace's fourth section consists of an approximate process to shew that the ring *may* be of the form obtained by revolving an ellipse about a straight line in its plane, outside it, and parallel to the minor axis; the minor axis being supposed very small compared with the major axis.

Take the centre of Saturn as the origin. Then the equation of relative equilibrium of the supposed fluid film will be

$$\text{constant} = \frac{1}{2}\omega r^2 + V + \frac{S}{\sqrt{(r^2+z^2)}} \quad \text{......................(3)},$$

where r and z are taken as in (2) of Art. 866.

Then since V is symmetrical with respect to the plane from which z is measured we shall have, if we expand V in powers of z, approximately

$$V = A + Bz^2;$$

where A and B are functions of r. That is there will be no term in V involving the first power of z, or indeed odd powers of z; and terms in z^4 and higher powers are rejected.

With this value of V we have from (2), by considering the terms independent of z,

$$B = -\frac{1}{2r}\frac{d}{dr}\left(\frac{dA}{dr}r\right).$$

And
$$\frac{1}{\sqrt{(r^2+z^2)}} = \frac{1}{r} - \frac{z^2}{2r^3} \text{ nearly},$$

so that (3) becomes

$$\text{constant} = A + \frac{1}{2}\omega r^2 + \frac{S}{r} - \frac{z^2}{2r}\left\{\frac{S}{r^2} + \frac{d}{dr}\left(\frac{dA}{dr}r\right)\right\} \quad \text{..........(4)}.$$

Next Laplace supposes $r = l - u$ where l is constant and u is small; so that l is a mean value of r.

Let Q be what A becomes when l is put for r.

Then
$$A = Q - u\frac{dQ}{dl} + \frac{u^2}{2}\frac{d^2Q}{dl^2} \text{ nearly},$$

$$\frac{S}{r} = \frac{S}{l} + \frac{Su}{l^2} + \frac{Su^2}{l^3} \text{ nearly},$$

so that (4) becomes

$$\text{constant} = Q + \frac{1}{2}\omega l^2 + \frac{S}{l} - u\left(\omega l - \frac{S}{l^2} + \frac{dQ}{dl}\right)$$

$$+ \frac{1}{2}u^2\left(\omega + \frac{2S}{l^3} + \frac{d^2Q}{dl^2}\right) - \frac{z^2}{2l}\left\{\frac{S}{l^2} + \frac{d}{dl}\left(\frac{dQ}{dl}\,l\right)\right\} \text{..............(5).}$$

Let l be found from $\omega l - \frac{S}{l^2} + \frac{dQ}{dl} = 0$,

and put c for $$\frac{1}{2}\omega + \frac{S}{l^3} + \frac{1}{2}\frac{d^2Q}{dl^2}:$$

thus (5) becomes

$$cu^2 + (\omega - c)z^2 = \text{constant.}$$

This gives an ellipse as the generating figure. The approximations are rather rude; and the process is not reproduced in the *Mécanique Céleste:* see Livre III. § 44.

869. In his fifth section Laplace finds approximately the attraction which the ring would exert at any point of its surface. He treats the ring in fact as if it were an ellipsoid having a principal section coincident with the section of the ring, and the axis at right angles to this section infinite. Then he can find the attraction by formulæ given in his fourth memoir. The result is the same as we have in the *Mécanique Céleste*, Livre III. at the end of § 44. Then Laplace finds the equation to the generating ellipse of the ring in the same form as in the *Mécanique Céleste*, Vol. II. page 161.

870. In his sixth section Laplace discusses the result obtained in the fifth section; and here we have the matter which substantially is reproduced in the *Mécanique Céleste*, Vol. II. page 162.

The memoir also contains a numerical illustration which is not reproduced in the *Mécanique Céleste;* it may be of interest to give it here.

Laplace supposes that the breadth of the interior part of the ring is $\frac{R}{4}$; this he says may be admitted without violence to the observations. Moreover he takes the inner radius of this ring to be $\frac{5}{3}R$ as in Art. 867. Thus the distance from the centre of Saturn to the middle of the ring, will be $\frac{5}{3}R+\frac{1}{8}R$, that is $\frac{43}{24}R$. Denote this by a.

The semiaxis major of the generating ellipse is thus $\frac{1}{8}R$; suppose the semiaxis minor to be $\frac{1}{10}$ of this, that is $\frac{1}{80}R$.

Now Laplace shews in the Memoir, and in the *Mécanique Céleste*, that

$$\frac{S}{4\pi a^3}=\frac{\lambda(\lambda-1)}{(\lambda+1)(3\lambda^2+1)}\text{..................(6)},$$

where λ is the ratio of the major to the minor axis of the generating ellipse; the density of the ring being taken as unity.

Let ρ be the mean density of Saturn; then (6) may be written

$$\frac{\rho R^3}{3a^3}=\frac{\lambda(\lambda-1)}{(\lambda+1)(3\lambda^2+1)}\text{..................(7)}.$$

Put 10 for λ, and $\frac{43}{24}$ for $\frac{a}{R}$, then (7) gives

$$\rho=\frac{3\times 90}{11\times 301}\left(\frac{43}{24}\right)^3,$$

and therefore

$$\frac{1}{\rho}=\frac{11\times 301}{3\times 90}\left(\frac{24}{43}\right)^3.$$

This is then the ratio of the density of the ring to the mean density of Saturn; the value will be found to be 2·13.

Then the volume of ring will be $2\pi a\times\pi\times\frac{R}{8}\times\frac{R}{80}$; that is $\frac{43\pi^2R^3}{12\times 8\times 80}$. Hence the ratio of the volume of the ring to the volume of Saturn is $\frac{43\pi}{10240}$. The ratio of the mass of the ring to the mass of Saturn is therefore $\frac{1}{\rho}\times\frac{43\pi}{10240}$; which is about $\frac{1}{36}$.

These results present according to Laplace nothing that is impossible.

I have introduced the example mainly on account of an application which I want to make of it to Art. 867. In the investigation which led to equation (6), Laplace treated the ring as an infinite cylinder in estimating the attraction at a point of its surface. In this way, for the attraction at the end of the major axis of the generating ellipse, he obtains $\frac{4\pi}{\lambda+1}\times$ semiaxis; thus in our example this becomes $\frac{4\pi}{11}\times\frac{R}{8}$.

This then is the p_1 or p_2 of Art. 867; for to this order of approximation they are equal. And $q=\frac{S}{R^2}=\frac{4\pi\rho R}{3}$. Hence we have

$$\frac{p_1}{q}=\frac{3}{88}\times\frac{1}{\rho}.$$

This is about $\frac{1}{14}$, with the value of $\frac{1}{\rho}$ found above.

This in fact so far agrees with what Laplace had stated, that it makes $\frac{p_1}{q}$ *small*; but it can hardly be said to make q *much greater* than p_1. However in taking the thickness of the ring to be $\frac{1}{80}$ of the diameter of Saturn, the thickness is probably exaggerated.

871. We will return to the remarks made by Plana, to which we adverted in Art. 867. Plana treats the ring as the difference of two circular cylinders of slender height; so that instead of a flat ellipse he supposes a narrow rectangle to generate the ring by revolution. He supposes the thickness of the ring to be $\frac{1}{18}$ of the diameter of Saturn.

Then according to his numerical calculations, using the notation of Art. 867,

$$p_1=\cdot 39722,$$
$$p_2=\cdot 397846.$$

. He takes his unit of distance such that $r_1 = 1$; then $r_2 = \frac{7}{5}$, and $R = \frac{3}{5}$.

Hence according to Laplace, if the density of the rings be supposed the same as the density of Saturn, we ought to have

$$p_1 + \frac{5}{7} p_2 \text{ greater than } \frac{4\pi}{3}\left(\frac{3}{5}\right)^3 \left\{1 - \left(\frac{5}{7}\right)^3\right\}.$$

The values of p_1 and p_2 make $p_1 + \frac{5}{7} p_2 = \cdot 681395$.

But $\frac{4\pi}{3}\left(\frac{3}{5}\right)^3 \left\{1 - \left(\frac{5}{7}\right)^3\right\} = \cdot 57505$; which is indeed less than $p_1 + \frac{5}{7} p_2$; but not much less.

It is possible that Plana's numerical values of p_1 and p_2 are not quite accurate; I think they are not: but still the result may be of the nature he indicates.

Plana adds that the thickness which he has ascribed to the ring is in truth greater than can be admitted. If it is diminished then p_1 and p_2 are diminished, and Laplace's inequality ceases to hold; and in order to restore it we must suppose the density of the ring greater than that of the Planet. Plana concludes thus on page 420 of the Turin *Memorie*, Vol. XXIV.:

Mais il ne me paraît pas que l'on puisse tirer de-là la division de l'anneau en plusieurs anneaux concentriques, d'après un raisonnement semblable à celui que M. Laplace a exposé à la page 250 de son Mémoire sur la figure de l'anneau de Saturne, imprimé dans les volumes de l'Académie des Sciences de Paris (année 1787).

We observe that according to Plana's figures we find that $\frac{\rho}{q} = \frac{\cdot 397}{2 \cdot 5132}$ supposing the ring and Saturn of the *same density*: this is between $\frac{1}{6}$ and $\frac{1}{7}$.

But I do not see the force of Plana's remark, that to restore the inequality we *must* increase the density of the ring. It might be said that to restore the inequality we should diminish the difference between r_1 and r_2. By Art. 867, we require that $p_1 + \omega r_1 - \frac{S}{r_1^2}$ should be positive, and also $p_2 + \frac{S}{r_2^2} - \omega r_2$, where r_2 is greater than r_1. Then it is obvious that if p_1 and p_2 are small, r_1 and r_2 cannot differ much.

It appears in the course of Laplace's investigations that $\omega a = \frac{S}{a^2}$, where $a = \frac{1}{2}(r_1 + r_2)$.

872. Laplace's seventh section contains the demonstration that if the ring were circular and perfectly alike in all its parts, its equilibrium would be unstable. The demonstration is reproduced in the *Mécanique Céleste*, Livre III. § 46. It involves the following properties of Laplace's coefficients when expressed as functions of $\cos\theta$:

if n be odd $$\int_0^\pi P_n \, d\theta = 0;$$

if n be even $$\int_0^\pi P_n \, d\theta = \pi \left\{\frac{1 \,.\, 3 \ldots (n-1)}{2 \,.\, 4 \ldots n}\right\}^2.$$

The mechanical problem discussed is equivalent to that of the resultant attraction of a circular ring on an internal particle. The method which Newton uses in discussing the attraction of a spherical shell on an internal particle may be easily used; thus it will appear that the only position of equilibrium is at the centre of the ring, and then the equilibrium is unstable.

873. Laplace's eighth section consists of two paragraphs. The first relates to the mutual action of the rings; it is reproduced in the *Mécanique Céleste* forming the last paragraph of the Chapter. The second paragraph in the memoir makes some statements as to the oscillations of the rings; these are not reproduced in the part of the *Mécanique Céleste* with which we are concerned: the motion of the rings round their centres of gravity is discussed however in Livre V. Chapitre III.

CHAPTER XXIV.

LEGENDRE'S THIRD MEMOIR.

874. In the Paris *Mémoires* for 1788, published in 1791, there is a memoir by Legendre, entitled *Mémoire sur les Intégrales Doubles:* it occupies pages 454...486 of the volume. The memoir was presented on the 12th of December, 1789.

875. Legendre thus states the object of his memoir in his first paragraph:

Je me propose d'indiquer dans ce Mémoire, un moyen de transformation auquel on n'a pas fait attention jusqu'à présent, et qui paroît très-propre à faciliter l'évaluation des intégrales doubles ou multiples, lesquelles servent à déterminer les solidités des corps, leurs surfaces courbes, la position de leurs centres de gravité, &c. L'objet que j'ai particulièrement en vue, est d'intégrer par ce moyen les formules qui donnent l'attraction d'un sphéroïde elliptique quelconque sur un point extérieur; d'où résultera la démonstration directe de ce théorème déjà connu: *Si deux sphéroïdes elliptiques ont leurs trois sections principales décrites des mêmes foyers, les attractions qu'ils exercent sur un même point extérieur, auront la même direction, et seront entr'elles comme leurs masses.*

876. Legendre then proceeds in his next paragraph to speak of this problem in attractions:

Cette proposition que j'avois démontrée rigoureusement pour les sphéroïdes de révolution (*Sav. étrang. Tom.* x), et qui devenoit infiniment probable pour ceux dont toutes les coupes sont elliptiques, a l'avantage de ramener le cas des points extérieurs à celui des points situés sur la surface du sphéroïde, et de réduire ainsi à une forme très-simple la valeur absolue de l'attraction. Mais si la verité de ce théorème peut être constatée assez facilement par l'induction et par une approximation poussée très-loin, il n'est pas aussi facile de s'en procurer une démonstration rigoureuse, et je ne crains pas de dire que cette question est une des plus épineuses de l'analyse. La seule solution qui en existe, est celle que M. de la Place a donnée dans les Mém. de l'Académie, année 1783; mais la méthode de ce savant géomètre quelque ingénieuse qu'elle soit, laisse à désirer un procédé plus direct, et ne répand d'ailleurs aucune lumière sur l'intégration indéfinie.

We may observe that instead of 1783 we ought to read 1782. It is curious to see how the difficulties of one age are removed by the labours of another; the question which Legendre regarded as one of the most difficult in analysis has since been solved in a very simple manner by what we call Ivory's theorem.

877. As to the transformation of double or multiple integrals Legendre recognises the priority of Lagrange. Legendre says:

Ce principe auquel j'étois parvenu par des considérations géométriques, et que j'ai examiné ensuite avec plus de soin, ne s'est point trouvé différent d'un moyen de transformation indiqué par M. de la Grange dans les Mémoires de Berlin, an 1773, *pag.* 125. La propriété en appartient donc à cet illustre géomètre; il ne me reste que la nouvelle forme sous laquelle j'ai présenté ce principe et l'usage que j'en ai indiqué, usage auquel il paroît que M. de la Grange n'a pas pensé, ou dont au moins il n'a fourni aucun exemple.

878. Legendre's memoir consists of four sections. The first section treats of the transformation of double or multiple integrals. Legendre gives the same unsatisfactory method as Lagrange had given previously: see Article 710.

It must however be observed that this section on the transformation of double or multiple integrals is *scarcely used* in the subsequent sections. Legendre does indeed employ the polar form of the element of mass; but he does not say that this is to be obtained by transformation from the rectangular form, and we know that it can be obtained independently. See Arts. 710 and 753.

When Legendre transforms his integrals in the course of the memoir he practically only transforms single integrals by the change of one independent variable; or at least his results may be easily obtained in this way. It is hard to see much force in the concluding words of the extract we have made in Art. 877. Legendre's page 470 gives the only case of an apparent double transformation. In fact the title of this memoir by Legendre is bad: it should have been *On the attraction of ellipsoids;* for that is really the subject discussed. In the well-known *Repertorium Commentationum* by J. D. Reuss there is no reference to this memoir in the sections where the titles of memoirs on attraction are recorded.

879. Legendre's second section gives the general formulæ for the attraction of an ellipsoid at an external point; these formulæ take the shape of double integrals. Legendre uses the method of polar coordinates which had been adopted by Lagrange in 1773, and which is now in all elementary books: see *Statics*, Art. 220.

880. Legendre's third section is devoted to the particular case in which the attracted particle is in a principal plane of the ellipsoid. In this case one integration can be effected by the ordinary process, that is without adopting any novel method of cutting up the ellipsoid into elements. Legendre says on his page 463:

...Ce cas est d'autant plus intéressant à développer, qu'il avoit échappé à tous ceux qui se sont occupés de cette matière, et que la théorie de l'attraction des sphéroïdes de révolution s'y trouve comprise dans toute sa généralité.

Legendre's treatment of this particular case is sound but very laborious; he leaves much work to be effected by the reader, the results being given, but many of the intermediate operations being omitted. For instance, he states on his page 465, a result which we may state in our own notation thus:

$$\int_{-\frac{\pi}{2}}^{\frac{\pi}{2}} \frac{(M + N\sin\phi)\,d\phi}{a + b\sin\phi + c\sin^2\phi} = \pi\,\frac{(\lambda + a + c)\,M - bN}{\lambda\sqrt{\{(\lambda + a)^2 - c^2\}}},$$

where $$\lambda = \sqrt{\{(a + c)^2 - b^2\}}.$$

It is supposed here that a, b, and c are such that

$$a + b \sin \phi + c \sin^2 \phi$$

cannot vanish. The student will find that this result is correct, but the verification will be tedious.

Plana has supplied the details of Legendre's operations in a memoir published in Crelle's *Journal für...Mathematik*, Vol. XXVI. pages 132...146.

881. Legendre's fourth section is devoted to the general problem of the attraction of an ellipsoid at an external point. By a change of variables Legendre effects one integration out of the two which are involved. The process is very laborious; much is left for the student to perform for himself, the results being rather indicated than worked out.

Plana has supplied the details of Legendre's operations in a memoir published in Crelle's *Journal für...Mathematik*, Vol. XX. pages 240...270.

It would be impossible to render Legendre's method intelligible within the limits of the space we can devote to the present memoir. We may however state the nature of the decomposition which he effects of the attracting ellipsoid. A series of conical surfaces is described after a certain law, each cone having its vertex at the attracted point; the outer cone touches the ellipsoid. Then the one integration which Legendre effects, amounts to determining the attraction exerted parallel to an axis by the portion of the ellipsoid which is comprised between two indefinitely close conical surfaces out of the series. The series of cones is obtained by varying a parameter ω which is zero for the tangent cone, and has its maximum value when the cone degenerates into a straight line.

Now the remarkable fact is that Legendre succeeds in obtaining an expression *free from the integral sign* which represents the resolved attraction of one of these portions of a conical shell: and when we look at the very laborious process by which the result is obtained, we may safely pronounce it one of the most extraordinary mathematical feats ever performed.

882. A point of some interest in the Integral Calculus presents itself in the course of Legendre's first integration: see his page 477. As usual we state the matter in our own notation. A certain definite integral, which we can see is necessarily finite, becomes by transformation

$$\int_{-\infty}^{\infty} \frac{(1+x^2)\,dx}{(a+2bx+cx^2)(a_1+2b_1x+c_1x^2)},$$

where the constants are such that the denominator of the expression under the integral sign never vanishes. To effect the integration the fraction is resolved into partial fractions, say that

$$\frac{1+x^2}{(a+2bx+cx^2)(a_1+2b_1x+c_1x^2)}=\frac{l+mx}{a+2bx+cx^2}+\frac{l_1+m_1x}{a_1+2b_1x+c_1x^2}.$$

And $$\int\frac{(l+mx)\,dx}{a+2bx+cx^2}=\int\frac{l-\frac{mb}{c}+m\left(x+\frac{b}{c}\right)}{c\left(x+\frac{b}{c}\right)^2+a-\frac{b^2}{c}}\,dx.$$

Legendre then implicitly states that between the limits $-\infty$ and ∞ the integral gives

$$\frac{(lc-mb)\,\pi}{c\sqrt{ac-b^2}},$$

so that he considers

$$\int_{-\infty}^{\infty}\frac{m\left(x+\frac{b}{c}\right)dx}{c\left(x+\frac{b}{c}\right)^2+a-\frac{b^2}{c}}=0.$$

This last integral may be asserted to be zero in this sense; it consists of a positive and a negative part each of which is *infinite*, which may be taken to balance each other: but this is hardly satisfactory.

The best method is to proceed thus:

$$\int\frac{m\left(x+\frac{b}{c}\right)dx}{c\left(x+\frac{b}{c}\right)^2+a-\frac{b^2}{c}}+\int\frac{m_1\left(x+\frac{b_1}{c_1}\right)dx}{c_1\left(x+\frac{b_1}{c_1}\right)^2+a_1-\frac{b_1^2}{c_1}}$$

$$=\frac{m}{2c}\log\left\{c\left(x+\frac{b}{c}\right)^2+a-\frac{b^2}{c}\right\}+\frac{m_1}{2c_1}\log\left\{c_1\left(x+\frac{b_1}{c_1}\right)^2+a_1-\frac{b_1^2}{c_1}\right\}.$$

Suppose we take this between the limits 0 and ξ, we obtain

$$\frac{m}{c}\log\xi + \frac{m}{2c}\log\frac{c\left(1+\frac{b}{c\xi}\right)^2 + \left(a-\frac{b^2}{c}\right)\frac{1}{\xi^2}}{a}$$

$$+\frac{m_1}{c_1}\log\xi + \frac{m_1}{2c_1}\log\frac{c_1\left(1+\frac{b_1}{c_1\xi}\right)^2 + \left(a_1-\frac{b_1^2}{c_1}\right)\frac{1}{\xi^2}}{a_1}.$$

Now by the theory of the decomposition of rational fractions we see that $\frac{m}{c}+\frac{m_1}{c_1}=0$; thus the term in $\log\xi$ disappears from the above result; and when ξ is made indefinitely great we obtain simply between the limits 0 and ∞

$$\frac{m}{2c}\log\frac{c}{a}+\frac{m_1}{2c_1}\log\frac{c_1}{a_1}.$$

In like manner between the limits $-\infty$ and 0 we obtain the same numerical result with the opposite sign. Thus the entire integral is zero.

883. In the course of his investigations Legendre arrives at the following result which he justly calls a remarkable theorem:

Si on imagine plusieurs sphéroïdes semblables, dont la densité soit la même et les axes situés dans la même direction, et que ces sphéroïdes agissent sur un même point extérieur, l'attraction du plus petit sphéroïde sera équivalente à celle d'une portion de chacun des autres, retranchée par la surface conique dans l'étendue de laquelle ϖ est égal au maximum de cette quantité dans le plus petit sphéroïde.

Legendre says that this proposition can be easily verified in the case of concentric spheres. On examination it will be found that in this case the proposition coincides with the result given in Art. 251.

It may be convenient to state explicitly by means of symbols the general result which constitutes this remarkable theorem.

Let a, b, c be the semiaxes of an ellipsoid; let f, g, h be the coordinates of an external point. Let $m=\frac{a^2}{b^2}$ and $n=\frac{a^2}{c^2}$.

Put u for $x^2+my^2+nz^2-a^2$,

v for $fx+mgy+nhz-a^2$,

r^2 for $(x-f)^2+(y-g)^2+(z-h)^2$,

ζ for $f^2+mg^2+nh^2-a^2$.

Then the attraction exerted at the external point by the body bounded by the ellipsoid

$$u=0,$$

and the cone

$$v^2-\zeta u=\omega^2 r^2$$

is independent of a; that is this attraction is a function of f, g, h, m, n, and ω.

If we transfer the origin to the external point, the equations to the ellipsoid and the cone become respectively

$$x^2+my^2+nz^2+2(fx+gmy+hnz)+\zeta=0 \dots\dots\dots (1),$$

$$(fx+gmy+hnz)^2-\zeta(x^2+my^2+nz^2)=\omega^2(x^2+y^2+z^2) \dots. (2).$$

884. Legendre we see arrived at his theorem incidentally as he was developing a new demonstration of Laplace's theorem; and the improvement subsequently effected by Ivory in the treatment of Laplace's theorem has probably much diminished the interest which would otherwise have continued to belong to Legendre's. Nevertheless it is to be wished that a simple investigation could be supplied of the remarkable result; and perhaps this may be attained in consequence of thus drawing attention to it. The nature of the theorem will become more obvious if we consider the particular case in which the external point is situated on the prolongation of an axis of the ellipsoid, which can be worked out without much difficulty.

Suppose then that with the notation of the preceding Article $g=0$ and $h=0$; and let us seek the attraction of the element comprised between the ellipsoid and two cones corresponding respectively to the parameters ω and $\omega+d\omega$.

With the usual polar notation the attraction will be equal to

$$\iiint dr\,d\theta\,d\phi \sin\theta\cos\theta.$$

The limits for r will be r_1 and r_2, which denote the two distances from the external point to the ellipsoid corresponding to an assigned direction determined by θ and ϕ.

The limits for θ will be θ_1 and θ_2, which differ infinitesimally, corresponding to the change of ω into $\omega + d\omega$, while other quantities are constant. The limits for ϕ will be 0 and 2π.

Thus our expression first becomes

$$\iint (r_2 - r_1) \sin\theta \cos\theta \, d\theta \, d\phi ;$$

and then it may be written

$$- d\omega \int (r_2 - r_1) \sin\theta \cos\theta \frac{d\theta}{d\omega} \, d\phi ;$$

the negative sign is used because $\frac{d\theta}{d\omega}$ is negative.

It would only remain to transform the expression under the integral sign into a function of ω and ϕ, and to integrate with respect to ϕ from 0 to 2π.

Let t stand for $\cos^2\theta + m \sin^2\theta \cos^2\phi + n \sin^2\theta \sin^2\phi$.

Then from equation (1) of Art. 883 we find that

$$r_2 - r_1 = \frac{2\sqrt{(f^2 \cos^2\theta - \zeta t)}}{t} ;$$

and equation (2) of Art. 883 becomes

$$f^2 \cos^2\theta - \zeta t = \omega^2 \text{ ....} (3);$$

thus

$$r_2 - r_1 = \frac{2\omega}{t} .$$

Again the value of $\frac{d\theta}{d\omega}$ is to be found from (3); this gives

$$[\zeta - f^2 - \zeta (m \cos^2\phi + n \sin^2\phi)] \sin\theta \cos\theta \frac{d\theta}{d\omega} = \omega .$$

Hence the expression for the attraction becomes

$$- 2\omega^2 d\omega \int \frac{d\phi}{t\,[\zeta - f^2 - \zeta (m \cos^2\phi + n \sin^2\phi)]} .$$

The values of $\sin^2\theta$ and $\cos^2\theta$ must be found from (3), and substituted in t; then our expression becomes

$$2\omega^2 d\omega \int_0^{2\pi} \frac{d\phi}{\omega^2 + (f^2 - \omega^2)(m\cos^2\phi + n\sin^2\phi)},$$

that is
$$\frac{4\pi\omega^2 d\omega}{\sqrt{\{(\omega^2 + mf^2 - m\omega^2)(\omega^2 + nf^2 - n\omega^2)\}}}.$$

Thus the expression for the attraction of the element is definitely found, and it is independent of a.

885. The following is the conclusion at which Legendre arrives respecting the attraction of an ellipsoid at an external point. Let a, b, c be the semiaxes of an ellipsoid; let f, g, h be the co-ordinates parallel to these semiaxes respectively of an external point; and let M be the mass of the ellipsoid. Then the attraction parallel to the semiaxis a is

$$\frac{3Mf}{k}\int_0^1 \frac{x^2 dx}{\sqrt{\{k^2 + (b^2 - a^2)x^2\}}\sqrt{\{k^2 + (c^2 - a^2)x^2\}}},$$

where k denotes the greatest root of the equation

$$1 = \frac{f^2}{k^2} + \frac{g^2}{k^2 + b^2 - a^2} + \frac{h^2}{k^2 + c^2 - a^2}.$$

Thus the attraction depends only on the mass and on $b^2 - a^2$ and $c^2 - a^2$; therefore we have Laplace's theorem, namely that if there be two confocal ellipsoids the attractions which they exert at the same point external to both are in the same direction and proportional to the masses.

The expression for the attraction was first given by Laplace in his *Théorie...de la Figure des Planètes*, being deduced by him from his theorem: here Legendre has obtained the expression independently, and deduces the theorem.

886. Since the attracted point is external to the ellipsoid we have the condition that $\frac{f^2}{a^2} + \frac{g^2}{b^2} + \frac{h^2}{c^2} - 1$ is positive. Legendre's demonstration is worked out on the supposition that something more than this holds, namely that $f^2 - a^2$ is positive. Legendre

himself draws attention to this; he justifies himself by asserting that there can be but one formula which represents the attraction, and that if the formula is obtained for the assumed case in which f is greater than a this formula must be the general formula. See his page 472.

This is undoubtedly a drawback from the value of Legendre's demonstration; but so far as I can perceive it is the only drawback.

It must however be admitted that the demonstration is extremely complicated; so that in fact it seems like a stupendous feat of mathematical athletics. By means of Ivory's theorem, as it is called, the difficult integrations which Legendre encountered are avoided; so that probably little more than an historical interest would now belong to the investigations of Legendre. Moreover, as we shall see, Poisson has obtained the formula of Art. 885 by an easier route.

We will briefly notice the opinions of Legendre's method expressed by subsequent writers.

887. Ivory's theorem, as it is called, was first published in the *Philosophical Transactions* for 1809. Ivory remarks on his page 347:

Le Gendre has given a direct demonstration of the theorem of La Place, by integrating the fluxional expressions of the attractive forces; a work of no small difficulty, and which is not accomplished without complicated calculations.

Legendre himself published a memoir on the attraction of homogeneous ellipsoids in the *Mémoires de l'Institut* for 1810. Here he speaks thus respecting the last section of his memoir of 1788:

J'ai ensuite considéré le problême dans toute sa généralité, et j'ai fait voir qu'on pouvait vaincre les difficultés de l'intégration, de manière à parvenir enfin au théorème desiré. J'avoue néanmoins que cette partie de mon Mémoire n'a que le mérite d'être directe, et de montrer, dès l'abord, la possibilité de la solution, mais que d'ailleurs l'analyse en est d'une extrême complication. Il était donc à desirer qu'on découvrit une route plus facile pour parvenir au même résultat.

A memoir by Poisson on the attraction of a homogeneous ellipsoid is published in the *Mémoires...de l'Institut*, Vol. XIII. 1835: the memoir was read to the Academy on the 7th of October, 1833. Poisson says on his page 499:

Dans le Mémoire que je présente aujourd'hui à l'Académie, je me propose d'envisager la question sous un nouveau point de vue, et de considérer directement et indépendamment l'une de l'autre, les intégrations relatives aux points intérieurs et aux points extérieurs, de sorte que le double problème de calcul intégral que présente l'attraction d'un ellipsoïde homogène, puisse être résolu d'une manière complète. C'est à quoi Legendre est parvenu dans le cas particulier où le point attiré appartient au plan de l'une des sections principales de l'ellipsoïde; mais quand ce point est extérieur et situé d'une manière quelconque, les calculs deviennent inextricables dans la méthode qu'il a suivie (Mémoires de l'Académie, année 1788, page 480); et Legendre s'est borné à en déduire une démonstration nouvelle du théorème de Maclaurin, sur la réduction du cas du point extérieur à celui du point intérieur; démonstration plus directe, mais encore plus compliquée que celle que Laplace avait donnée auparavant, qu'il a reproduite dans le IIIe livre de la *Mécanique céleste*, et que Burckhardt a commentée dans sa traduction allemande de cet ouvrage.

Pontécoulant in the Supplement to the fifth Book of his *Théorie analytique du Système du Monde*, reproduces the substance of the memoir of Poisson, which has just been noticed. Having arrived at formulæ which correspond to that of Art. 865, Pontécoulant adds in a note:

Ces formules correspondent à celles qu'avait obtenues Legendre dans ses savantes recherches sur les attractions des sphéroïdes elliptiques (*Mémoires de l'Académie des Sciences*, 1788); mais ce n'est qu'à travers une série de calculs inextricables, et en altérant même les expressions primitives des attractions par des considérations qu'il justifie, il est vrai, mais qui laissent toujours quelques doutes dans les esprits, qu'il y est parvenu.

Chasles in the *Mémoires...par divers Savants*, Vol. IX. page 637, says of Legendre's investigation, "...nécessitaient d'autres calculs qui parurent inextricables." Chasles gives the following reference on his page 635:

Voir l'excellent mémoire de M. le baron Maurice, sur les travaux et les écrits de Legendre (*Bibliothèque universelle de Genève*; janvier 1833).

Poinsot appears to have entertained a very favourable view of Legendre's investigation. See the Paris *Comptes Rendus*, Vol. VI. page 660. Poinsot insists that Legendre's solution was the first direct solution, though he admits that it was long and complicated.

888. I will now make some remarks suggested by the preceding extracts.

What Poisson calls Maclaurin's theorem, I call Laplace's theorem; the propriety of my appellation is sufficiently obvious from what has been already said: see Art. 234.

I may observe that Poisson repeats his opinion of Legendre's investigation in nearly the same words in the Paris *Comptes Rendus*, Vol. VI. page 838: he says, "...mais l'analyse...était vraiment inextricable." To my satisfaction Poisson there has *théorème de Laplace* instead of *théorème de Maclaurin*.

It will be seen that Poisson, Pontécoulant and Chasles all use the word *inextricables* with respect to Legendre's investigations. Poisson says, in the first extract, that Legendre's investigations *become* inextricable; this probably means simply that Legendre could not extract from his definite integral the result which he wanted by any direct process, and so was obliged to adopt an indirect process. With Pontécoulant and Chasles the word seems used merely as equivalent to *complicated*. At all events I do not consider that any objection holds against the soundness of Legendre's process; though this word might perhaps appear to imply such a suggestion.

I do not feel quite certain as to what Pontécoulant means by saying that Legendre alters the primitive expressions; I suppose it refers to the indirect considerations which are introduced by Legendre on his page 480: but I should be at a loss to point out the precise step which appears doubtful to Pontécoulant.

889. We may observe that Poisson's own investigations of the attraction of an ellipsoid, to which we refer in Art. 887, are conducted by decomposing the ellipsoid into similar infinitesimal shells. Legendre expressed rather incautiously the opinion that

his own method of decomposition appeared the only one that was applicable; he says on page 486 of his memoir:

...il ne paroît pas, qu'il y ait d'autre moyen que de décomposer, comme nous avons fait, le sphéroïde en couches ou enveloppes coniques dans lesquelles ω est constant:..

Poisson has drawn attention to the incautious remark: see the Paris *Comptes Rendus*, Vol. VII. page 2.

Poisson, as we see in Art. 887, refers specially to page 480 of Legendre's memoir; and he repeats the reference in the Paris *Comptes Rendus*, Vol. VI. page 838, and Vol. VII. page 2.

The passage is to this effect: Legendre has arrived at an expression which denotes the attraction of one of his conical elements, and which must be integrated in order to obtain the attraction of the whole ellipsoid. Then he states that the matter looks hopeless, but nevertheless, by a particular consideration, he attains his end. Poisson seems to me to lay too much stress on the passage. Legendre's words are:

Quoique la difficulté se trouve ainsi considérablement diminuée, elle n'est cependant pas réduite au point où elle doit être pour faire sortir du résultat le théorème que nous avons en vue. Sans doute qu'une substitution ultérieure réduiroit les choses à leur dernier état de simplicité; mais cette substitution ne se présente pas naturellement, et faute de l'appercevoir, il n'y auroit presque aucune conclusion à tirer de tant de calculs. Heureusement une considération particulière sur la forme de l'expression (g'), nous dispense d'attaquer de front cette difficulté algébrique, et va nous conduire au résultat d'une manière très-simple.

When Poisson says that Legendre confines himself to giving a new demonstration of Laplace's theorem, it would be natural to reply that this was his sole object, and also a very important object.

890. An account of Legendre's investigation of the attraction of an ellipsoid at an external point, by Professor Cayley, will be found in the fourth volume of the *Cambridge and Dublin Mathematical Journal*, 1849. Legendre's investigation is said to be "one of the earliest and (notwithstanding its complexity) most elegant solutions of the problem."

CHAPTER XXV.

LEGENDRE'S FOURTH MEMOIR.

891. In the Paris *Mémoires* for 1789, published in 1793, there are two very important memoirs on our subject, one by Laplace, and one by Legendre. The memoir by Legendre occupies pages 372...454; it is entitled *Suite des Recherches sur la Figure des Planètes*. The following note is given at the foot of page 372:

On trouve dans un Mémoire de M. de la Place, imprimé à la tête de ce volume, des recherches analogues aux miennes. Sur quoi j'observe que mon Mémoire a été remis le 28 août 1790, et que la date de celui de M. de la Place est postérieure.

892. Legendre begins thus:

J'ai déja considéré le cas de l'homogénéité dans les Mémoires de l'Académie, année 1784, et j'ai fait voir *à priori*, que la figure elliptique est la seule qui convienne à l'équilibre. On savoit bien auparavant que cette figure satisfaisoit rigoureusement; mais il n'étoit point démontré que ce fût la seule, et même plusieurs Géomètres penchoient en faveur de la proposition contraire. Je crois avoir fondé ma démonstration sur une analyse rigoureuse, et dont il n'existoit aucune trace dans les auteurs qui m'ont précédé. Il est vrai qu'on trouve dans le volume de l'Académie de 1782, un très-beau Mémoire de M. de la Place, où la proposition dont je parle est démontrée, ainsi que plusieurs autres du même genre, en négligeant le quarré et les autres puissances de la force centrifuge. Mais quoique je ne sois pas cité dans cet ouvrage, j'ai déja observé dans une note, à la tête de mon Mémoire de 1784, que mon travail est le premier en date, et qu'il a donné lieu à M. de la Place de suivre ses idées sur le même objet, et de généraliser mes résultats.

It must be remembered that in the researches to which Legendre here refers, he assumed the figure to be one of re-

volution; Laplace's demonstration is free from this restriction, although it assumes that the figure differs but little from a sphere.

893. On his page 374 Legendre says:

Pour parvenir aux nouvelles formules de l'attraction, il a fallu démontrer avant tout plusieurs théorèmes très-intéressans, sur une espèce de fonctions que M. de la Place a considérées le premier dans son Mémoire imprimé en 1785, et qui sont une généralisation de celles dont j'avois détaillé les propriétés dans mon Mémoire de 1784. On verra qu'en adoptant le fondement des démonstrations de M. de la Place, j'ai traité cette matière avec plus d'étendue, et je suis parvenu à des résultats entièrement nouveaux.

The generalisation effected by Laplace consisted in treating the functions as functions of *two* independent variables; Legendre had formerly treated them as functions of *one* variable.

894. Up to page 426 of his memoir Legendre confines himself to figures of revolution. He first investigates general formulæ of attraction; and then discusses three different hypotheses as to the nature of the body.

Legendre's notation is not inviting; I shall not preserve it completely, but must retain as much as possible for the sake of comparison with him.

895. Legendre employs V to denote the sum of every element of the attracting body, divided by its distance from the attracted point; that is, V is what we now call the *Potential*.

896. Legendre says that there is a difficulty as to this subject which ought to be mentioned: see his page 370. If a particle be within the hollow part of a shell, whose surfaces are homothetical ellipsoids, it experiences no attraction. Therefore, the potential *must be zero*. But the potential cannot be zero, since it is the sum of a number of positive elements. This contradiction forms his difficulty.

There is, however, no difficulty, but only an extraordinary error involved in the words which I have put in Italics. All that is necessary is that the potential should be *constant*: it is not necessary that this constant should be zero.

897. The function P_n, which in Legendre's first memoir had presented itself only for *even* values of n, now presents itself for both even and odd values. We have given the general form of P_n in Art. 780. It may be convenient to notice the form which Legendre uses, and which is equivalent to that we have given. His expressions for the first seven functions are these, supposing the variable to be x:

$$P_1 = x,$$

$$P_2 = \frac{3}{2}x^2 - \frac{1}{2},$$

$$P_3 = \frac{5}{2}x^3 - \frac{3}{2}x,$$

$$P_4 = \frac{5.7}{2.4}x^4 - \frac{3.5}{2.4}2x^2 + \frac{1.3}{2.4},$$

$$P_5 = \frac{7.9}{2.4}x^5 - \frac{5.7}{2.4}2x^3 + \frac{3.5}{2.4}x,$$

$$P_6 = \frac{7.9.11}{2.4.6}x^6 - \frac{5.7.9}{2.4.6}3x^4 + \frac{3.5.7}{2.4.6}3x^2 - \frac{1.3.5}{2.4.6},$$

$$P_7 = \frac{9.11.13}{2.4.6}x^7 - \frac{7.9.11}{2.4.6}3x^5 + \frac{5.7.9}{2.4.6}3x^3 - \frac{3.5.7}{2.4.6}x.$$

898. On his page 378 Legendre gives the theorem which we noticed in Art. 787. It may be easily verified for the first two or three functions. He says that a general proposition which includes this will be found in the memoir; there is a blank as to the Article in the memoir to which he here refers: he means his Article 41.

899. On his page 379 Legendre gives the theorem which we have noticed in Art. 791.

900. Legendre requires the value of the potential for any point within the mass or on its surface. Let r, θ, ϕ be the polar coordinates of the point; let r', θ', ϕ' be the polar coordinates

of any element of the mass; let $\mu' = \cos\theta'$; let ρ be the density. Then the potential, which we will denote by V,

$$= \iiint \frac{\rho r'^2 \, dr' \, d\mu' \, d\phi'}{\sqrt{(r'^2 - 2rr't + r^2)}},$$

where t stands for $\cos\theta \cos\theta' + \sin\theta \sin\theta' \cos(\phi - \phi')$.

The expression $\dfrac{1}{\sqrt{(r'^2 - 2rr't + r^2)}}$ is expanded in a series; and to ensure a convergent series we must expand in ascending powers of that one of the two quantities $\dfrac{r'}{r}$ and $\dfrac{r}{r'}$, which is less than unity. In this expansion we shall denote by Y_n the coefficient, which we call Laplace's coefficient of the nth order.

901. Legendre's formulæ for V may be said to be substantially equivalent to Laplace's, as given in the fourth memoir, and reproduced in the *Mécanique Céleste*; there are two cases, namely, when the point considered is on the surface of the body, and when the point considered is within the body.

The method of obtaining these formulæ, however, is not quite satisfactory, as I have already remarked in Art. 792.

902. On his pages 382...394 Legendre discusses the first of his three hypotheses: see Art. 894. He proposes to determine the figure of a planet of which the interior is solid and composed of strata similar to the surface; the superficial stratum is supposed to be fluid. This problem was discussed by Laplace in his fourth memoir. Laplace does not assume that the strata are all similar. Laplace takes for the radius vector of any point of a stratum $a(1 + \alpha y')$, where a is the parameter of the stratum, and α is very small. If we assume that y' is independent of a we in effect suppose that all the strata are similar. Laplace does not make this assumption, which however would have but little effect on the solution of the problem.

903. Since there are no external forces supposed to act we have as the condition of relative equilibrium of the stratum of fluid

$$V + \frac{\omega^2 r^2}{2} \sin^2\theta = \text{constant},$$

where ω is the angular velocity.

Adopt for V the expression given in Art. 900; and let M denote the mass of the fluid: thus

$$\frac{M}{r} + \frac{U_1}{r^2} + \frac{U_2}{r^3} + \dots + \frac{\omega^2 r^2 \sin^2\theta}{2} = \text{constant} \dots\dots\dots\dots (1),$$

where U_n is put for $\iiint \rho r'^{n+2}\, Y'_n\, dr'\, d\mu'\, d\phi'$.

Now as the strata are supposed to be figures of revolution round the common axis, r' is independent of ϕ; thus in U_n the integration with respect to ϕ' may be effected by Art. 898. Therefore

$$U_n = 2\pi P_n \iint \rho r'^{n+2}\, P'_n\, dr'\, d\mu' \dots\dots\dots\dots\dots\dots (2),$$

where P_n has the meaning of Art. 897 with μ, that is $\cos\theta$, substituted for x; and P'_n is obtained from P_n by changing μ to μ'.

Assume $r = bu$ where b is the polar semiaxis of the body, and u is a function of θ. And similarly let $r' = \beta u'$, where u' is the same function of θ' that u is of θ, and β is a parameter which belongs to the stratum considered: hence β varies from 0 to b as we pass from the centre to the surface.

904. Let a_n stand for $\dfrac{\int_0^b \rho \beta^{n+2}\, d\beta}{\int_0^b \rho \beta^2\, d\beta}$,

and let ζ_n stand for $\dfrac{a_n \int_{-1}^{1} u'^{n+3}\, P'_n\, d\mu'}{\int_{-1}^{1} u'^3\, d\mu'}$;

then we see that

$$M = 2\pi \int_0^b \rho \beta^2\, d\beta \,.\, \int_{-1}^{1} u'^3\, d\mu'$$

and that $$\frac{U_n}{M} = \zeta_n P_n.$$

Hence dividing (1) by M we obtain

$$\frac{1}{r} + \frac{\zeta_1 P_1}{r^2} + \frac{\zeta_2 P_2}{r^3} + \ldots + \frac{\omega^2}{2M} r^2 \sin^2\theta = \text{const.} \quad \ldots\ldots\ldots (3).$$

Suppose that $u = 1 + v$, where v is so small that its square may be neglected. Hence $\frac{1}{r} = \frac{1}{b}(1 - v)$; and (3) gives

$$v = \frac{b\zeta_1 P_1}{r^2} + \frac{b\zeta_2 P_2}{r^3} + \frac{b\zeta_3 P_3}{r^4} + \ldots + \frac{b\omega^2}{2M} r^2 \sin^2\theta - \text{constant} \quad \ldots\ldots (4).$$

905. Now Legendre shews on his page 334 that if m and n are different positive integers we have

$$\int_{-1}^{1} P'_m P'_n \, d\mu' = 0 \quad \ldots\ldots\ldots\ldots\ldots\ldots\ldots\ldots (5),$$

and as a particular case of this

$$\int_{-1}^{1} P'_n \, d\mu' = 0 \quad \ldots\ldots\ldots\ldots\ldots\ldots\ldots\ldots (6).$$

Also he shews that

$$\int_{-1}^{1} (P'_n)^2 \, d\mu' = \frac{2}{2n+1} \quad \ldots\ldots\ldots\ldots\ldots\ldots\ldots\ldots (7).$$

Legendre had formerly established these results for the case in which m and n are *even* integers: see Art. 827.

906. From the value of ζ_n given in Art. 904, and the value of v furnished by (4) we may infer that ζ_n is of the first order. Hence to the order we wish to retain we may change r into b in the denominators of (4). And we also suppose the centrifugal force to be small when compared with the attraction at the equator or at the pole; hence $b\omega^2$ is small compared with $\frac{M}{b^2}$. Therefore we have approximately

$$\frac{b\omega^2 r^2 \sin^2\theta}{2M} = \frac{b^3\omega^2}{2M}(1 - \cos^2\theta) = \frac{b^3\omega^2}{2M}\left(\frac{2}{3} + \frac{1}{3} - \cos^2\theta\right)$$

$$= \frac{b^3\omega^2}{3M} - \frac{b^3\omega^2}{M} \cdot \frac{P_2}{3}.$$

The term $\frac{b^2\omega^2}{3M}$ can be connected with the constant of equation (4). We will put κ for $\frac{b^3\omega^2}{M}$; thus (4) becomes

$$v = \text{constant} + \frac{\zeta_1 P_1}{b} + \left(\frac{\zeta_2}{b^2} - \frac{\kappa}{3}\right) P_2 + \frac{\zeta_3 P_3}{b^3} + \dots \quad \dots\dots\dots (8).$$

907. We may observe that a_n and ζ_n have the same meaning with us as with Legendre. He uses n for what we call κ. He uses b as we do; but as he proceeds he supposes b equal to unity.

908. Thus by the equation (8) Legendre shews that v must be equal to a series of what we call Legendre's or Laplace's coefficients. That is, since $b(1+v)$ is the radius vector, we infer that, neglecting the squares of small quantities, the radius vector of any body which will *satisfy our problem* must be expressible in such a series. Laplace, in his treatment of the subject in his fourth memoir and in the *Mécanique Céleste*, undertakes to demonstrate that *any function* whatever can be so expressed.

We do not assert that these demonstrations by Legendre and Laplace are quite satisfactory.

909. Since $u = 1 + v$ we have in like manner $u' = 1 + v'$, where the value of v' is to be obtained from (8) by changing P_n into P'_n.

Hence we find that to our order of approximation

$$\zeta_n = \frac{(n+3)\,a_n}{2}\int_{-1}^{1} P'_n v'\,d\mu'.$$

Substitute for v'; then for any value of n except 2 we have by (5) and (7)

$$\zeta_n = \frac{(n+3)\,a_n}{2}\int_{-1}^{1} \frac{\zeta_n (P'_n)^2}{b^n}\,d\mu' = \frac{n+3}{2n+1}\,\frac{a_n}{b^n}\,\zeta_n \quad \dots\dots\dots\dots (9).$$

For $n = 2$ we have

$$\zeta_2 = a_2\left(\frac{\zeta_2}{b^2} - \frac{\kappa}{3}\right) \dots\dots\dots\dots\dots\dots\dots\dots (10).$$

From (9) we shall be able to shew that ζ_n is always zero when n is greater than 2.

For
$$\frac{a_n}{b^n} = \frac{\int_0^b \rho\beta^{n+3}\,d\beta}{b^n\int_0^b \rho\beta^3\,d\beta} = \frac{\int_0^b \rho\beta^{n+3}\,d\beta}{\int_0^b \rho b^n\beta^3\,d\beta},$$

and as β is less than b, except at the limit, the numerator is less than the denominator; so that $\frac{a_n}{b^n}$ is less than unity. And $\frac{n+3}{2n+1}$ is less than unity if n is greater than 2. Hence from (9) we must have $\zeta_n = 0$ if n is greater than 2.

Hence equation (8) reduces to

$$v = \text{constant} + \frac{\zeta_1 P_1}{b} + \left(\frac{\zeta_2}{b^2} - \frac{\kappa}{3}\right) P_2 \text{..............(11)}.$$

The term in v which involves P_1 might be removed by having the origin of the radii vectores suitably fixed; in fact by shifting this origin through a space ζ_1 along the axis.

Then in v there remains only the term which involves P_2 besides the constant; and by (10) we have

$$\zeta_2 = -\frac{\kappa a_2 b^2}{3(b^2 - a_2)}.$$

Therefore $v = \text{constant} - \frac{\kappa b^2 P_2}{3(b^2 - a_2)} = \text{constant} + \frac{\kappa b^2}{3(b^2 - a_2)}(1 - P_2)$

$$= \text{constant} + \frac{\kappa b^2}{2(b^2 - a_2)} \sin^2\theta.$$

The constant in the last expression vanishes because by supposition $v = 0$ when $\theta = 0$; therefore

$$v = \frac{\kappa b^2}{2(b^2 - a_2)} \sin^2\theta \text{ ....(12)}.$$

910. We may observe that if the density diminishes from the centre to the surface we shall have $\frac{a_n}{b^n}$ less than $\frac{3}{n+3}$.

For, by integration by parts

$$\int_0^b \rho\beta^{n+2}\,d\beta = \frac{\rho_1 b^{n+3}}{n+3} - \frac{1}{n+3}\int_{\rho_0}^{\rho_1}\beta^{n+3}\,d\rho,$$

where ρ_1 denotes the density at the surface, and ρ_0 at the centre.

Hence
$$\frac{a_0}{b^2} = \frac{3}{n+3}\,\frac{\rho_1 b^{n+3} - \int_{\rho_0}^{\rho_1}\beta^{n+3}\,d\rho}{\rho_1 b^{n+3} - \int_{\rho_0}^{\rho_1} b^n\beta^3\,d\rho};$$

the multiplier of $\frac{3}{n+3}$ is less than unity, for $-\int_{\rho_0}^{\rho_1}\beta^{n+3}\,d\rho$ and $-\int_{\rho_0}^{\rho_1} b^n\beta^3\,d\rho$ are both positive, but the latter is the greater.

911. When the body is homogeneous

$$\frac{a_0}{b^2} = \frac{\int_0^b \beta^4\,d\beta}{\int_0^b b^2\beta^2\,d\beta} = \frac{3}{5},$$

so that when the body is homogeneous we have from (12)

$$v = \frac{5\kappa}{4}\sin^2\theta.$$

912. Thus far we have really no more than Laplace had already given in substance in his fourth memoir; but Legendre proceeds to a *second approximation*. This is a great addition to previous investigations, and it is for the sake of this process that I have adopted much of Legendre's own notation.

913. Put, as Legendre does, e for $-\frac{b^2\kappa}{3(b^2-a_0)}$, so that $v = e(P_2 - 1)$. Suppose that $u = 1 + v + w$, where v denotes the term of the first order already determined, and w a term of the second order which is now to be determined. Then instead of (4) we shall now have to the second order

$$w + v - v^2 = \frac{b\zeta_1 P_1}{r^2} + \frac{b\zeta_2 P_2}{r^3} + \ldots + \frac{\kappa r^2\sin^2\theta}{2b^2} + \text{constant} \ldots\ldots (13).$$

We know that to the first order ζ_n is zero if n is greater than 2; hence in (13) we may put b for r in the corresponding terms; the same remark also holds with respect to ζ_2.

For the term $\frac{b\zeta_2 P_2}{r^4}$ we may put $\frac{\zeta_2}{b^3}(1-3v)P_2$.

And
$$\frac{\kappa r^2 \sin^2\theta}{2b^3} = -\frac{\kappa r^2}{3b^3}(P_2-1) = -\frac{\kappa}{3}[1+2e(P_2-1)](P_2-1)$$
$$= -\frac{\kappa}{3}(P_2-1) - \frac{2e\kappa}{3}(P_2-1)^2.$$

Then (13) becomes

$$w = \text{constant} + \frac{\zeta_1 P_1}{b} + \frac{\zeta_3 P_3}{b^4} + \frac{\zeta_4 P_4}{b^5} + \dots + \frac{\zeta_2}{b^3} P_2 [1 - 3e(P_2-1)]$$
$$-\left(e + \frac{\kappa}{3}\right)(P_2-1) + \left(e^2 - \frac{2e\kappa}{3}\right)(P_2-1)^2 \dots\dots(14).$$

Now the general expression for ζ_n in Art. 904 shews that to our order of approximation we have for any value of n greater than 2,

$$\zeta_n = \frac{a_n}{2}\int_{-1}^{1}\left\{(n+3)w' + \frac{(n+3)(n+2)}{2}e^2(P_2'-1)^2\right\}P_n' d\mu' \dots\dots(15).$$

The integration is facilitated by the aid of the following formula which readily follows from the expressions in Art 897,

$$(P_2-1)^2 = \frac{18P_4 - 60P_2 + 42}{35}.$$

Moreover this gives

$$P_2^2 = \frac{18P_4}{35} + \frac{10P_2}{35} + \frac{7}{35}.$$

Hence from (15), as from the corresponding equation (9), we may shew that $\zeta_n = 0$, except when $n = 1$ or 2 or 4. The case of $n = 1$ we need not consider: see Art. 909.

The equation for finding ζ_2 is

$$\zeta_2 = \frac{a_2 \int_{-1}^{1} P_2' (1 + v' + w')^5 \, d\mu'}{\int_{-1}^{1} (1 + v' + w')^3 \, d\mu'}$$

$$= \frac{a_2 \int_{-1}^{1} [5e (P_2' - 1) + 5w' + 10e^2 (P_2' - 1)^2] \, P_2' \, d\mu'}{\int_{-1}^{1} [1 + 3e (P_2' - 1)] \, d\mu'} .$$

914. As we have thus shewn that we need only consider the value of ζ_n for the cases of $n = 2$ and $n = 4$, we may write (14) for shortness thus:

$$w = f + gP_2 + hP_4,$$

where f, g, and h are certain constants.

Hence we shall find that the numerator of ζ_2 reduces to

$$a_2 \int_{-1}^{1} \left(5e + 5g - \frac{600}{35} e^2\right) (P_2')^2 \, d\mu',$$

that is to
$$2a_2 \left(e + g - \frac{120e^2}{35}\right).$$

The denominator of ζ_2 reduces to $2(1 - 3e)$.

Hence
$$\zeta_2 = a_2 \left(e + g - \frac{3}{7} e^2\right).$$

Also we find that

$$\zeta_4 = a_4 \left(\frac{7}{9} h + \frac{8}{5} e^2\right).$$

Substitute these values in the expressions for g and h, which are

$$g = \frac{\zeta_2}{b^2} - e - \frac{\kappa}{3} - \frac{12}{7} \left(e^2 - \frac{2e\kappa}{3}\right) + \frac{15}{7} e \frac{\zeta_2}{b^2} ,$$

$$h = \frac{18}{35} \left(e^2 - \frac{2e\kappa}{3}\right) - \frac{54}{35} e \frac{\zeta_2}{b^2} + \frac{\zeta_4}{b^4} ,$$

then there will remain only g and h to determine.

We shall find that to our order

$$g = -\frac{36}{7} e^2,$$

$$h = \frac{54}{35} e^2 \frac{9b^7 - 15a_2 b^2 + 7a_4}{9b^7 - 7a_4}.$$

The value of f may be determined from the relation $f + g + h = 0$ which holds by reason of the supposition that $v + w$ vanishes with θ: see Art. 909.

For abbreviation put k for $\dfrac{9b^7 - 15a_2 b^2 + 7a_4}{9b^7 - 7a_4}$.

Thus $\quad w = f + gP_2 + hP_4 = g(P_2 - 1) + h(P_4 - 1)$

$$= -\frac{36}{7} e^2 (P_2 - 1) + \frac{54}{35} e^2 k (P_4 - 1);$$

therefore to the second order we have

$$r = b\left\{1 + \left(e - \frac{36e^2}{7}\right)(P_2 - 1) + \frac{54}{35} e^2 k (P_4 - 1)\right\}.$$

Put for P_2 and P_4 their values, and $-\dfrac{b^5\kappa}{3(b^5 - a_2)}$ for e; thus we find that

$$r = b\left\{1 + \frac{\kappa b^5}{2(b^5 - a_2)} \sin^2\theta + \frac{3\kappa^2 b^{10}}{28(b^5 - a_2)^2} \sin^2\theta (8 - k - 7k\cos^2\theta)\right\} \ldots (16).$$

Let $b(1 + \epsilon)$ denote the radius vector at the equator; thus

$$\epsilon = \frac{\kappa b^5}{2(b^5 - a_2)} + \frac{3\kappa^2 b^{10}(8 - k)}{28(b^5 - a_2)^2} \ldots\ldots\ldots\ldots\ldots (17).$$

If we introduce the expression for ϵ in the above value of r we shall find that to our order

$$r = b[1 + \epsilon \sin^2\theta - 3k\epsilon^2 \sin^2\theta \cos^2\theta].$$

If the body is homogeneous, we have

$$a_2 = \frac{3}{5} b^5, \quad a_4 = \frac{3}{7} b^7; \text{ hence } k = \frac{1}{2};$$

and then $\quad r = b\left\{1 + \epsilon \sin^2\theta - \dfrac{3\epsilon^2}{2} \sin^2\theta \cos^2\theta\right\}.$

We know that when the body is homogeneous, an oblatum is a rigorous solution; and it will be found that the value of r just obtained agrees to the second order with that which we should derive from

$$\frac{r^2\cos^2\theta}{b^2}+\frac{r^2\sin^2\theta}{b^2(1+\epsilon)^2}=1.$$

915. The connexion between e and ϵ should be noticed.

We have $$\epsilon=-\frac{3e}{2}+\frac{27}{28}(6-k)e^2;$$

therefore $$e=-\frac{2\epsilon}{3}+\frac{9}{14}(6-k)e^2$$

$$=-\frac{2\epsilon}{3}+\frac{2}{7}(6-k)\epsilon^2,$$

to our order of approximation.

916. We may also notice that

$$a_4=\frac{9b^4(k-1)+15a_2b^2}{7(k+1)};$$

$$\zeta_4=a_4\left(\frac{7}{9}h+\frac{6}{5}e^2\right)=\frac{6}{5}a_4e^2(k+1)$$

$$=\frac{6}{35}\left[9b^4(k-1)+15a_2b^2\right]e^2=\frac{6b^2}{3.5.7}\left[9b^2(k-1)+15a_2\right]e^2.$$

This expression for ζ_4 will be found useful in verifying the result which will be given in Art. 921.

917. Legendre expresses the value of the ellipticity in terms of the ratio of centrifugal force to gravity.

Let X be the attraction resolved parallel to the polar axis, and Y the attraction resolved parallel to the equatorial axis, at a point whose coordinates are x and y. Then we know that

$$X=-\frac{dV}{dx},\quad Y=-\frac{dV}{dy},$$

so that $$dV=-Xdx-Ydy.$$

But $x = r\cos\theta$, and $y = r\sin\theta$; therefore

$$dV = -(X\cos\theta + Y\sin\theta)\,dr + (X\sin\theta - Y\cos\theta)\,r\,d\theta.$$

Now we see, by Arts. 903 and 904, that

$$V = \frac{M}{r}\left\{1 + \frac{\zeta_1 P_1}{r} + \frac{\zeta_2 P_2}{r^2} + \dots\right\}.$$

Thus, $$X\cos\theta + Y\sin\theta = \frac{M}{r^2}\left\{1 + \frac{2\zeta_1 P_1}{r} + \frac{3\zeta_2 P_2}{r^2} + \dots\right\},$$

$$X\sin\theta - Y\cos\theta = -\frac{M\sin\theta}{r^2}\left\{\frac{\zeta_1}{r}\frac{dP_1}{d\mu} + \frac{\zeta_2}{r^2}\frac{dP_2}{d\mu} + \dots\right\}.$$

Put $\theta = 90°$, and for r put a the radius of the equator. Then the first equation gives for the attraction at the equator *along the radius*

$$\frac{M}{a^2}\left\{1 + \frac{2\zeta_1}{a}P_1 + \frac{3\zeta_2 P_2}{a^2} + \dots\right\}.$$

But when $\theta = 90°$, we have by Art. 897,

$$P_1 = 0,\quad P_2 = -\frac{1}{2},\quad P_3 = 0,\quad P_4 = \frac{1.3}{2.4}, \dots$$

so that this attraction becomes

$$\frac{M}{a^2}\left\{1 - \frac{3}{2}\frac{\zeta_2}{a^2} + \frac{3.5}{2.4}\frac{\zeta_4}{a^4} - \dots\right\} \dots\dots\dots\dots\dots (18).$$

This attraction is the whole attraction at the equator, provided the resolved attraction parallel to the polar axis vanishes there. On examining the value of X we see that it vanishes when $\theta = 90°$, provided the coefficients of an *odd order* $\zeta_1, \zeta_3, \zeta_5 \dots$ vanish then. This will certainly be the case if the plane of the equator divides the body *symmetrically*. We will suppose this to be the case; and then (18) represents the whole attraction at the equator.

Let A represent this attraction, and Φ the centrifugal force at the equator; and suppose that at the equator the centrifugal force is i times gravity.

Then $\Phi = i(A - \Phi)$; therefore $\Phi = \dfrac{iA}{1+i}$.

But $\Phi = a\omega^2 = \frac{Ma\kappa}{b^3}$; therefore

$$\frac{Ma\kappa}{b^3} = \frac{i}{1+i}\frac{M}{a^2}\left\{1 - \frac{3}{2}\frac{\zeta}{a^2} + \frac{3 \cdot 5}{2 \cdot 4}\frac{\zeta_1}{a^4} - \dots\right\};$$

therefore $$\kappa = \frac{i}{1+i}\frac{b^3}{a^3}\left\{1 - \frac{3}{2}\frac{\zeta}{a^2} + \frac{3 \cdot 5}{2 \cdot 4}\frac{\zeta_1}{a^4} - \dots\right\}.$$

Restricting ourselves to terms of the second order we have

$$\kappa = (i - i^2)(1 - 3\epsilon)\left(1 - \frac{3}{2}\frac{a_2\epsilon}{b^5}\right)$$

$$= (i - i^2)(1 - 3\epsilon)\left\{1 + \frac{\kappa a_2}{2(b^5 - a_2)}\right\}$$

$$= (i - i^2)\left\{1 - \frac{3\kappa b^5}{2(b^5 - a_2)} + \frac{\kappa a_2}{2(b^5 - a_2)}\right\}.$$

Hence, to the second order,

$$\kappa = i + \frac{3a_2 - 5b^5}{2(b^5 - a_2)}i^2.$$

If we substitute this value in the expression for the ellipticity, we obtain

$$\epsilon = \frac{ib^5}{2(b^5 - a_2)} + \frac{i^2b^5}{28(b^5 - a_2)^2}(21a_2 - 11b^5 - 3kb^5).$$

In the case of a homogeneous body $a_2 = \frac{3}{5}b^5$, and $k = \frac{1}{2}$; then $\epsilon = \frac{5}{4}i + \frac{5}{224}i^2$.

918. Legendre now finds an expression for the force of gravity at any point; this requires some preliminary analysis: the processes are carried on so far as to make the results true to the second order.

Let L denote the latitude at any point. Then

$$\tan L = -\frac{d(r\sin\theta)}{d(r\cos\theta)};$$

this gives to the second order

$$\tan L = \cot \theta \, [1 + 2\epsilon + \epsilon^2 + (3 - 6k) \, \epsilon^2 \cos 2\theta] \, ;$$

whence we get

$$\theta = \frac{\pi}{2} - L + \epsilon \sin 2L - \frac{\epsilon^2}{2} \sin 2L + \frac{6k - 5}{4} \epsilon^2 \sin 4L.$$

Substitute this value in that of r, and we obtain

$$r = b \, [1 + \epsilon \cos^2 L + (4 - 3k) \, \epsilon^2 \sin^2 L \cos^2 L].$$

Let s be the arc of the meridian measured from the equator to the latitude L. Then

$$ds^2 = dr^2 + r^2 d\theta^2.$$

Hence after substitution we find that

$$\frac{ds}{dL} = b \, \{1 + \epsilon \, (3 \sin^2 L - 1) + (2 - 3k) \, \epsilon^2 \, (2 - 15 \sin^2 L \cos^2 L)\}.$$

Now $\dfrac{ds}{dL}$ is equal to the radius of curvature of the meridian. Therefore if D be the length of the degree of the meridian which has its middle point at the equator, the length of the degree of the meridian which has its middle point at the latitude L, is

$$D \, [1 + 3\epsilon \sin^2 L + 3\epsilon^2 \sin^2 L - 15\epsilon^2 \, (2 - 3k) \sin^2 L \cos^2 L].$$

919. Eliminate Y from the equations of Art. 917; thus we get

$$X = \frac{M}{r^2} \cos \theta \left\{1 + \frac{2\zeta_1}{r} P_1 + \frac{3\zeta_2}{r^2} P_2 + \ldots\right\},$$

$$- \frac{M}{r^2} \sin^2 \theta \left\{\frac{\zeta_1}{r} \frac{dP_1}{d\mu} + \frac{\zeta_2}{r^2} \frac{dP_2}{d\mu} + \ldots\right\}.$$

The two series may be incorporated by the aid of the following general theorem:

$$P_n \, (n + 1) \cos \theta - \sin^2 \theta \, \frac{dP_n}{d\mu} = (n + 1) \, P_{n+1} \ldots\ldots\ldots\ldots \, (19),$$

so that we get

$$X = \frac{M}{r^2} \left\{P_1 + \frac{2\zeta_1}{r} P_2 + \frac{3\zeta_2}{r^2} P_3 + \frac{4\zeta_3}{r^3} P_4 + \ldots\right\}.$$

If we assume as before, in Art. 917, that ζ_1, ζ_3, ζ_5, ... vanish, this reduces to

$$X = \frac{M}{r^2}\left\{P_1 + \frac{3\zeta_2}{r^2}P_3 + \frac{5\zeta_4}{r^4}P_5 + \dots\right\}.$$

920. Legendre says nothing about this general theorem; though I presume he must have known it: but it would be sufficient for his purpose here to verify the truth of the theorem for the simple cases of $n = 0, 2, 4$. I do not perceive the theorem in the work of Heine already cited.

The theorem may be established in various ways. We may use the general expression for P_n given in Art. 786, and verify the theorem by examining the coefficients of the various powers of x. Or we may use the general expression for P_n first given by Rodrigues; namely,

$$P_n = \frac{1}{2^n \lfloor\underline{n}}\,\frac{d^n T^n}{dx^n},$$

where T stands for $x^2 - 1$: see Heine's *Handbuch der Kugelfunctionen*, page 10. Here x takes the place of our former μ.

For thus the expression on the left-hand side of (19) becomes

$$\frac{(n+1)}{2^n \lfloor\underline{n}}\,x\,\frac{d^n T^n}{dx^n} + \frac{T}{2^n \lfloor\underline{n}}\,\frac{d^{n+1} T^n}{dx^{n+1}};$$

and the expression on the right-hand becomes $\dfrac{1}{2^{n+1}\lfloor\underline{n}}\,\dfrac{d^{n+1}T^{n+1}}{dx^{n+1}}$, which is equal to

$$\frac{1}{2^{n+1}\lfloor\underline{n}}\left\{T\frac{d^{n+1}T^n}{dx^{n+1}} + 2(n+1)\,x\,\frac{d^n T^n}{dx^n} + n(n+1)\,\frac{d^{n-1}T^n}{dx^{n-1}}\right\},$$

and therefore to establish the theorem we have only to show that

$$T\frac{d^{n+1}T^n}{dx^{n+1}} = n\,(n+1)\,\frac{d^{n-1}T^n}{dx^{n-1}} \quad \text{.................(20).}$$

Now this may be established by comparing the coefficients of the various powers of x. Or more simply thus. It is obvious that

$$\frac{d^{n+1}T^{n+1}}{dx^{n+1}} = 2\,(n+1)\,\frac{d^n(xT^n)}{dx^n};$$

hence, developing each member, we get

$$T\frac{d^{n+1}T^n}{dx^{n+1}}+2(n+1)x\frac{d^nT^n}{dx^n}+n(n+1)\frac{d^{n-1}T^n}{dx^{n-1}}$$

$$=2(n+1)\left\{x\frac{d^nT^n}{dx^n}+n\frac{d^{n-1}T^n}{dx^{n-1}}\right\}:$$

thus (20) is established.

921. Let Π denote the gravity at the latitude L. Then

$$\Pi \sin L = X.$$

For when there is relative equilibrium Π is the *whole force* at the point considered, and its direction is that of the normal. Hence $\Pi \sin L$ must be equal to the force resolved parallel to the polar axis.

Thus, $$\Pi=\frac{X}{\sin L}=\frac{M}{r^2\sin L}\left\{P_1+\frac{3\zeta_1}{r^2}P_3+\frac{5\zeta_2}{r^4}P_5+\ldots\right\}.$$

Legendre evaluates this expression to the second order. I have verified his result which may be thus expressed:

$$\Pi=\frac{M}{b^2}\left\{1+\left(\frac{3a_1}{b^2}-4\right)\epsilon+\left(4-\frac{5a_1}{b^2}\right)\epsilon\sin^2 L\right.$$
$$\left.+(\gamma\sin^4 L+\gamma_1\sin^2 L+\gamma_2)\epsilon^2\right\},$$

where $$\gamma=9k-\frac{5a_2}{b^4},$$

$$\gamma_1=-10-6k+\frac{116}{7}\frac{a_1}{b^2}-\frac{15}{7}\frac{a_1}{b^2}k,$$

$$\gamma_2=\frac{40}{7}+\frac{3k}{7}-\frac{45}{7}\frac{a_1}{b^2}+\frac{9}{7}\frac{a_1}{b^2}k.$$

922. Let Π_e denote the gravity at the equator so that

$$\Pi_e=\frac{M}{b^2}\left\{1+\left(\frac{3a_1}{b^2}-4\right)\epsilon+\gamma_2\epsilon^2\right\}.$$

Hence we find that

$$\Pi=\Pi_e\left\{1+\left(4-\frac{5a_1}{b^2}\right)\epsilon\sin^2 L+\left(9k-\frac{5a_2}{b^4}\right)\epsilon^2\sin^4 L+\gamma_1\epsilon^2\sin^2 L\right\}.$$

where γ_2 stands for $\gamma_1 - \left(\frac{3a_2}{b^2} - 4\right)\left(4 - \frac{5a_2}{b^2}\right)$,

that is for $0 - 6k - \frac{108}{7}\frac{a_2}{b^2} + \frac{15a_2^{\,2}}{b^4} - \frac{15}{7}\frac{a_2}{b^2}k$.

Therefore the gravity at the pole is

$$\Pi_e\left\{1 + \left(4 - \frac{5a_2}{b^2}\right)\epsilon + \left(9k - \frac{5a_2}{b^2} + \gamma_2\right)\epsilon^2\right\}.$$

In the case of a homogeneous fluid *oblatum* we know that the gravity at the pole is exactly $(1 + \epsilon)$ times the gravity at the equator. This is accordant with our result: for if we put $a_2 = \frac{3}{5}b^2$ and $k = \frac{1}{2}$ we have

$$4 - \frac{5a_2}{b^2} = 1, \text{ and } 9k - \frac{5a_2}{b^2} + \gamma_2 = 0.$$

923. In the case of a variable density we no longer have *Clairaut's fraction* exactly equal to ϵ: see Art. 171. This fraction is now equal to

$$\left(4 - \frac{5a_2}{b^2}\right)\epsilon + \left(9k - \frac{5a_2}{b^2} + \gamma_2\right)\epsilon^2;$$

denoting this by ϖ we have

$$\varpi + \epsilon = 5\left(1 - \frac{a_2}{b^2}\right)\epsilon + \left(9k - \frac{5a_2}{b^2} + \gamma_2\right)\epsilon^2.$$

Put for ϵ on the right-hand side its value in terms of i from Art. 917; thus we obtain

$$\varpi + \epsilon = \frac{5i}{2} + \frac{i^2}{28}\cdot\frac{17a_2 - 13b^2 + 6kb^2}{(b^2 - a_2)^2}b^2 \text{ .... (21).}$$

If we restrict ourselves to the first term on the right-hand side we have Clairaut's theorem.

If the body is homogeneous (21) becomes

$$\varpi + \epsilon = \frac{5i}{2} + \frac{5i^2}{112}.$$

Denoting the expression on the right-hand side of this equation by Δ we shall find that (21) becomes

$$\varpi + \epsilon = \Delta + \frac{i^2}{112} \frac{(5a_1 - 3b^2)(15b^2 - a_1) + 12(2k-1)b^4}{(b^2 - a_1)^2}.$$

924. Thus the solution has been carried to the second order inclusive. Legendre says that it would not be difficult to push the approximations further; and he states what will be the general form of the expression for the radius vector: see his page 394.

Legendre remarks that the formulæ shew that the augmentation of the length of a degree of the meridian, and the augmentation of gravity both vary approximately as the square of the sine of the latitude in passing from the equator to the pole. Thus it is impossible to admit the truth of a law suggested by Bouguer, namely that the augmentation of the length of a degree varies as the fourth power of the sine of the latitude: see Art. 363.

925. On his pages 395...420, Legendre discusses the second of his three hypotheses: see Art. 894. He proposes to determine the figure of a planet considered in a fluid state. This problem had not been discussed before, except by Clairaut on the assumption that the strata were ellipsoidal.

Here we require the value of the potential for an internal point. Accordingly V is now taken to be equal to the sum of two series; the general terms of these are

$$\frac{2\pi}{r^{n+1}} P_n \iint \rho r'^{n+2} P_n' \, dr' \, d\mu',$$

and

$$2\pi r^n P_n \iint \rho \frac{P_n'}{r'^{n-1}} \, dr' \, d\mu'.$$

This expression will be *accurately* true if we suppose the former integral to extend over those parts of the body for which r' is less than r, and the latter over those parts of the body for which r' is greater than r. But Legendre is not sufficiently careful. He makes the former integral extend over those strata of the body which are *beneath* the stratum on which the point (r, θ) is situated; and the latter integral over those strata which are *beyond* this stratum. This value of the potential had been given by Laplace in his fourth memoir, page 179. Poisson first shewed that the formula of Legendre and Laplace was really true, though it had not previously been strictly established: see Art. 792.

926. The strata are now not to be assumed similar; so that when we put $r' = \beta u'$ the value of u' must not be assumed to be independent of β. We shall denote by ϵ the value of β corresponding to the stratum on which the point (r, θ) is situated so that $r = \epsilon u$. And as before the value of β at the surface will be denoted by b.

Let
$$\lambda_n = \int_0^{\epsilon u'} \rho r'^{n+2} dr',$$

and
$$\nu_n = \int_{\epsilon u'}^{bu'} \frac{\rho}{r'^{n-1}} dr'.$$

Let
$$2a = \int_{-1}^{1} \lambda_0 d\mu';$$

let
$$\zeta_n = \frac{1}{2a} \int_{-1}^{1} \lambda_n P'_n d\mu';$$

and let
$$\xi_n = \frac{1}{2} \int_{-1}^{1} \nu_n P'_n d\mu'.$$

The equation for relative equilibrium is
$$V + \frac{r^2 \omega^2}{2} \sin^2\theta = \text{constant}.$$

Hence dividing by $4\pi a$, we obtain
$$\frac{1}{r} + \frac{\zeta_1 P_1}{r^2} + \frac{\zeta_2 P_2}{r^3} + \ldots + \frac{1}{a} \{\xi_1 P_1 r + \xi_2 P_2 r^2 + \ldots\}$$
$$+ \frac{\epsilon a_1}{3ab^3} r^2 (1 - P_2) = \text{constant} \ldots\ldots\ldots\ldots (22).$$

Here a_1 is the value of a at the surface; so that if M denote the whole mass we have $M = 4\pi a_1$. The term $\frac{\xi_0 P_0}{a}$ is not expressed because it does not involve r or θ explicitly, and so may be supposed comprised in the constant.

927. Suppose that $u = 1 + v$, where v is so small that its square may be neglected; then we have from (22), to the first order of small quantities,

$$v = \text{constant} + \left(\frac{\zeta_1}{b} + \frac{\xi_1 b^3}{a}\right) P_1 + \left(\frac{\zeta_2}{b^2} + \frac{\xi_2 b^4}{a} - \frac{\kappa \eta_1 b^5}{3ab^3}\right) P_2$$

$$+ \left(\frac{\zeta_3}{b^3} + \frac{\xi_3 b^5}{a}\right) P_3 + \ldots$$

Thus we may put

$$v = A_0 + A_1 P_1 + A_2 P_2 + A_3 P_3 + \ldots\ldots;$$

and we shall have

$$A_0 + A_1 + A_2 + \ldots = 0,$$

because $r = b$ when $\theta = 0$.

928. Now

$$\lambda_n = \int_0^{a'} \rho r'^{n+2} dr' = \frac{1}{n+3} \int_0^{a'} \rho \frac{d}{dr'} r'^{n+3} dr'$$

$$= \frac{1}{n+3} \int_0^a \rho \frac{d}{d\beta} (r'^{n+3}) \, d\beta$$

$$= \frac{1}{n+3} \int_0^a \rho \frac{d}{d\beta} [1 + (n+3) v'] \beta^{n+3} d\beta$$

$$= \int_0^a \rho \beta^{n+2} d\beta + \int_0^a \rho \frac{d(\beta^{n+3} v')}{d\beta} d\beta.$$

When we substitute for v' the second of the two expressions on the right-hand side gives rise to a series of which the general term is

$$P_n \int_0^a \rho \frac{d}{d\beta} (\beta^{n+3} A_n) \, d\beta.$$

Substitute the value of λ_n in the expression for ζ_n given in Art. 920; then, by the aid of Art. 905, we obtain

$$\zeta_n = \frac{1}{a(2n+1)} \int_0^a \rho \frac{d}{d\beta} (\beta^{n+3} A_n) \, d\beta.$$

In like manner we find that

$$2a = \int_{-1}^{1} \left\{ \int_0^c \rho \beta^2 \, d\beta + \int_0^c \rho \frac{d}{d\beta} (\beta^3 A_0) \, d\beta \right\} d\mu',$$

so that
$$a = \int_0^c \rho \beta^2 \, d\beta + \int_0^c \rho \frac{d}{d\beta} (\beta^3 A_0) \, d\beta.$$

We will denote the first of the two expressions on the right-hand side by σ; so that when we neglect the small quantity of the first order we may put $a = \sigma$.

In the same manner we find that

$$\xi_n = \frac{1}{2n+1} \int_0^b \rho \frac{d}{d\beta} \left(\frac{A_n}{\beta^{n-2}} \right) d\beta,$$

which we may express thus

$$\xi_n = \frac{1}{2n+1} \left\{ N_n - \int_0^c \rho \frac{d}{d\beta} \left(\frac{A_n}{\beta^{n-2}} \right) d\beta \right\},$$

where N_n is a constant, namely

$$\int_0^b \rho \frac{d}{d\beta} \left(\frac{A_n}{\beta^{n-2}} \right) d\beta.$$

929. Now for any value of n except 2 we have

$$A_n = \frac{\zeta_n}{c^n} + \frac{c^{n+1} \xi_n}{a}.$$

Substitute for ζ_n and ξ_n; thus we obtain

$$(2n+1) \sigma c^n A_n = \int_0^c \rho \frac{d}{d\beta} (\beta^{n+3} A_n) \, d\beta$$
$$+ c^{2n+1} \left\{ N_n - \int_0^c \rho \frac{d}{d\beta} \left(\frac{A_n}{\beta^{n-2}} \right) d\beta \right\} \ldots\ldots (23).$$

In the case of $n = 2$ we have

$$5\sigma c^2 A_2 = \int_0^c \rho \frac{d}{d\beta} (\beta^5 A_2) \, d\beta + c^5 \left\{ N_2 - \int_0^c \rho \frac{dA_2}{d\beta} d\beta - \frac{5\pi q_2}{3b^3} \right\} \ldots (24).$$

But this is of substantially the same form as the general equation, for $N_2 - \frac{5\pi q_2}{3b^3}$ is a constant.

930. This is the first appearance of these important equations for all values of n. Clairaut had substantially arrived at the equation for the case of $n = 2$; and D'Alembert in addition at the equation for the cases of $n = 1$, and $n = 3$. See Art. 444.

931. In the particular case of $n = 1$, we can shew that A_1 must be zero. For then we have

$$3\sigma\epsilon A_1 = \int_0^\epsilon \rho \frac{d}{d\beta}(\beta^3 A_1) + \epsilon^3 \left\{N_1 - \int_0^\epsilon \rho \frac{d}{d\beta}(\beta A_1)\, d\beta\right\}.$$

Here by A_1 when free from the integral sign we mean the value corresponding to the value ϵ of the parameter; and the same remark applies to ρ when it occurs free from the integral sign. Differentiate with respect to ϵ, observing that $\frac{d\sigma}{d\epsilon} = \rho\epsilon^2$; thus we get

$$\frac{d}{d\epsilon}(\epsilon A_1) \int_0^\epsilon \rho\beta^2 d\beta = \epsilon^2 \left\{N_1 - \int_0^\epsilon \rho \frac{d}{d\beta}(\beta A_1) d\beta\right\}.$$

Integrate the last expression with respect to ϵ; thus

$$\left\{N_1 - \int_0^\epsilon \rho \frac{d}{d\beta}(\beta A_1)\, d\beta\right\} \int_0^\epsilon \rho\beta^2 d\beta = \text{constant}.$$

But the left-hand member vanishes when $\epsilon = b$; hence the constant must be zero; therefore

$$N_1 = \int_0^\epsilon \rho \frac{d}{d\beta}(\beta A_1)\, d\beta.$$

Differentiate with respect to ϵ; hence $\frac{d}{d\epsilon}(\epsilon A_1) = 0$; therefore $A_1 = \frac{C}{\epsilon}$, where C is a constant. But C must be zero, or A_1 would be infinite at the centre. Hence A_1 is always zero.

932. Take the general equation (23) and differentiate with respect to ϵ; thus

$$\sigma \frac{d(\epsilon^n A_n)}{d\epsilon} = \epsilon^{2n} \left\{N_n - \int_0^\epsilon \rho \frac{d}{d\beta}\left(\frac{A_n}{\beta^{n-3}}\right) d\beta\right\};$$

therefore

$$\sigma\left\{\varepsilon^{n}\frac{dA_n}{d\varepsilon}+n\varepsilon^{n-1}A_n\right\}=N_n-\int_0^{\varepsilon}\rho\frac{d}{d\beta}\left(\frac{A_n}{\beta^{n-1}}\right)d\beta \text{ .....(25).}$$

Differentiate again; thus

$$\sigma\left\{\frac{d^2A_n}{d\varepsilon^2}-n(n+1)\frac{A_n}{\varepsilon^2}\right\}+2\rho\varepsilon^2\left(\frac{dA_n}{d\varepsilon}+\frac{A_n}{\varepsilon}\right)=0 \text{ ......(26).}$$

This equation is a little simplified by putting $\frac{Q_n}{\sigma}$ for A_n; for thus we get

$$\frac{d^2Q_n}{d\varepsilon^2}-n(n+1)\frac{Q_n}{\varepsilon^2}-\frac{\varepsilon^2}{\sigma}\frac{d\rho}{d\varepsilon}Q_n=0 \text{ .... (27).}$$

933. Legendre now proposes to demonstrate that A_n must vanish for every value of n greater than 2. The demonstration rests on the following principles: A_n must satisfy the equation (26); also A_n must always be a small quantity; and, moreover, it is assumed that the density diminishes from the centre to the surface: see his pages 399...403.

Legendre's demonstration bears a general resemblance to that which Laplace afterwards used: see the *Mécanique Céleste*, Livre III. § 30. But the two demonstrations are not identical. I have discussed the matter in a memoir published in the *Cambridge Philosophical Transactions*, Vol. XII.

934. Let us now take the first step of the demonstration. The distinction between β and ε need not be retained hereafter, when we shall be free from integral signs.

The solution of the differential equation (26) will give A_n in the form

$$A_n=C_1f_1(\beta)+C_2f_2(\beta),$$

where C_1 and C_2 are arbitrary constants, and $f_1(\beta)$ and $f_2(\beta)$ are definite functions of β. Now Legendre and Laplace shew in effect that one of the two functions $f_1(\beta)$ and $f_2(\beta)$ will be infinite when $\beta=0$; suppose that this is $f_2(\beta)$. Then since A_n is always to be a small quantity, we must have $C_2=0$.

We will now give the method by which Legendre shews that $f_2(\beta)$ will be infinite when $\beta=0$. Since the density decreases from the centre to the surface, whatever be the law of density, we may assume that when β is very small, $\rho = g\beta^{-m}$, where m is positive or zero, and g is a constant. If m is not zero the density will be infinite at the centre; but still the hypothesis is admissible, provided the mass included within a finite volume is finite. But this mass $= 4\pi\int \rho\beta^2 d\beta = \frac{4\pi g}{3-m}\beta^{3-m}$. Hence, provided m is not greater than 3, there is nothing inadmissible in our law of density. With this law of density we shall have

$$\frac{\beta^2}{\sigma}\frac{d\rho}{d\beta} = -\frac{m(3-m)}{\beta^2}.$$

Hence (27) becomes

$$\frac{d^2 Q_n}{d\beta^2} = \{n(n+1) - m(3-m)\}\frac{Q_n}{\beta^2}.$$

The solution of this differential equation is

$$Q_n = C_1\beta^c + C_2\beta^{1-c},$$

where
$$c = \frac{1}{2} + \sqrt{\left\{\left(n+\frac{1}{2}\right)^2 - m(3-m)\right\}}.$$

Now it is obvious that if C_2 is not zero, Q_n will be infinite when $\beta = 0$; for c is greater than unity, since n is not less than 2, and m not greater than 3; and *à fortiori* Δ_n will be infinite, because $\Delta_n = \frac{Q_n}{\sigma}$. Hence C_2 must be zero.

935. We may observe that an investigation resembling the preceding was given by Clairaut: see the pages 277...281 of his *Figure de la Terre*. A peculiarity in Legendre's investigation is the admission of a possible infinite density at the centre. What Legendre says on this point appears to me satisfactory. Laplace, however, holds that the density must be finite at the centre.

Laplace treats this first step of the demonstration in a different manner, as we shall see hereafter.

936. There will then be only one arbitrary constant in the value of Δ_n; for we have $\Delta_n = \frac{C_1\psi_1(\beta)}{\sigma}$. To determine this

constant we use equation (23). Suppose that $\beta = b$; then the second part of the right-hand member vanishes: hence we must have

$$(2n+1)\,\sigma b^n A_n = \int_0^b \rho \frac{d}{d\beta}\left(\beta^{n+3} A_n\right) d\beta,$$

where σ and A_n on the left-hand side denote the values of these quantities when β is equal to b.

One very obvious way to satisfy this condition is to suppose $C_1 = 0$.

937. But it remains to shew that $C_1 = 0$ is the only way to satisfy the relation just given. To this Legendre proceeds; he first shews that A_n must increase from the centre to the surface, and from this be deduces the required result. This is true when $n = 2$ as well as for other values; the demonstration applying as well to (24) as to (23). Laplace's process rests on the same principles as Legendre's, but is rather simpler.

938. Thus we have only left the coefficient A_0. This cannot be explicitly determined until some law of density is assumed. But without assuming any particular law we arrive at the result that the strata are ellipsoidal, and that the excentricity increases continually from the centre to the surface.

Legendre finds also the law of gravity, and shews that Clairaut's theorem holds: see his pages 404, 405.

939. Legendre gives three examples of laws of density in which the equation for finding A_n can be solved: see his pages 406...412.

940. The first example is that of a homogeneous mass. We may take $\rho = 1$. Thus $\sigma = \frac{\beta^3}{3}$. Hence equation (27) becomes

$$\frac{d^2 Q_n}{d\beta^2} = n(n+1)\frac{Q_n}{\beta^2};$$

whence $$Q_n = C_1 \beta^{n+1} + C_2 \beta^{-n}.$$

Then in order that Q_n and A_n may not be infinite at the centre, we must have $C_2 = 0$. Hence

$$A_n = 3 C_1 \beta^{n-2}.$$

Then to determine C_1 we employ the equation obtained by differentiating (23): see Art. 932. Thus $\frac{d}{d\beta}(3C_1\beta^{2n-2})$ must vanish when $\beta = b$; therefore $C_1(2n-2) = 0$.

Hence $C_1 = 0$; except when $n = 1$, and then it would not follow necessarily from this result that $C_1 = 0$. In the case of $n = 1$, however, we must have $C_1 = 0$, in order that Δ_1 may not be infinite at the centre.

When $n = 2$ we have $A_2 = 3C_1$; and we may find the value of C_1 by differentiating (24) and putting b for β in the result. Thus,

$$\frac{1}{b^4}\frac{d}{db}(3C_1b^5) = -\frac{5\kappa}{3b^2}, \text{ so that } 3C_1 = -\frac{5\kappa}{6}.$$

941. For the next example Legendre supposes that

$$\rho = g\beta^{-m} + h\beta^{m-3}.$$

This, as we have seen in Art. 934, is admissible if m is not greater than 3. Legendre says that m is *greater* than $\frac{3}{2}$. It is obvious that we may without loss of generality suppose that m is either less than $\frac{3}{2}$ or greater than $\frac{3}{2}$; and towards the end of the discussion Legendre really supposes m to be less than $\frac{3}{2}$. But we shall make neither supposition as we can proceed as well without.

The constants g and h are not necessarily both positive; but ρ must always be positive. Thus if we suppose h negative we must have m greater than $m - 3$, so that ρ may be positive at the centre; and also hb^{m-3} must be numerically less than gb^{-m}, so that ρ may be positive at the surface.

With this value of ρ we have

$$\sigma = \frac{g}{3-m}\beta^{-m} + \frac{h}{m}\beta^{m-3},$$

and

$$\frac{\beta^2}{\sigma}\frac{d\rho}{d\beta} = -\frac{m(3-m)}{\beta^2}.$$

Hence, as in Art. 934, we have

$$Q_n = C_1 \beta^n,$$

$$A_n = \frac{C_1 \beta^n}{\sigma}.$$

To determine the constant C_1 we employ the equation obtained by differentiating (23): see Art. 932. This leads to

$$C_1 \left[(n+c) \left(\frac{g}{3-m} b^{3-m} + \frac{h}{m} b^m \right) - (gb^{3-m} + hb^m) \right] = 0.$$

This is of course satisfied by $C_1 = 0$. It may indeed also be satisfied by supposing

$$(n+c) \left(\frac{g}{3-m} b^{3-m} + \frac{h}{m} b^m \right) - (gb^{3-m} + hb^m) = 0 \ldots\ldots (28).$$

But it will be found that if (28) is supposed to hold the condition that the density diminishes from the centre to the surface is not satisfied. For

$$\frac{d\rho}{d\beta} = -\left[mg\beta^{-m-1} + (3-m) h\beta^{m-4} \right];$$

and if we use (28) we shall find that $\frac{d\rho}{d\beta}$ vanishes and changes sign when

$$\frac{n+c-m}{n+c-3+m} = \left(\frac{\beta}{b} \right)^{2m-3},$$

and unless $m = \frac{3}{2}$ some value of β less than b will satisfy this equation.

We may observe that by comparing the values of σ and $\frac{d\rho}{d\beta}$ it follows that if $\frac{d\rho}{d\beta}$ could vanish and change sign, we should have σ vanishing also; but this is quite inadmissible.

We may also shew in the following way that equation (28) cannot subsist. From this equation it would follow that $\sigma = \frac{\rho \beta^n}{n+c}$, when $\beta = b$; but $\sigma = \frac{\rho \beta^3}{3} - \frac{1}{3} \int \beta^3 \frac{d\rho}{d\beta} d\beta$, which is greater than $\frac{\rho\beta^n}{3}$ if $\frac{d\rho}{d\beta}$ is negative. And $\frac{\rho\beta^n}{3}$ is greater than $\frac{\rho\beta^n}{n+c}$; for c is greater than unity, and n is not less than 2.

In the case of $n=2$ we find the value of the constant C_1 by differentiating (24), and putting b for β in the result. Thus

$$\left\{\frac{1}{\beta^4}\frac{d}{d\beta}\left(\frac{C_1\beta^{c+2}}{\sigma}\right)\right\}_1 = -\frac{5\kappa}{3b^3};$$

where the subscript 1 indicates that b is to be put for β after the differentiation. Thus

$$C_1\left\{\frac{(c+2)\,\sigma-\beta\frac{d\sigma}{d\beta}}{\sigma^2}\right\}_1 b^{c-3} = -\frac{5\kappa}{3b^3};$$

therefore $$C_1 = -\frac{\frac{5\kappa}{3}\left\{\frac{g}{3-m}b^{3-m}+\frac{h}{m}b^m\right\}^2 b^{-c}}{(c+2)\left(\frac{g}{3-m}b^{3-m}+\frac{h}{m}b^m\right)-(gb^{3-m}+hb^m)}.$$

The ellipticity of any stratum is $-\frac{3}{2}A_2$, as in Arts. 909 and 914, that is $-\frac{3C_1}{2\sigma}\beta^c$; this may be expressed thus

$$\epsilon = -\frac{\frac{3C_1}{2}\beta^{c-m}}{\frac{h}{m}+\frac{g}{3-m}\beta^{3-2m}} \quad \text{..................(29).}$$

Let ϵ_1 denote the ellipticity at the surface; then

$$\epsilon_1 = \frac{\frac{5\kappa}{2}\left\{\frac{g}{3-m}b^{3-m}+\frac{h}{m}b^m\right\}}{(c+2)\left(\frac{g}{3-m}b^{3-m}+\frac{h}{m}b^m\right)-(gb^{3-m}+hb^m)}$$

$$= \frac{\frac{5\kappa}{2}}{2+c-(3-m)\,m\,\frac{gb^{3-m}+hb^m}{mgb^{3-m}+(3-m)\,hb^m}}.$$

Hence we see that ϵ_1 is less than $\frac{5\pi}{4}$. For now

$$c = \frac{1}{2} + \sqrt{\left\{\frac{25}{4} - m(3-m)\right\}};$$

and hence it will be found that c is greater than m and also greater than $3-m$. Hence $c-(3-m) = \frac{gb^{c-m} + hb^m}{mgb^{c-m} + (3-m)hb^m}$ is positive; for its sign is the same as the sign of

$$m\,[c-(3-m)]\,gb^{c-m} + (3-m)(c-m)\,hb^m;$$

and this is positive, even if h is negative.

From (29) we see that the ellipticity increases continually from the centre to the surface. For since C_1 is negative, it will be found that the sign of $\frac{d\epsilon}{d\beta}$ is the same as the sign of

$$(c-m)\frac{h}{m} + \frac{g(c+m-3)}{3-m}\beta^{c-2m};$$

and this is positive for its sign is the same as the sign of

$$m\,[c-(3-m)]\,g\beta^{c-m} + (3-m)(c-m)\,h\beta^m.$$

942. For the next example Legendre supposes that

$$\rho = \frac{\sin\frac{m\beta}{b}}{\frac{\beta}{b}}.$$

I will here, though with some reluctance, follow him in putting $b=1$, as the formulæ thus become simpler. Hence we take $\rho = \frac{\sin m\beta}{\beta}$. If m have any constant value less than π, we thus obtain a density which is always positive and which diminishes continually from the centre to the surface. The density at the centre is denoted by m; and the density at the surface by $\sin m$.

With this value of ρ we have

$$\sigma = \frac{\sin m\beta - m\beta\cos m\beta}{m^2},$$

and

$$\frac{\beta^2}{\sigma}\frac{d\rho}{d\beta} = -m^2.$$

Hence equation (27) becomes

$$\frac{d^2 Q_n}{d\beta^2} - n(n+1)\frac{Q_n}{\beta^2} + m^2 Q_n = 0.$$

Legendre states without demonstration the integral of this equation; namely Q_n

$$= (C_1 \sin m\beta + C_2 \cos m\beta)\left\{1 - \frac{n(n^2-1)(n+2)}{2.4m^2\beta^2} + \frac{n(n^2-1)(n^2-4)(n^2-9)(n+4)}{2.4.6.8\,m^4\beta^4} - \dots\right\}$$

$$+ (C_1 \cos m\beta - C_2 \sin m\beta)\left\{\frac{n(n+1)}{2m\beta} - \frac{n(n^2-1)(n^2-4)(n+3)}{2.4.6\,m^3\beta^3} + \dots\right\}.$$

Since n is supposed an integer the series are finite.

The integral may also be exhibited as the sum of two infinite series; namely Q_n

$$= C_3\beta^{n+1}\left\{1 - \frac{m^2\beta^2}{2(2n+3)} + \frac{m^4\beta^4}{2.4(2n+3)(2n+5)} - \dots\right\}$$

$$+ C_4\beta^{-n}\left\{1 + \frac{m^2\beta^2}{2(2n-1)} + \frac{m^4\beta^4}{2.4.(2n-1)(2n-3)} + \dots\right\}.$$

It is easy to verify these statements.

Laplace has given some of the details of the process of integrating the equation in the *Mécanique Celeste*, Livre XI. § 9.

The solution, it is now known, can be put into the following compact symbolical form

$$Q_n = \frac{C}{\beta^n}\frac{d^n}{da^n}\frac{\sin(\beta\sqrt{a} + B)}{\sqrt{a}},$$

where after the differentiations we put m^2 for a. See the *Cambridge Mathematical Journal*, Vol. II. page 105.

We shall confine ourselves to the case of $n = 2$.

Thus

$$Q_2 = C_1\left\{\left(1-\frac{3}{m^2\beta^2}\right)\sin m\beta+\frac{3}{m\beta}\cos m\beta\right\}$$

$$+ C_2\left\{\left(1-\frac{3}{m^2\beta^2}\right)\cos m\beta-\frac{3}{m\beta}\sin m\beta\right\}.$$

Here C_2 must be zero in order that Q_2 may not be infinite at the centre. Thus

$$Q_2 = C_1\left\{\left(1-\frac{3}{m^2\beta^2}\right)\sin m\beta+\frac{3}{m\beta}\cos m\beta\right\}.$$

The constant C_1 must be determined by differentiating equation (24) and putting $\beta = 1$ after the differentiation.

Hence we find that

$$C_1 = \frac{\frac{5}{3}\kappa\left(\frac{\sin m}{m}-\cos m\right)^2}{m^2-2\sin^2 m+m\sin m\cos m}.$$

The ellipticity of any stratum is $-\frac{3}{2}A_2$; hence denoting the ellipticity by ϵ we have

$$\epsilon = -\frac{\frac{3}{2}C_1\left\{\left(1-\frac{3}{m^2\beta^2}\right)\sin m\beta+\frac{3}{m\beta}\cos m\beta\right\}}{\frac{\sin m\beta - m\beta\cos m\beta}{m^2}}.$$

Let ϵ_1 denote the ellipticity at the surface and ϵ_0 the ellipticity at the centre. Then

$$\epsilon_1 = \frac{5\kappa(\sin m - m\cos m)\{(3-m^2)\sin m - 3m\cos m\}}{2m^2(m^2-2\sin^2 m+m\sin m\cos m)},$$

$$\epsilon_0 = \frac{\kappa(\sin m - m\cos m)^2}{2(m^2-2\sin^2 m+m\sin m\cos m)}.$$

Legendre states the numerical results which will be found corresponding to various values of m. We collect them in the following table; taking $\kappa = \frac{1}{288}$.

m	ϵ_0	ϵ_1
$\frac{\pi}{2}$	$\frac{1}{269}$	$\frac{1}{250}$
$\frac{2\pi}{3}$	$\frac{1}{312}$	$\frac{1}{269}$
$\frac{7\pi}{8}$	$\frac{1}{424}$	$\frac{1}{319}$
π	$\frac{1}{576}$	$\frac{1}{379}$

In the second case Legendre says that the density at the centre is to that at the surface in the ratio of $\frac{2\pi}{3}$ to $\frac{1}{2}$; but it should of course be in the ratio of $\frac{2\pi}{3}$ to $\frac{\sqrt{3}}{2}$.

In the third case which he gives he considers the ellipticity very nearly equal to the actual ellipticity of the earth's surface; he says that the mean density is three times that at the surface: this may be easily verified.

943. In his pages 412...420 Legendre proceeds to a second approximation for the case of his second hypothesis, that is of a planet in a state of fluidity. He obtains the formulæ for this purpose; for an application of the formulæ he supposes the fluid homogeneous.

944. In his pages 420...426 Legendre discusses the third of his three hypotheses: see Art. 894. He proposes to determine the figure of a planet of which the interior is solid and composed of ellipsoidal strata in which the ellipticities follow any law.

Let ϵ be the ellipticity of any stratum, ϵ_1 the ellipticity of the surface. The only equation which must be satisfied is (24) applied to the surface. And as $\epsilon = -\frac{3}{2}A_2$, we thus get

$$\epsilon_1 - \frac{1}{2}\kappa = \frac{\int_0^b \rho \frac{d}{d\beta}(\epsilon\beta^5)\, d\beta}{5b^2 \int_0^b \rho\beta^2\, d\beta}.$$

The expression for the value of gravity at the surface may be found, and it may be shewn that Clairaut's theorem holds.

945. Legendre makes some remarks as to the numerical values of the quantities.

Let ϖ denote Clairaut's fraction; then if the earth were homogeneous and fluid, we should have $\varpi = \frac{1}{230}$. Pendulum observations shew, however, that ϖ is greater than $\frac{1}{230}$; on this point Legendre refers to Laplace's fifth memoir. Since then ϖ is greater than $\frac{1}{230}$ it follows by Clairaut's theorem that ϵ_1 must be less than $\frac{1}{230}$.

The result found by experience that ϖ is greater than $\frac{1}{230}$, is in agreement, Legendre says, with the theory for the case of entire fluidity. I am not certain as to what he has here in view. Perhaps he alludes to the values obtained in Art. 942. Or perhaps he means that assuming ρ to diminish from the centre to the surface, we can shew by the formula of Art. 944, which will hold here, that ϵ_1 is less than $\frac{5}{4}\kappa$, so that ϖ is greater than $\frac{5}{4}\kappa$. The theorem that ϵ_1 is less than $\frac{5}{4}\kappa$ is easily deduced from the formula; indeed Clairaut gives this: see page 227 of his *Figure de la Terre*. The result holds if ϵ increases from the centre to the surface, or even if only $\beta^2\epsilon$ does: see Art. 329. Hence the result holds if ϵ is constant. But this particular case Legendre himself treats. We have then

$$\epsilon_1 - \frac{1}{2}\kappa = \frac{\epsilon_1 \int \rho\beta^4 d\beta}{b^2 \int \rho\beta^2 d\beta}.$$

Now he says that we know the coefficient of ϵ_1 on the right-hand side to be less than $\frac{3}{5}$. This is true, *assuming that the density decreases from the centre to the surface:* see Art. 910. Therefore $\epsilon_1 - \frac{1}{2}\kappa$ is less than $\frac{3}{5}\epsilon_1$; and hence ϵ_1 is less than $\frac{5}{4}\kappa$.

946. Legendre considers that the pendulum observations make $\varpi = \frac{1}{180}$ very nearly; and thus $\epsilon_1 = \frac{1}{318}$ very nearly. He says it is easy to imagine hypotheses respecting the density and the ellipticity of the strata which will produce this value of ϵ_1. For example, suppose that the densities along a radius increase in arithmetical progression from the surface to the centre. Let 1 be the density at the surface where the radius is 1; let m be the density at the middle of the radius; then $\rho = 2m - 1 - 2\beta(m-1)$. Let all the strata be similar, so that the ellipticity is constant. Then

$$\frac{\int_0^1 \rho\beta^4 d\beta}{\int_0^1 \rho\beta^2 d\beta} = \frac{2}{5}\,\frac{m+2}{m+1};$$

whence
$$\epsilon_1 = \frac{5}{4}\kappa\,\frac{2m+2}{3m+1}.$$

If we suppose $\kappa = \frac{1}{288}$, and make $m = 8$, we find that $\epsilon_1 = \frac{1}{320}$. In this case the mean density on a radius is about eight times that at the surface. But the mean density of the Earth is $\frac{m+1}{2}$ times that at the surface, that is $4\frac{1}{2}$ times; which appears quite admissible. But other hypotheses might give the same value of ϵ_1 with a much less value of the mean density; this appears in Art. 942.

947. Legendre adverts to the subject of precession and nutation. The expression $\dfrac{\int \rho \frac{d}{d\beta}(\beta^5\epsilon)\,d\beta}{\int \rho \frac{d\beta^5}{d\beta}\,d\beta}$ occurs as a coefficient in the values of these quantities found by theory. Hence, comparing

the values found by observation we may determine this coefficient, and thus obtain information as to the Figure of the Earth. We have already seen that this idea was used by D'Alembert: see Art. 385.

Legendre considers that the comparison confirms the value of ϵ_1 which he had adopted, namely about $\frac{1}{318}$.

948. Legendre says on his page 425 that the solutions hitherto given have been restricted to the case of figures of revolution, but we might desire an investigation of a more general character, so that the figure of revolution, if it must of necessity hold, should be a result of investigation and not an hypothesis.

But he does not think it possible to obtain suitable formulæ for the attraction of bodies of any figure. But still a form may be given to the radius vector which shall be applicable to a large number of figures.

This in fact leads Legendre to consider the properties of what we call Laplace's coefficients, and accordingly pages 420...442 are devoted to the demonstration of various theorems of analysis. See also Art. 783.

949. With respect to these theorems Legendre says on his page 426:

Plusieurs de ces théorèmes sont dûs à M. de la Place, qui en a donné la démonstration dans son Mémoire de 1782, fondée sur une équation aux différences partielles à laquelle les fonctions doivent satisfaire. J'adopterai ici le fondement de ces démonstrations, mais on verra que j'ai considéré cet objet sous un point de vue différent, et que je suis parvenu à des résultats entièrement nouveaux.

950. The first thing Legendre does is to find an expression for what we now call Laplace's coefficient of the nth order: see his pages 420...432.

Let $(1-2zt+z^2)^{-\frac{1}{2}}$ be expanded in ascending powers of z, where $t = \cos\theta\cos\theta' + \sin\theta\sin\theta'\cos(\phi-\phi')$; then the coefficient of z^n will be called Y_n. We shall put μ for $\cos\theta$, μ' for $\cos\theta'$, and ψ for $\phi-\phi'$.

Legendre shews that

$$Y_n = P_n(\mu)\, P_n(\mu')$$

$$+ \frac{2}{n(n+1)} \frac{dP_n(\mu)}{d\mu} \frac{dP_n(\mu')}{d\mu'} \sin\theta \sin\theta' \cos\psi$$

$$+ \frac{2}{(n-1)\,n\,(n+1)\,(n+2)} \frac{d^2P_n(\mu)}{d\mu^2} \frac{d^2P_n(\mu')}{d\mu'^2} \sin^2\theta \sin^2\theta' \cos 2\psi$$

$$+ \frac{2}{(n-2)\,(n-1)\,n \ldots (n+3)} \frac{d^3P_n(\mu)}{d\mu^3} \frac{d^3P_n(\mu')}{d\mu'^3} \sin^3\theta \sin^3\theta' \cos 3\psi$$

$$+ \ldots\ldots\ldots$$

Here $P_n(\mu)$ has the value assigned in Art. 780.

Legendre's investigation is better than that given by Laplace in his fourth memoir; and an important error of Laplace's is corrected. Laplace had omitted the terms involving $\cos r\psi$ in the case in which $n+r$ is odd: see Art. 831. The investigation in the *Mécanique Céleste*, Livre III. is correct and much resembles Legendre's.

951. Let Z_n and Z_m be two Laplace's functions, of the nth and m^{th} order respectively. Then n and m being different

$$\int_{-1}^{1}\int_{0}^{2\pi} Z_n Z_m \, d\mu\, d\phi = 0.$$

This had been established by Laplace in his fourth memoir: Legendre demonstrates this in his pages 433...435.

Laplace's demonstration depends on the fact that Z_n and Z_m satisfy the partial differential equation of Art. 851. Legendre assumes for Z_n and Z_m expressions of the same *form* as Y_n and Y_m, but with arbitrary constants as the coefficients of the various terms.

Legendre also indicates the form which the value of the double integral will take when $m = n$. In this case the following result is interesting and important,

$$\int_{-1}^{1}\int_{0}^{2\pi} Z_n Y_n \, d\mu\, d\phi = \frac{4\pi}{2n+1} Z_n,$$

where Z'_n indicates what Z_n becomes when θ and ϕ are changed to θ' and ϕ' respectively. Also Y_n is Laplace's n^{th} *coefficient.* This important result was first formally given by Legendre. Laplace however had really obtained it. For it forms equation (1) on page 44 of the *Mécanique Céleste,* Vol. II.: and this equation is implicitly involved on the page 152 of Laplace's fourth memoir, but he does not there bring it into special notice.

As a particular example suppose that $Z_n = Y_n$; then $Z'_n = 1$; so that

$$\int_{-1}^{1}\int_{0}^{2\pi}(Y_n)^2\,d\mu\,d\phi = \frac{4\pi}{2n+1}.$$

952. A part of Legendre's investigation may be usefully presented here.

Let P_m and P_n be Legendre's coefficients of the mth and n^{th} order respectively, the variable being denoted by x; then will

$$\int_{-1}^{1}\frac{d^rP_m}{dx^r}\,\frac{d^rP_n}{dx^r}\,(1-x^2)^r\,dx$$
$$= (n+r)(n-r+1)\int_{-1}^{1}\frac{d^{r-1}P_m}{dx^{r-1}}\,\frac{d^{r-1}P_n}{dx^{r-1}}\,(1-x^2)^{r-1}\,dx\,\ldots\ldots(30).$$

For by integration by parts we have

$$\int_{-1}^{1}\frac{d^rP_m}{dx^r}\,\frac{d^rP_n}{dx^r}\,(1-x^2)^r\,dx$$
$$= -\int_{-1}^{1}\frac{d^{r-1}P_m}{dx^{r-1}}\,\frac{d}{dx}\left\{\frac{d^rP_n}{dx^r}\,(1-x^2)^r\right\}dx\,\ldots\ldots\ldots\ldots(31).$$

But by the fundamental differential equation of Art. 851,

$$\frac{d}{dx}\left\{(1-x^2)\,\frac{dP_n}{dx}\right\} = -n(n+1)\,P_n.$$

Thus $$(1-x^2)\,\frac{d^2P_n}{dx^2} - 2x\,\frac{dP_n}{dx} = -n(n+1)\,P_n.$$

Differentiate both sides $r-1$ times: thus

$$(1-x^2)\,\frac{d^{r+1}P_n}{dx^{r+1}} - 2xr\,\frac{d^rP_n}{dx^r} - (r^2-r)\,\frac{d^{r-1}P_n}{dx^{r-1}} = -n(n+1)\,\frac{d^{r-1}P_n}{dx^{r-1}};$$

therefore

$$(1-x^2)\frac{d^{r+1}P_n}{dx^{r+1}} - 2xr\frac{d^rP_n}{dx^r} = (r^2 - r - n^2 - n)\frac{d^{r-1}P_n}{dx^{r-1}},$$

which we may put thus

$$\frac{d}{dx}\left\{(1-x^2)^r\frac{d^rP_n}{dx^r}\right\} = (r^2 - r - n^2 - n)(1-x^2)^{r-1}\frac{d^{r-1}P_n}{dx^{r-1}}.$$

Hence substituting in (31) we obtain the required result.

In the same manner as (30) was established we may obtain a similar result with $(m+r)(m-r+1)$ instead of $(n+r)(n-r+1)$ as the coefficient.

Hence if m and n are different, we must have

$$\int_{-1}^{1}\frac{d^rP_m}{dx^r}\frac{d^rP_n}{dx^r}(1-x^2)^r\,dx = 0.$$

If $m = n$ we have from (30)

$$\int_{-1}^{1}\left(\frac{d^rP_n}{dx^r}\right)^2(1-x^2)^r\,dx = (n+r)(n-r+1)\int_{-1}^{1}\left(\frac{d^{r-1}P_n}{dx^{r-1}}\right)^2(1-x^2)^{r-1}\,dx.$$

By successive applications of the formula we can obtain the value of the integral. For instance, if $r = 1$ we have

$$\int_{-1}^{1}\left(\frac{dP_n}{dx}\right)^2(1-x^2)\,dx = n(n+1)\int_{-1}^{1}(P_n)^2\,dx$$
$$= \frac{2}{2n+1}\,n(n+1)$$

by Art. 905.

953. For an example of the formulæ Legendre proposes this problem: to determine the solids, homogeneous or heterogeneous, for which every axis passing through the centre of gravity is a *principal* axis. Laplace had considered this problem for the case of a homogeneous solid in his fifth memoir: see Art. 863.

If we take x, y, z for the rectangular coordinates of the element of mass dM, we must have

$$\int xy\,dM = 0, \qquad \int yz\,dM = 0, \qquad \int zx\,dM = 0,$$

the origin being at the centre of gravity, and the integrations extending over the whole body.

Let r be the radius vector from the origin, and suppose that the integral $\int \rho r^4 dr$ is taken from the centre to the surface along any radius, and that the result assumes the form of a series of Laplace's functions

$$U_0 + U_1 + U_2 + \ldots\ldots$$

Then, as in the *Mécanique Céleste*, Livre III. § 32, the above three conditions determine to some extent the nature of U_2. The general form of U_2 being

$$H\left(\mu^2 - \frac{1}{3}\right) + H_1\mu\sqrt{(1-\mu^2)}\sin\phi + H_2\mu\sqrt{(1-\mu^2)}\cos\phi$$
$$+ H_3(1-\mu^2)\sin 2\phi + H_4(1-\mu^2)\cos 2\phi,$$

the conditions shew that we must have

$$H_1 = 0, \quad H_2 = 0, \quad H_3 = 0.$$

Legendre's treatment is in substance the same as that which was adopted by Laplace.

The next step in the problem is to observe that we must also have

$$\int x^2 dM = \int y^2 dM = \int z^2 dM.$$

Legendre treats this by estimating the value of

$$\int (\alpha x^2 + \beta y^2 + \gamma z^2)\, dM,$$

where α, β, γ are constants.

We have

$$\alpha x^2 + \beta y^2 + \gamma z^2 = \alpha r^2 \cos^2\theta + r^2 \sin^2\theta\,(\beta\cos^2\phi + \gamma\sin^2\phi)$$
$$= \left\{\frac{\alpha+\beta+\gamma}{3} + \frac{2\alpha-\beta-\gamma}{3}\left(\frac{3}{2}\cos^2\theta - \frac{1}{2}\right) + \frac{\beta-\gamma}{2}\sin^2\theta\cos 2\phi\right\} r^2.$$

If we use the same supposition as before relative to $\int \rho r^4 dr$, we shall find that our integral reduces to

$$\frac{4\pi}{3}(\alpha+\beta+\gamma)\,U_0 + \frac{8\pi}{45}(2\alpha-\beta-\gamma)\,H + \frac{8\pi}{15}(\beta-\gamma)\,H_4$$

For the problem we have in hand this expression must retain the same value when out of the three constants α, β, γ any two are made zero, and the third unity.

Therefore we must have $H = 0$ and $H_1 = 0$. Thus we shew, in fact, that U_2 must vanish completely.

Again, since by supposition the centre of gravity is the origin, we must have

$$\int x\,dM = 0, \quad \int y\,dM = 0, \quad \int z\,dM = 0.$$

Treat these in the same manner as the other conditions. Hence we shall find that if $\int \rho r^3 dr$ be supposed developed in the form of a series of Laplace's functions,

$$Z_0 + Z_1 + Z_2 + \ldots\ldots$$

then Z_1 must vanish.

This fact Laplace forgot to notice when he discussed the problem in his fifth memoir: that is, supposing the solid to be homogeneous, he ought to have excluded also the function of the *first* order from the fourth power of the radius vector. In the *Mécanique Céleste* he does not solve the problem; but only so much of it as leads to $H_0 = 0$, $H_2 = 0$, $H_4 = 0$.

Legendre says that it is easy to satisfy simultaneously the conditions which have been obtained. As a general example for a homogeneous solid be suppose that we exclude from r^4 all terms in which $\cos\theta$ and $\cos\phi$ are raised to odd powers; and so take

$$r^4 = A_0 + B_1 P_4 + B_2 \frac{d^2 P_4}{d\mu^2} \sin^2\theta \cos 2\phi + B_3 \frac{d^4 P_4}{d\mu^4} \sin^4\theta \cos 4\phi$$
$$+ C_1 P_6 + C_2 \frac{d^2 P_6}{d\mu^2} \sin^2\theta \cos 2\phi + C_3 \frac{d^4 P_6}{d\mu^4} \sin^4\theta \cos 4\phi$$
$$+ C_4 \frac{d^6 P_6}{d\mu^6} \sin^6\theta \cos 6\phi$$
$$+ \ldots\ldots\ldots$$

See Art. 950.

By reason of the exclusion of the odd powers the centre of gravity is at the origin.

As a very simple example we may take

$$r^4 = A_0 + B_4 P_4,$$

which is equivalent to

$$r^4 = a^4 + b^4 (7 \cos^4 \theta - 6 \cos^2 \theta).$$

Thus if a solid be generated by the revolution of this curve round the initial line, the moment of inertia will have the same value for any axis which passes through the origin.

954. Legendre devotes his pages 443...445 to formulæ of attraction applicable to an infinite number of figures which are not solids of revolution. He supposes the solid to be composed of strata, and that any assigned power of the radius vector of a stratum can be expressed in terms of a series of Laplace's functions. Then he gives an expression for the potential at any point. Laplace had already obtained results substantially equivalent in his fourth memoir.

955. On his pages 445...447, Legendre considers the figure of a planet supposed entirely fluid. He now assumes that the radius vector of any stratum is of the form of a series of Laplace's coefficients,

$$\beta \{1 + Z_0 + Z_1 + Z_2 + \ldots\ldots\}.$$

Thus he obtains equations of the same form as those in Art. 929; but instead of A_n we have now Z_n.

Hence he concludes that Z_n must vanish for values of n greater than 2.

Therefore the form of the planet must be that of an oblatum; this of course is only shewn under the assumption just stated as to the radius vector. Laplace had obtained this result for the case of a homogeneous mass in his fourth memoir; now Legendre extends it to a mass of variable density.

Legendre's investigations are substantially the same as those subsequently given by Laplace in the *Mécanique Céleste*, Livre III. § 29 and § 30.

956. On his pages 447...454, Legendre considers the case of a solid planet covered by a very thin stratum of fluid.

The equations to which we have referred in Articles 920 and 955 now hold, not generally, but at the surface; thus we cannot now shew, as in Art. 955, that Z_n must be zero when n is greater than 2. This part of Legendre's memoir had been substantially given by Laplace in his fourth memoir; and is reproduced in the *Mécanique Céleste*, Livre III. § 31, § 32, and § 33.

957. It will be seen from our account of Legendre's memoir, that it occupies an important position in the history of our subject. The most striking addition which is here made to previous researches consists in the treatment of a planet supposed entirely fluid; the general equation for the form of a stratum is given for the first time and discussed: see Art. 929. The investigation carried on to the second order of small quantities, which we have reproduced in Arts. 913...923, is also deserving of notice. Moreover, here for the first time we have a correct and convenient expression for Laplace's n^{th} coefficient: see Art. 950.

As we have stated in our analysis, Laplace adopted in his *Mécanique Céleste* the substance of much of Legendre's memoir, which has thus become permanently incorporated in our subject.

CHAPTER XXVI.

LAPLACE'S SEVENTH MEMOIR.

958. LAPLACE'S seventh memoir on our subject is contained in the Paris *Mémoires* for 1789, published in 1793, being the same volume as contained Legendre's fourth memoir. In the first 176 pages of the memoirs in this volume the word *Royale* occurs as part of the heading of the left-hand pages; but this word is omitted in the remainder of the volume. The explanation is furnished by the announcement on the back of the title page: "Les vingt-deux premières feuilles des mémoires de ce volume, étoient imprimées avant l'époque du 10 août 1792."

Laplace's memoir is entitled *Sur quelques points du Système du monde;* it occupies pages 1...87 of the volume: we are concerned only with pages 18...55.

959. The pages 18...43 constitute one section which is entitled *Sur les degrés mesurés des méridiens, et sur les longueurs observées du pendule.*

960. The pages 18...21 of the memoir consist of general remarks which are reproduced in the beginning of § 38 of Livre III. of the *Mécanique Céleste;* the rest of this long section of the *Mécanique Céleste* does not occur in the memoir.

The pages 21...27 of the memoir contain an account of a mode of treating the measured lengths of degrees, so as to determine from them, if possible, the elements of an elliptic figure. Laplace says on his page 21:

Cependant avant que de renoncer entièrement à la figure elliptique, il faut déterminer celle dans laquelle le plus grand écart des degrés mesurés est plus petit que dans toute autre figure elliptique, et voir si cet écart est dans les limites des erreurs des observations. J'ai donné dans nos Mémoires de 1783, une méthode pour résoudre ce problême, et je l'ai appliquée aux quatre mesures des degrés du nord, de France, du cap de Bonne-Espérance et du Pérou; mais cette méthode devient très-pénible, lorsque l'on considère à la fois un grand nombre de degrés. La méthode suivante est beaucoup plus simple.

These pages of the memoir constitute that part of § 39 of Livre III. of the *Mécanique Céleste*, which follows the first three pages. It will be observed that Laplace proposes to obtain from observations the same kind of result as in his fifth memoir, namely, that in which the greatest deviation of the observations is the least possible: but he now expounds another mode of obtaining the result.

In the *Mécanique Céleste*, Livre III. § 39, both methods are given; namely, in the first three pages the method of the fifth memoir, and in the remaining pages the method of the seventh memoir.

961. On pages 29...32 of the memoir the method is applied to nine measured lengths of degrees. In the corresponding part of the *Mécanique Céleste*, namely § 41, only seven measured lengths of degrees are used. In the memoir two French degrees are used. One is in the latitude 45° 43', "...que M. l'abbé de la Caille, dans nos Mémoires de 1758, a fixé à 57034 toises." The other is in the latitude 49° 23', "...et qu'après plusieurs vérifications, on a fixé enfin à 57074·5 toises." In the *Mécanique Céleste* only one French degree is used, namely, the mean length of the degree of France as determined by Delambre and Méchain.

One of the degrees of the memoir is not used in the *Mécanique Céleste*, namely, one which he thus describes:

Le degré de Hollande, par 52° 41' de latitude, mesuré primitivement par Snellius, et ensuite rectifié par MM. de Cassini, qui l'ont fixé à 57145 toises. La grandeur de ce degré vient d'être confirmée par les nouvelles mesures que l'on a faites en Angleterre, et avec lesquelles elle est à fort peu près d'accord.

The conclusion from the numerical calculations is thus stated:

Ainsi, de quelque manière que l'on combine les neuf degrés précédens, quelque rapport que l'on choisisse pour celui des deux axes de la terre, il est impossible d'éviter dans l'ellipse, une erreur de 108^{r}; et comme cette erreur étant la limite de celles qui peuvent être admises, elle est par cela même infiniment peu probable; il faudroit, pour admettre une figure elliptique, supposer des erreurs plus grandes encore que 108^{r}, dans quelques uns de ces degrés.

La valeur que nous venons de trouver pour y, donne une ellipse dont les axes sont dans le rapport de 249 à 250. Dans cette ellipse, les trois plus grandes erreurs tomberoient sur les degrés de Pensilvanie, du cap de Bonne-Espérance, et du Nord. En considérant avec attention les mesures de ces trois degrés, il me semble impossible qu'il se soit glissé dans chacun d'eux une erreur de 108^{r}, sur-tout après les réductions que j'ai déja faites au degré du nord. Il me paroît donc prouvé par les mesures précédentes, que la variation des degrés des méridiens terrestres s'écarte sensiblement de la loi du carré du sinus de la latitude, qui résulte d'une figure elliptique.

This conclusion may be compared with the corresponding passage in the *Mécanique Céleste*. There instead of the 108 toises we have 48·6 double toises, that is 97·2 toises; the degree in the *Mécanique Céleste* is taken in the *centesimal* scale. The signification of y is the same in the two places; in fact, if y be divided by the mean length of a degree of the meridian, the result is three times the ellipticity. In the memoir y is found to be 684·73 toises; in the *Mécanique Céleste* it is 616·404 toises. The ellipticity in the *Mécanique Céleste* is found to be $\frac{1}{277}$.

In the *Mécanique Céleste* the inference as to the inadmissibility of the elliptic law of variation of the length of the degrees is stated less confidently than in the memoir.

962. In pages 32...35 of the memoir Laplace expounds another principle which may guide us in treating the observations; this part of the memoir forms § 40 of Livre III. of the *Mécanique Céleste*. Laplace here proposes to find an ellipse such that (1) the sum of the errors should be zero, and (2) the sum of the errors taken with the positive sign should be a mini-

mum. He calls such an ellipse the *most probable ellipse* in the *Mécanique Céleste:* see page 140 of Vol. II. The following sentence respecting the author of the principle is given in the memoir, but not in the *Mécanique Céleste*.

M. Boscovich a donné pour cet objet, une méthode ingénieuse qui est exposée à la fin de l'édition française de son Voyage astronomique et géographique; mais comme il l'a inutilement compliquée de la considération des figures, je vais le présenter ici sous la forme analytique la plus simple.

Boscovich had previously expounded his principle in his supplementary annotations to Stay's poem *Philosophiae Recentioris*. See Arts. 511 and 514.

963. On page 36 of the memoir Boscovich's method is applied numerically to the nine measured lengths already adopted. In like manner in § 41 of the *Mécanique Céleste*, Boscovich's method is applied numerically to the seven measured lengths adopted in that work. In the memoir Laplace thus finds, for the length in toises of a degree of the meridian at the latitude θ, the expression

$$56753 + 613 \cdot 1 \sin^2 \theta.$$

He states his conclusion thus:

Le rapport des axes de la terre est alors celui de 278 à 279; mais l'expression précédente donne une erreur en plus, de $137^t.7$ dans le degré du Nord, et une erreur en moins, de $109^t.9$ dans celui de Pensilvanie, ce qui ne peut pas être admis. On voit ainsi qu'il n'est pas possible de concilier avec une figure elliptique, les degrés du méridien.

This may be compared with the corresponding passage in the *Mécanique Céleste:* see Vol. II. page 141. The length in toises of a *centesimal* degree of the meridian at the *centesimal* latitude θ is found to be

$$51077 \cdot 7 + 499 \cdot 86 \sin^2 \theta.$$

964. The memoir now passes to the subject of the length of the seconds pendulum; on pages 37 and 38 we have thirteen such measures, with references to the authorities. The following table gives in the first column the place, in the second the latitude, in the third the length of the pendulum in lines, and in the fourth the name of the observer on whom the result depends.

Equator	0° 0'	439·21	Bouguer.
Portobello...........	9°34'	439·30	Bouguer.
Little Goave.........	18°27'	439·47	Bouguer.
Cape of Good Hope	33°18'	440·14	La Caille.
Rome	41°54'	440·38	Jacquer and Sueur.
Vienna	48°12'30"	440·56	Liesganig.
Paris..................	48°50'	440·67	Bouguer.
London...............	51°31'	440·75	Graham.
Arensbourg	58°15'	441·07	Griscow.
Pernavia	58°20'	441·10	Griscow.
Petersburg	59°56'	441·21	Griscow.
Pello..................	66°48'	441·27	Maupertuis.
Ponoi	67°4'30"	441·41	Mallet.

Laplace says that the lengths have been reduced to a vacuum, to the level of the sea, and to the temperature of about 14° of Réaumur. With respect to the length at London, Laplace says that it has been determined by assuming with Maupertuis that the Paris seconds pendulum, if transported to London, makes 7·7 more oscillations in a day. I have put Graham's name to this length because he obtained the result which Maupertuis adopted; see Maupertuis's *La Figure de la Terre*...page 172.

For the observations of Griscow Laplace refers to the *Nouveaux Mémoires de Pétersbourg*, Vol. VII.; and for the observation of Mallet to the *Nouveaux Mémoires de Pétersbourg*, Vol. XIV. part II.

965. In the corresponding part of the *Mécanique Céleste* fifteen lengths of the seconds pendulum are used; the observations at Rome and Pernavia are omitted; but observations at Pondicherry, Jamaica, Toulouse and Gotha are introduced.

Moreover the length at Petersburg is now stated to rest on observations made by Mallet, and is put at a smaller value. The length at Paris is taken for the unit, and that of Petersburg is put at 1·00074.

It will be seen in the memoir that although the difference between the latitudes of London and Paris is *more* than four times as great as the difference between the latitudes of Paris and Vienna, yet the difference between the lengths of the pendulum is *less* in the former case than in the latter. In the *Mécanique Céleste* this anomaly is reduced by putting the length of the pendulum at Vienna greater than in the memoir, namely at ·99987, that of Paris being unity. But no reason is assigned for the change.

There is however a curious mistake in the *Mécanique Céleste.* According to Laplace's words the last two lengths which he uses stand thus, the latitudes being in the *centesimal* scale:

Place.	Latitude.	Length of Pendulum.
Ponoi............	74°22	1·00137
Pello	74°33	1·00148

There can be no doubt that Ponoi and Pello must interchange places. The centesimal latitude 74°22 corresponds to the ordinary 66°48′, which is the latitude of Pello according to Maupertuis; and the 1·00137 corresponds to the 1·0014 of Maupertuis; see pages 162 and 179 of *La Figure de la Terre....*

The mistake is of course reproduced in the national edition of Laplace's works; it escaped the notice of the accurate Bowditch: the table occurs on page 470 of the second volume of his translation of the *Mécanique Céleste.*

966. The lengths of the seconds pendulum adopted in the memoir are subjected to the two methods of treatment which were applied to the lengths of degrees: see pages 38...43 of the memoir. The corresponding investigation in the *Mécanique Céleste* will be found in § 42 of Livre III.

The length of the seconds pendulum in the latitude θ, expressed in lines, is found in the memoir to be by the first method

$$439·3090 + 2·4286 \sin^2 \theta,$$

and by the second method

$$439·2110 + 2·5827 \sin^2 \theta.$$

From the latter expression, by using Clairaut's theorem, Laplace finds $\frac{1}{359}$ for the ellipticity. Although this value is much smaller than that which results from the measures of degrees, yet Laplace thinks that it is not too small. Besides the evidence from pendulum observations, Laplace appeals to the phenomena of precession and nutation; and also to the value of the ellipticity of Jupiter as an analogy in favour even of a still smaller ellipticity.

967. The words which Laplace uses with respect to Clairaut's theorem should be noticed; he says on page 42 of the memoir:

Ce résultat a généralement lieu, quelle que soit la figure de la terre, pourvu que les variations des longueurs du pendule suivent, à fort peu près, la loi du carré du sinus de la latitude; ce qui, comme on vient de le voir, est le cas de la nature (*Voyez nos Mémoires pour l'année* 1783).

The language seems too strong; for I presume Laplace really intends to assume that the strata are nearly spherical. His reference to the memoirs of 1783 is not very precise. I think he means to direct attention to the use he makes in that memoir of formulæ which he had obtained in his memoir of 1782, and which are reproduced in the *Mécanique Céleste*, Livre III. § 33.

968. The pages 44...55 of the memoir constitute a section which is entitled *Sur la Figure de la Terre*; it begins thus:

J'ai fait voir dans nos Mémoires pour l'année 1782, que si l'on suppose la Terre fluide et homogène, sa figure ne peut être que celle d'un ellipsoïde de révolution. Je me propose ici d'étendre ce résultat, au cas où la Terre ayant été primitivement fluide, elle seroit formée de couches de densités variables. M. Clairaut a déjà fait voir que la figure elliptique remplit dans ce cas, les conditions de l'équilibre; mais il s'agit de prouver qu'elle est la seule qui satisfasse à ces conditions. Pour cela, je vais rappeller quelques propositions que j'ai démontrées dans les Mémoires cités.

These pages of the memoir treat then of the figure of the Earth considered as a heterogeneous fluid; they occur in the *Mécanique Céleste*, Livre III. §§ 29, 30, and 31. Laplace however makes no substantial addition to the results which Legendre had obtained on the subject in his fourth memoir.

CHAPTER XXVII.

MISCELLANEOUS INVESTIGATIONS BETWEEN THE YEARS 1781 AND 1800.

969. THE present Chapter will contain an account of various miscellaneous investigations between the years 1781 and 1800.

970. We have first to notice a memoir by Euler, entitled *Enodatio difficultatis super Figura Terrae a vi centrifuga oriunda.*

This memoir occurs in the *Nova Acta...*, St Petersburg, Vol. II.; the volume is for the year 1784, and was published in 1787. The memoir occupies pages 121...130 of the volume.

The memoir was presented on the 2nd of Nov. 1775.

If the Earth is considered as fluid and nearly spherical, and the fluid is acted on by a force tending towards the centre, then whatever be the law of force, if the centrifugal force be small compared with the attractive force, the ellipticity is about $\frac{1}{578}$; see Art. 57. But Euler says the measures of degrees give the ellipticity about $\frac{1}{200}$. This is the difficulty to be resolved.

Instead of adopting the theory of universal gravitation and attempting to determine the figure of the Earth from that, Euler supposes a force tending to the centre which is some function of the distance, and also a transversal force. This transversal force he arbitrarily assumes to be proportional to $\frac{\sin\theta\cos\theta}{r}$, where r and θ are the usual polar coordinates. By taking the magnitude of this transversal force such that its greatest value at the surface is $\frac{1}{300}$ of the attraction there, Euler manages to arrive at the ellipticity which he wants, namely $\frac{1}{200}$.

Such a memoir at the beginning of the eighteenth century would not have caused any surprise; but it is certainly remarkable that it should have appeared towards the end of the century. The memoir is quite destitute of value, and it is difficult to see on what ground it could have been published nearly fifty years after Maclaurin had established the relative equilibrium of rotating fluid in the form of an oblatum; and also after Laplace had produced his work on the *Figure of the Planets.*

971. We have next a memoir by W. L. Krafft, entitled *Essay relatif aux recherches de M. De La Grange sur l'attraction des sphéroïdes elliptiques.*

This memoir occurs in the *Nova Acta*...St Petersburg, Vol. II.; the volume is for the year 1784, and was published in 1787. The memoir occupies pages 148...160 of the volume. The memoir was read on the 8th of March, 1787. The author is, I presume, the same as that of the essay noticed in Art. 687.

Krafft's object is to obtain by the aid of *rectangular* coordinates what Lagrange obtained in his memoir of 1773 by the aid of *polar* coordinates; namely the attraction of an oblatum at any point of the axis or of the axis produced, and at any point of the equator.

Krafft's memoir then adds nothing to preceding results, but constitutes an example in the Integral Calculus which might be used for the exercise of students.

If the major axis of the oblatum is to the minor axis as 101 is to 100, Krafft says that the attraction at the Pole is to the attraction at the Equator as 1 is to ·99773. Krafft refers to Euler's memoir of 1738, where the ratio is stated to be as 1 is to ·99803. See Arts. 229 and 693.

At this time the Academy of St Petersburg was under the direction of a lady; and the historical part of the volume in which these two memoirs are contained makes frequent reference to *Madame la Princesse de Daschkaw.*

972. The celebrity of Bernardin de Saint-Pierre may justify a short notice, though all which he contributed to our subject was the revival of an antiquated blunder. In his *Études de la Nature* he maintained that the fact of the increase of the length of a degree of the meridian, in passing from the equator to the pole, established the oblong form of the Earth. The work was published in 1784; more details on this point seem however to have been given in a subsequent edition; but I have had access only to the collected works of Saint-Pierre, published in 12 volumes at Paris in 1818, in which the matter occurs in Vol. III. pages vii...xii, and in Vol. V. pages 413...417. The nature of the error is the same as we have pointed out in the case of Keill: see Art. 76. Such an eminent example might be cited as some excuse, but we must remember that owing to the steady advancement of knowledge, even a novelist at one epoch may be fairly expected to understand elementary principles which puzzled a professor of an earlier century. The author of *Paul and Virginia*, however, seems to have had no qualifications for the pursuit of exact science; besides his error as to the Figure of the Earth, he advocated an absurd hypothesis of his own to account for the phenomena of the Tides.

973. Cousin. A work entitled *Introduction à l'étude de l'Astronomie Physique*, was published in 1787 at Paris by Cousin. The author styles himself *Lecteur et Professeur royal, de l'Académie royale des Sciences.* The work is in quarto; the title, dedication, and preliminary discourse occupy xvi pages; then the text follows on 323 pages: there are two plates.

974. The part of the work with which we are concerned is the fourth Chapter on pages 135...176, which is entitled *De l'action mutuelle des corps, lorsqu'elle résulte des attractions de toutes les parties qui les composent.*

975. Cousin finds the attraction of an ellipsoid of revolution, oblate or oblong, on an internal particle; he follows the method of Lagrange given in the Berlin Memoirs for 1773: see Art. 707.

With respect to an external point, Cousin refers to the first and second memoirs by Legendre; and contents himself with working out the case in which the attracted particle is on the prolongation of the axis of revolution. He gives the result for an oblatum and for an oblongum.

976. Cousin passes on to the equations of fluid equilibrium. He begins very unfortunately, on his page 141, by confounding equality of pressure with equable transmission of pressure. He asserts that a fluid will not remain at rest unless all points of its surface are acted on by *equal* normal forces. This of course is untrue.

However, he obtains the correct equations of fluid equilibrium; and says he will make some applications of them. He refers to Clairaut's *Figure de la Terre*, to Euler's memoir of 1755, and to D'Alembert's *Opuscules Mathématiques*, Vols. V. and VI.

Accordingly he applies the equations of fluid equilibrium to the relative equilibrium of rotating fluid; and arrives at the accurate equation connecting the angular velocity with the excentricity. He shews that there cannot be more than two values of the excentricity of an oblatum for a given angular velocity. He proceeds in a manner which would be naturally suggested by pages 47...67 of the sixth volume of D'Alembert's *Opuscules Mathématiques*. As Cousin gives no reference, I presume that we may attribute to himself this demonstration, that only two values are possible. Laplace gave the first demonstration in his *Figure des Planetes*; see Arts. 657 and 811. Cousin's demonstration is perhaps a little simpler than Laplace's of 1784; but inferior to that adopted in the *Mécanique Céleste*. Cousin's is founded on D'Alembert's, but avoids the errors in it: see the *Opuscules Mathématiques*, Vol. VIII. pages 292 and 293.

There are, however, a few words inserted in his process by Cousin which should be noticed. He says on his page 148: "On parviendroit au même résultat en supposant le demi-axe plus grand que le rayon de l'équateur;..." It would seem that Cousin was not aware that an *oblongum* could not be a possible form of relative equilibrium; yet Laplace had drawn attention to this fact in the *Figure des Planetes*. See Art. 812.

977. Cousin establishes the theorem due to Maclaurin which Lagrange has discussed in the Berlin *Mémoires* for 1775: see Art. 720.

Cousin does not, like other French mathematicians, assert that Maclaurin only enunciated the theorem: Cousin says more cautiously on his page 148:

Maclaurin a cherché (Traité *des fluxions*, n°. 653) s'il n'y auroit pas quelque analogie semblable entre des sphéroïdes homogenes qui ne seroient pas des solides de révolution...

Cousin's demonstration is somewhat simpler than Lagrange's; I suppose it is Cousin's own, for no reference is given.

978. Certain approximate formulæ which Cousin gives may be reproduced; his notation is rather different from that which is common, and thus his results may be usefully recorded.

Let c be the smaller semiaxis, and $c(1+\alpha)$ the larger semiaxis of an ellipsoid of revolution. Then for an oblatum:

the attraction at the pole

$$= \frac{4\pi\rho c}{3}\left\{1 + \frac{4\alpha}{5} - \frac{2\alpha^2}{7} + \frac{8\alpha^3}{105} - \ldots\right\};$$

the attraction at the equator

$$= \frac{4\pi\rho c}{3}\left\{1 + \frac{3\alpha}{5} - \frac{9\alpha^2}{35} + \frac{11\alpha^3}{105} - \ldots\right\}.$$

And for an oblongum:

the attraction at the pole

$$= \frac{4\pi\rho c}{3}\left\{1 + \frac{\alpha}{3} - \frac{2\alpha^2}{7} + \frac{22\alpha^3}{105} - \ldots\right\};$$

the attraction at the equator

$$= \frac{4\pi\rho c}{3}\left\{1 + \frac{2\alpha}{5} - \frac{9\alpha^2}{35} + \frac{16\alpha^3}{105} - \ldots\right\}.$$

Then if an oblatum of rotating fluid is in relative equilibrium and β denote the ratio of centrifugal force at the equator to the *gravity* there, not the *attraction* there, we must have

$$\beta = \frac{4\alpha}{5} - \frac{2\alpha^2}{5.35} + \frac{8\alpha^3}{25.35} - \ldots$$

whence
$$\alpha = \frac{5\beta}{4} + \frac{5\beta^2}{224} - \frac{135\beta^3}{6272} - \ldots$$

Here $\beta = \frac{j}{1-j}$, where j denotes the ratio of centrifugal force at the equator to *attraction* there.

979. Cousin investigates approximate expressions for the attraction of an ellipsoid, not of revolution, at a point on its surface. He applies them to determine the form of relative equilibrium of the moon, supposed homogeneous and fluid. The investigation is of the same character as Laplace had given in his *Figure des Planetes;* but Cousin does not refer to any preceding author for it: see Art. 809.

980. Cousin proposes on his page 156 to pass to the case in which the fluid is not homogeneous but composed of ellipsoidal shells. He says on his page 158 that he has tried to develop and generalise what Clairaut had said in the second and third chapters of the second part of his *Figure de la Terre.* The attempt at generalisation consists in discussing the relative equilibrium of a revolving ellipsoid which is nearly spherical. See Cousin's pages 156...163. The process is long and tedious. Cousin arrives at an equation connecting the angular velocity with the ellipticities; and at a result which is analogous to Clairaut's theorem. But the investigation is a failure. Cousin makes out that the attraction on a particle at the surface of an ellipsoid is exerted in the *meridian plane,* which is not true to the order of approximation he requires. The fact is that he takes the particle on the surface to be in a *principal plane,* and then he forgets this restriction.

We now know from the discussions on Jacobi's theorem that the relative equilibrium of a rotating ellipsoid of fluid is indeed possible, but in that case the ellipsoid is not nearly spherical.

If Cousin's investigation had been accurate, he might have drawn from his equation (K) on page 163, the inference that the nearly spherical ellipsoid could not be in relative equilibrium.

For in this equation $\sin\beta$ is variable, and so we must have $H-I=0$, and this renders the ellipsoid a figure of revolution. It is curious that he makes no remarks on this fact, which presents itself so naturally in his investigation. But he gives no adequate account throughout of what he wishes to prove or of what he has proved.

981. Cousin then considers the special case of an oblatum; his result is correct, and exactly corresponds with that which is given by Clairaut on his page 217, and which we have reproduced in Art. 323. Cousin then proceeds to urge the same objection to another formula of Clairaut's which D'Alembert brought forward in the sixth volume of the *Opuscules Mathématiques*, and elsewhere: see Arts. 328 and 377. The objection seems to me to be of no importance.

Cousin arrives at what I call Clairaut's derived equation: see Art. 343.

982. Cousin's pages 167...176 are taken from Laplace's third memoir, to which, and to the second memoir, Cousin refers. The main result is that which we have noticed in Art. 765, and which is embodied in the second equation of Art. 773.

The error or misprint which occurred for a moment in Laplace seems to be seriously adopted by Cousin: see Art. 769.

Cousin, in fact, allows himself to use the integrals $\int\frac{dy}{\cos y}$ and $\int\tan y dy$, between the limits 0 and π; but the expression to be integrated becomes infinite, and so we cannot trust the process.

However, Cousin's final results are correct, as they can be obtained without this suspicious step: Laplace himself obtained them correctly.

983. On the whole, although I consider that the design of such a work as Cousin's is excellent, I cannot praise his performance. He presents Clairaut's main results, substituting more analytical work for Clairaut's, which has a geometrical character; and he gives the substance of Laplace's third memoir. He adds

nothing of his own; nor does he effect any improvement which renders the investigations more simple or more interesting. As we have seen he is not uniformly accurate; and his work is rendered repulsive by the want of distinct statements as to what he is about to investigate. There is a meagre summary on pages 314 and 315 of the contents of the chapter; but it is far too brief.

Cousin does not introduce the Potential function, nor Laplace's functions, though both of these had already been brought under the notice of mathematicians. And he never refers to the work of Laplace on the *Figure of the Planets*, of which we have given an account in Chapter XXI.

984. In the *Philosophical Transactions* for 1785, published in that year, we have a memoir entitled *An Account of the Measurement of a Base on Hounslow-Heath. By Major-General William Roy.*

The measurement of this base may be considered to be the foundation of the important Trigonometrical Survey of Great Britain. Other memoirs relating to the progress of the survey are given in the *Philosophical Transactions* for 1790, 1795, 1797, 1800 and 1803.

The memoirs are substantially reproduced in the *Account of the Operations carried on for accomplishing a Trigonometrical Survey of England and Wales....* This work consists of three quarto volumes published in 1799, 1801 and 1811 respectively: in the prefaces to the first and third volumes will be found notices of the differences between the original memoirs and the republication. We shall not need to give any notice of the original memoirs.

The account of the measurement of the base seems to have been translated into French: see Voiron's *Histoire de l'Astronomie*, page 228: he calls the locality *Houslowwheat*.

985. In the Paris *Mémoires* for 1785, published in 1788, there is a memoir by La Lande, entitled *Mémoire sur la quantité de l'aplatissement de la Terre*. The memoir occupies pages 1...6 of the volume.

La Lande refers to observations of the length of the seconds pendulum at Spitzbergen, made by Lyons in 1773. From this, and Bouguer's determination of the length of the pendulum at the equator, La Lande obtains $\frac{1}{185}$ as the value of Clairaut's fraction. Then Clairaut's theorem gives $\frac{1}{302}$ for the ellipticity, so that $\frac{1}{300}$ may be conveniently adopted. These values differ very little from those at present received.

We see on page 7 that La Lande now possessed the toise which formerly belonged to Mairan, and considered it to be $\frac{1}{11}$ of a line shorter than the toise of Peru.

986. In the *Philosophical Transactions* for 1787, published in that year, there is a memoir entitled *An Account of the Mode proposed to be followed in determining the relative Situation of the Royal Observatories of Greenwich and Paris. By Major-General William Roy.* The memoir occupies pages 188...228 of the volume; with an Appendix on pages 465...470.

The memoir begins by referring in these words to the operation which we noticed in Art. 984:

Two years have nearly elapsed since an account of the measurement of a base on Hounslow-Heath was laid before the Royal Society, being the first part of an operation ordered by his Majesty to be executed for the immediate purpose of ascertaining the relative situations of the Royal Observatories of Greenwich and Paris; but whose chief and ultimate object has always been considered of a still more important nature, namely, the laying the foundation of a general survey of the British Islands.

The memoir points out the stations which would be suitable for determining the relative situation of the two Observatories. There is an account of the execution of the proposed design in the *Philosophical Transactions* for 1790, which is reproduced in the work cited in Art. 984.

On pages 224 and 225 of the memoir it is suggested that trigonometrical surveys might be undertaken with advantage, in the East

Indies, near the mouth of the Amazon, and in Russia. The first and the last of these operations have since been conducted on a very extensive scale; let us hope that Brazil will soon undertake the other.

The memoir contains some elaborate numerical calculations which are more closely connected with our subject than the details of the proposed survey. A table is given in which seven numerical results taken from the great French arc of the meridian are compared with the values which would be obtained from certain assumed forms of the Earth. Ten such assumed forms are considered, namely, a sphere, seven ellipsoids of revolution, and two other spheroids. The differences between the observed and the calculated values are much the least for the spheroid which represents Bouguer's hypothesis, that the increment of the radius of curvature varies as the fourth power of the sine of the latitude. General Roy gives a decided preference to this hypothesis; he is, I think, the only follower of Bouguer in this respect. We read in a note on page 211 of the memoir:

...... when the comparison is fairly drawn between this and every other system that has hitherto been submitted to the consideration of the public, M. Bouguer's will be found to be justly entitled to the preference, which I have here endeavoured to give it. His works shew, that he was a man of very superior abilities, eminent as a mathematician, and perhaps the best practical one that ever existed.

I was glad to find the high opinion which I had previously formed of Bouguer confirmed by the testimony of General Roy, which I had not seen when my Article 363 was written.

987. There are some points which require notice in General Roy's memoir.

The first ellipsoid which he considers is one in which the ratio of the axes is nearly that of 179 to 178. This ratio has been assumed on the authority of pendulum experiments; it is not quite clear to me how the ratio was deduced. General Roy says:

With regard to the first ellipsoid, supposing the earth to be homogeneous, it is well known, that the ratio of its semidiameters may be found, by comparing with each other the lengths of the pendulums that vibrate seconds in different latitudes.

What I think General Roy did was simply to make use of the theorem given in Art. 39; but this is very strange, because we know that if the Earth be considered as a homogeneous fluid, we have also the theorem given in Art. 28, namely, that $\epsilon = \frac{5j}{4}$; and it is a very arbitrary process to adopt one of these two results of theory and reject the other.

The last ellipsoid considered is one in which the ratio of the axes is that of 540 to 539: we are not told what suggested the assumption of this ratio.

General Roy also gives a table of the lengths of degrees of the meridian, and of degrees of great circles perpendicular to the meridian, and of degrees of great circles oblique to the meridian, calculated for Bouguer's spheroid which seemed to agree so well with the observations. But a formula given by Bouguer presented obvious difficulty, and General Roy made some arbitrary changes in consequence. Maskelyne however pointed out that there was a misprint in Bouguer's formula; this had led to the difficulty and caused some error in General Roy's calculations: the *Appendix* to the memoir relates to this matter. I may state that the misprint in Bouguer's formula would seem to be sufficiently obvious: I had corrected it in my copy of the book before I saw General Roy's memoir.

Some other errors in the present memoir are corrected in the *Philosophical Transactions* for 1790; see page 201 and a page of *Errata*.

988. A few words may be given to two works by one author, which profess to treat on our subject, but are quite worthless.

The first is entitled, *An entire new work, and method of proceeding to discover the variation of the Earth's diameters,... by Thomas Williams, Inventor*....London...1786.

The second is entitled, *Method to discover the difference of the Earth's diameters;... by Thomas Williams.* London...1788.

The first is in quarto and consists of a Title and text on 16 pages, and 4 pages of Tables. The second is in octavo, and consists of viii + 75 pages, besides Errata and Tables on 14 pages, and two Plates.

The pamphlet in quarto gives an outline of the author's notions, which are exhibited at greater length in the octavo volume. He was obviously an illiterate and unscientific person, and his publications consist of arbitrary hypotheses and assertions; they are, moreover, so obscure as to be almost unintelligible. In modern language we may say that he *assumes* the formula $a + bn^2$ to represent the length of a degree of the meridian; where n is the number of degrees in the latitude, and a and b are constants. He determines the constants by the lengths of the degrees in Peru and Lapland, and he maintains that the formula will then agree reasonably well with the other measured degrees.

Moreover he asserts that the ratio of the lengths of the two extreme degrees of the meridian is also the ratio of the diameters. Thus he concludes that the equatorial diameter is to the polar as 46 is to 45; and consequently that the polar diameter is 174 miles shorter than the equatorial.

The author touches on the subject of a universal standard for weights and measures. He suggests that the English foot should be defined, such that 365472 feet will be the length of the degree of the meridian in the latitude of London. For a unit of weight he suggests the weight of a cubic foot of sea-water.

Bound up with the copy of the octavo volume which I have examined, there is a printed document by the author, entitled, *Proposals for defraying, by subscription, the expences attending the making experiments for ascertaining whether the Earth be a solid body, as at present supposed, or only a shell.* The nature of the experiments is not stated; the author thinks that from some calculations and experiments which he has already made, "the Thickness of Matter composing the Shell is not above 50 Miles."

The copy of the quarto pamphlet which I have examined contains the following note in manuscript, which is probably due to the author himself:

If any part of this work should seem ambigious and not fully comprehended in so short a space the Author is ready to illusterate any such part.

And it is also the Authors earnest request that if any person acquainted with the subject Judge it to contain any Error to point out the Error and send it me as my object in view above althings is the Truth of the thing Asserted and which all the World ought to be acquainted with, as the Error arising in a misconception of this Matter is a Source for many others.

988. In the Paris *Mémoires* for 1787, published in 1789, will be found some articles bearing indirectly on our subject, which we will briefly notice.

On pages 216...222 there is a memoir by La Lande, *Sur la mesure de la Terre, que Fernel publia en* 1528.

This memoir is important in connexion with the history of the measurement of the length of a degree; but this is a matter which we do not profess to treat upon with any detail. Fernel observed the sun's meridian altitude at Paris; and then proceeded northwards for one degree. The length was determined from the number of revolutions of the wheel of a carriage by which he returned to Paris. The length of a degree thus deduced is usually given as 56746 toises. La Lande however considers that allowance should be made for a change in the length of the toise, thus bringing Fernel's result to 57070 toises, which differs by only a toise from the received value at the date of La Lande's memoir. But there seems to be a serious error as to what was really the length of a foot according to Fernel, which completely changes La Lande's conclusion: see *Penny Cyclopædia*, article *Weights and Measures*.

990. On pages 352...383 there is a memoir by Legendre entitled, *Mémoire sur les Opérations trigonométriques, dont les résultats dépendent de la figure de la Terre.*

This memoir investigates formulæ which are necessary for the reduction and the calculation of triangles on the surface of a spheroid. The formulæ are applied to the triangles formed between Dunkirk and Greenwich.

The memoir contains, I presume for the first time, the theorem which we now call *Legendre's theorem;* it is not demonstrated, but only enunciated in these words:

Théorème concernant les triangles sphériques, dont les côtés sont très-petits par rapport au rayon de la sphère.

Si la somme des trois angles d'un triangle sphérique infiniment petit, est supposée $180^\circ + \omega$, et que de chaque angle on retranche $\frac{1}{3}\,\omega$, afin que la somme des angles restans soit précisément de 180°, les sinus de ces angles seront entr'eux comme les côtés opposés; de sorte que le triangle, avec les angles ainsi diminués, pourra être considéré et résolu comme s'il étoit parfaitement rectiligne.

901. On pages 506...529 there is a memoir by Monge entitled, *Mémoire sur quelques effets d'attraction ou de répulsion apparente entre les molécules de matière.*

This is an account of experiments relating to phenomena of the nature of what is called capillary attraction.

902. In the Paris *Mémoires* for 1788, published in 1791, will be found some articles bearing indirectly on our subject: we will give the titles:

I. A report made to the Academy on the choice of a unit of measures, by Borda, Lagrange, Laplace, Monge, and Condorcet. The report is dated 19 March, 1791: it occupies pages 7...16 of the historical portion of the volume.

II. An account of the labours of the Academy on the project of uniformity in measures and weights: it occupies pages 17...20 of the historical portion of the volume.

III. A memoir by Cassini on the connexion of the Observatories of Paris and Greenwich, with a sketch of the antecedent geographical operations in France: it occupies pages 706...717 of the volume.

IV. A memoir by Brisson on the uniformity of measures of length, volume, and weight; and on a new method of constructing toises which were to serve as standards: it occupies pages 722...727 of the volume.

V. A memoir by Legendre on the series of triangles which serve to determine the difference of longitude between the observatories of Paris and Greenwich. This memoir occupies pages 747...754 of the volume: it is a continuation of a memoir in the *Mémoires* for 1787.

993. A very important theorem with respect to attractions occurs in a memoir on Electricity, by Coulomb, in the Paris *Mémoires* for 1788, published in 1791: see page 677 of the volume. The theorem may be thus enunciated: Let there be a closed film of matter which attracts according to the ordinary law; let the form of the film be any whatever, provided that the resultant attraction at an internal point is zero: then the resultant attraction at an external point which is indefinitely near any part of the suface is $4\pi\rho$, where ρ is such that $\rho\omega$ is the quantity of matter in an element ω of the film close to the attracted point.

The proposition is not formally enunciated in this manner by Coulomb; but it is substantially involved in his demonstration. The principle of his demonstration is the same as had been used by Lagrange in 1759: see Art. 561.

Suppose two points P and P' indefinitely close to the film, on a common normal to the surface; let P be inside and P' outside. Through P draw a plane at right angles to the normal; this will divide the film into two parts, an infinitesimal part, say S, and the remainder of the film, say S'.

Now consider S as an infinitesimal plane circular area; its action at P' will be $2\pi\rho$, along the normal: this follows from elementary investigations on attraction. The action of S at P will be ultimately equal in amount to the action of S at P', though in an opposite direction; thus this action is also $2\pi\rho$. Then since a particle at P would be in equilibrium, the action of S' at P must also be equal to $2\pi\rho$, and be in the opposite direction to the action of S at P, that is in the same direction as the action of S at P'. Finally, the action of S' at P' will be ultimately equal to the action of S' at P, that is to $2\pi\rho$. Thus the joint action of S and S' at P' is $4\pi\rho$.

In the particular case in which the film is bounded by concentric spherical surfaces, the proposition is an immediate result of theorems given by Newton. If I understand the matter rightly, the order of Coulomb's investigations in electricity with respect to the proposition was the following: the result given by Newton's theory for the spherical film was verified experimentally; then experiment shewed that the result was true for films of other forms: and finally the theoretical demonstration of the general proposition presented itself.

Coulomb fell into a slight error in the application of his theorem: see the *Cambridge and Dublin Mathematical Journal*, Vol. I. page 93.

994. The first edition of a famous work by Lagrange, appeared in 1788 in one volume, entitled, *Méchanique Analitique*. There is nothing in this edition which bears explicitly on our subject. But on his page 474 Lagrange gives, in fact, an integral in the form of a series of the partial differential equation

$$\frac{d^2V}{da^2}+\frac{d^2V}{db^2}+\frac{d^2V}{dc^2}=0;$$

and from this integral, as we shall see hereafter, Biot drew important inferences with respect to the attraction of a body.

995. We next consider a memoir entitled, *On the Resolution of Attractive Powers*. By *Edward Waring*, M.D., F.R.S., *and Lucasian Professor of Mathematics at Cambridge.*

This memoir is contained in the *Philosophical Transactions* for 1789, published in that year. It occupies pages 185...198: it was read May 28, 1789.

This memoir investigates differential expressions for the attraction of a straight line, a plane area, and a solid, the law of attraction being expressed by any function of the distance. It then passes to other subjects, as for example the differential expression for the surface of any solid. I cannot understand on what ground this memoir was published, for it does not appear to contain the slightest novelty.

996. In the *Ephemerides Astronomicæ* for 1791, published at Vienna in 1790, there is a memoir entitled *Dissertatio de Figura Telluris e Solis Eclipsibus deducta, a Francisco de Paula Triesnecker*; the memoir occupies pages 387...412 of the volume.

This memoir belongs to practical astronomy rather than to the subject of which we are tracing the history; so that a very brief indication of its nature is all that need be given here.

If the circumstances of an eclipse of the sun are carefully observed we may obtain the value of the errors in the moon's longitude and latitude which are recorded in the tables; but this supposes that we know the figure of the earth, which is required in order to allow for parallax. If we assume that the *ratio* of the error in latitude to the error in longitude is sufficiently known, we can apply our observations of an eclipse to yield information as to the ellipticity of the earth.

Sixteen eclipses of the sun are finally used in the memoir; these occurred at various dates between 1706 and 1788: the ellipticity deduced is $\frac{1}{329}$. But the process is very unsatisfactory; for it is plain that the observations are not of sufficient accuracy to warrant any strong reliance; and they are treated in a very arbitrary manner to make them yield a result. The chief practical obstacles consist in the difficulty of determining the instants of initial and final contact of the sun and moon, and the uncertainty as to the values of the apparent diameters of these bodies. The remarks made on these points may be of some interest to astronomical observers; but regarded as a contribution to our subject the memoir may be safely pronounced of no value.

997. A memoir is contained in the *Philosophical Transactions* for 1791, published in that year, entitled *Considerations on the Convenience of measuring an Arch of the Meridian, and of the Parallel of Longitude, having the Observatory of Geneva for their common Intersection. By Mark Augustus Pictet, Professor of Philosophy in the Academy of Geneva.*

The memoir occupies pages 106...127 of the volume, and is accompanied by a map.

Pictet considered that an arc of meridian of about 1° 24′, and an arc of longitude of about 2°, intersecting at Geneva could be very advantageously measured; and he wished the Royal Society to undertake the operation. He indicates on the map suitable places for the various stations.

998. A memoir by Waring *On Infinite Series* is contained in the *Philosophical Transactions* for 1791, published in that year. On pages 161...164 of the memoir Waring touches on the subject of Attraction. He shews how to calculate the attraction of a solid of revolution at any point of the axis, when the attraction varies as any power of the distance; he considers especially the case in which the solid is a sphere, and gives the approximate result when the solid deviates but little from a sphere.

The investigations contain nothing new or important.

999. In the *Philosophical Transactions* for 1791, published in that year, we have a memoir entitled *The Longitudes of Dunkirk and Paris from Greenwich, deduced from the Triangular Measurement in 1787, 1788, supposing the Earth to be an Ellipsoid. By Mr Isaac Dalby.* The memoir occupies pages 236...245 of the volume.

The memoir consists chiefly of numerical results. Lengths are *assumed* for the major and minor axes of the generating ellipse, which are nearly in Newton's proportion of 230 to 229; and it is shewn that the various measured arcs, neglecting Beccaria's, agree remarkably well with the values they would have on the assumed oblatum.

On page 240, a theorem is used which is the same with respect to the *major* axis of an ellipse as that enunciated in Art. 479 is to the *minor* axis.

1000. A volume was published in 1791 entitled *Exposé des opérations faites en France en 1787, pour la jonction des observatoires de Paris et de Greenwich, par M. M. Cassini, Méchain et Le Gendre.* See La Lande's *Bibliographie Astronomique*, page 618.

I have not been able to consult this volume; it is the French contribution corresponding to that made by the English under the superintendence of General Roy to the determination of the relative situations of the two great observatories: see Art. 980. I presume this French work embodies the memoirs on the subject which we have noticed in Art. 992.

The Cassini here named was the son of Cassini de Thury; and is often distinguished from the other members of his illustrious family as Cassini IV.

1001. We pass to a memoir entitled *Nuovo e sicuro mezzo per riconoscere la figura della terra. Del Sig. Antonio Cagnoli.* This memoir is contained in the *Memorie di Matematica...della Società Italiana*, Vol. 6, Verona 1792. It occupies pages 227...235 of the volume.

Observations are to be made of the duration of occultations of fixed stars by the moon; and this duration is to be compared with that calculated on the supposition that the Earth is spherical. The difference in the duration will, according to Cagnoli, amount to 86 seconds under favourable circumstances, or even to 130 seconds. He considers that in this way the Figure of the Earth may be ascertained.

He says on his page 234, he has elsewhere shown that the discordant results obtained in measuring degrees on the Earth's surface may be explained by irregularity in the density of the upper strata. I do not know to what publication he here alludes.

The most interesting circumstance connected with Cagnoli's memoir is the attention which it received from a very eminent English astronomer. A pamphlet was printed for private circulation entitled, *Memoir on a new and certain method of ascertaining the Figure of the Earth by means of occultations of the fixed Stars. By A. Cagnoli. With notes and an appendix by Francis Baily, London*, 1819.

This is a very interesting production; in consists of 44 octavo pages, besides the Title and Advertisement.

Baily urges private observers to attend to the suggestions for ascertaining the Figure of the Earth. He also strongly recommends the formation of an *Astronomical Society*; and this was soon afterwards carried into effect.

Baily alludes to the dissertation on the Figure of the Earth by Triesnecker, which we have noticed in Art. 906. But Baily had not been able to procure a sight of the dissertation.

Baily says in a note on page 8:

I find it difficult here to give a faithful translation of the author's words: the original runs thus, Senza che ci possiamo attenere alle osservazioni più sicure: ed una sola fase, &c. &c.

He translates this,

"Can we, indeed, expect to obtain more certain observations? since a single observation..."

The fault is that Baily throws into an interrogative form what Cagnoli gives as a statement. Cagnoli has just been considering an extremely unfavourable case, and shews that even there his method maintains its credit; and then he proceeds: Besides we may obtain more certain observations; since a single observation...

1002. In the *Philosophical Transactions* for 1792, published in that year, we have an *Account of the Measurement of a Base Line upon the Sea Beach, near Porto Novo, on the Coast of Coromandel*, by Michael Topping; this account occupies pages 99 ...114 of the volume.

The base was to serve for a series of triangles carried from Madras down the coast of Coromandel. The base did not form one straight line, but consisted of six portions, involving slight changes of direction at five points. The total length deduced for the distance between the extreme stations was 11636 yards. It is plain from the account that the operations were of a rather rude kind; and the result is not of any importance in the history of our subject. The great Indian arc which has since been measured does not pass through the locality of this early base, but some degrees to the west of it.

1003. The third edition of La Lande's *Astronomie* was published in 1792, in three volumes quarto. The pages 1...47 of the third volume form the 15th Book, entitled *De la Grandeur et de la Figure de la Terre.*

The pages are not very correctly printed, and contribute nothing new to the subject. But they collect useful information and references, especially concerning the historically famous toises of the North and of Peru, and that which had belonged to Mairan. There is also a table of the observed lengths of the seconds pendulum with references.

1004. We have now to consider a memoir by Lagrange, entitled, *Sur les Sphéroïdes elliptiques.*

This memoir is contained in the volume for 1792 and 1793 of the Berlin *Mémoires*, published in 1798: the memoir occupies pages 258...270 of the volume.

1005. Let there be an ellipsoid whose equation is

$$\frac{x^2}{a^2}+\frac{y^2}{b^2}+\frac{z^2}{c^2}=1 \quad \text{.........................(1)};$$

Lagrange first finds the value of

$$\iiint x^{2m} y^{2n} z^{2l}\, dx\, dy\, dz,$$

where m, n, l are positive integers, and the integration is extended over all the elements of the ellipsoid.

Lagrange shews that the value of this definite integral is

$$\frac{1.3.5...(2m-1)\,1.3.5...(2n-1)\,1.3.5...(2l-1)}{5...(2m+2n+2l+3)}\,a^{2m}b^{2n}c^{2l}M,$$

where
$$M=\frac{4\pi}{3}abc.$$

This might now be treated as an obvious example of Dirichlet's theorem in definite integrals: see *Integral Calculus*, Chapter XII.

Lagrange's own method is very ingenious; it may be understood from considering a particular case. Suppose that we require

$\iiint x^4\,dx\,dy\,dz$: denote it by U. Transform to polar coordinates by the usual substitutions,

$$z = r\cos\theta, \quad y = r\sin\theta\sin\phi, \quad x = r\sin\theta\cos\phi.$$

Thus we obtain $\iiint r^6 \sin^5\theta \cos^4\phi\,d\theta\,d\phi\,dr$. The integration with respect to r can be immediately effected; and this gives

$$U = \frac{1}{7}\iint r_1^7 \sin^5\theta \cos^4\phi\,d\theta\,d\phi,$$

where r_1 is given by the following equation which is deduced from (1)

$$r_1^2\left(\frac{\sin^2\theta\cos^2\phi}{a^2} + \frac{\sin^2\theta\sin^2\phi}{b^2} + \frac{\cos^2\theta}{c^2}\right) = 1.$$

The limits of the integrations are 0 and π for θ, and 0 and 2π for ϕ.

Let $\quad a^2 = \frac{1}{\alpha}, \quad b^2 = \frac{1}{\beta}, \quad c^2 = \frac{1}{\gamma}$, and $r_1^2 = \frac{1}{R}$;

then
$$U = \frac{1}{7}\iint \frac{\sin^5\theta\cos^4\phi\,d\theta\,d\phi}{R^{\frac{7}{2}}} \quad\text{..............}\quad (2),$$

where $\quad R = \alpha\sin^2\theta\cos^2\phi + \beta\sin^2\theta\sin^2\phi + \gamma\cos^2\theta.$

In like manner if M denote the volume of the ellipsoid, we shall find that

$$M = \frac{1}{3}\iint \frac{\sin\theta\,d\theta\,d\phi}{R^{\frac{3}{2}}} \quad\text{....................}\quad (3).$$

From (2) and (3) we find that

$$U = \frac{2}{5}\cdot\frac{2}{7}\cdot\frac{d^2M}{d\alpha^2} \quad\text{........................}\quad (4).$$

Then as we can easily shew that

$$M = \frac{4\pi}{3\sqrt{(\alpha\beta\gamma)}},$$

we have from (4)

$$U = \frac{1\,.\,3}{5\,.\,7}\frac{M}{\alpha^2} = \frac{3}{5\,.\,7}Ma^4.$$

1006. It will be observed that (4) is the most important equation in the preceding Article. In the same manner as (4) was established we can shew that

$$\iiint x^{2m} y^{2n} z^{2l}\, dx\, dy\, dz = (-1)^{l+m+n} \frac{2}{5} \cdot \frac{2}{7} \cdot \frac{2}{9} \cdots \frac{2}{2m+2n+2l+3} \frac{d^{l+m+n} M}{d\alpha^m d\beta^n d\gamma^l}.$$

1007. Lagrange now proceeds to consider the attraction of the ellipsoid on an external particle. He introduces what we call the potential function, and denotes it by V. If f, g, h denote the coordinates of the attracted particle, the attractions in the corresponding directions are $\frac{dV}{df}$, $\frac{dV}{dg}$, $\frac{dV}{dh}$. Lagrange does not claim these expressions for himself; and we know that they are really due to Laplace: see Art. 789.

1008. Lagrange says on his page 263:

La recherche de l'attraction du sphéroïde dépend donc simplement de la détermination de la quantité V en fonction de a, b, c, f, g, h. Dans le mémoire déjà cité sur l'attraction des sphéroïdes, j'ai résolu la question pour le cas où le point attiré est dans l'intérieur ou à la surface; et dans une addition à ce mémoire, imprimée dans le volume de l'Année 1775, je l'ai résolue aussi pour le cas où le point attiré est sur le prolongement d'un des trois axes. Les autres cas ont été résolus d'abord par Le Gendre pour les seuls sphéroïdes de révolution, ensuite par La Place et Le Gendre pour des sphéroïdes quelconques. On ne peut regarder leurs solutions que comme des chef-d'œuvres d'analyse, mais on peut désirer encore une solution plus directe et plus simple; et les progrès continuels de l'analyse donnent lieu de l'espérer. En attendant, voici l'usage qu'on pourroit faire des formules précédentes dans cette recherche.

By the memoir already cited, Lagrange means the memoir of 1773; the attraction is there investigated without any use of the function V. The anticipation which Lagrange expresses respecting the progress of analysis has been completely fulfilled; for by Ivory's method a most direct and simple solution of the problem is furnished.

1009. The function V is given by

$$V = \iiint \frac{dx\, dy\, dz}{\sqrt{\{(f-x)^2 + (g-y)^2 + (h-z)^2\}}}.$$

Let $h = \rho \cos\lambda$, $g = \rho \sin\lambda \sin\mu$, $f = \rho \sin\lambda \cos\mu$; and suppose the radical in V expanded in the form

$$\frac{1}{\rho} + \frac{P_1}{\rho^2} + \frac{P_2}{\rho^3} + \frac{P_3}{\rho^4} + \ldots:$$

then P_n will be a homogeneous function of x, y, z of the degree n.

It is obvious by the symmetry of the ellipsoid that since the integrations extend over the whole ellipsoid, all the terms which involve odd values of n will disappear in the expression for V; so that we shall have

$$V = \frac{M}{\rho} + \frac{1}{\rho^3}\iiint P_2\, dx\, dy\, dz + \frac{1}{\rho^5}\iiint P_4\, dx\, dy\, dz + \ldots$$

1010. Lagrange then considers in detail the terms in V which arise from P_2, P_4, and P_6.

For instance, we must have

$$P_2 = Ax^2 + By^2 + Cz^2 + Eyz + Fzx + Gxy,$$

where $A, B, C, E, F, G,$ are certain quantities which are constants with respect to x, y, z. Hence by the general formula of Art. 1006, we have

$$\iiint P_2\, dx\, dy\, dz = \frac{M}{5}(Aa^2 + Bb^2 + Cc^2);$$

for the terms which depend on E, F, G obviously vanish.

But the expression which is under the integral sign in V satisfies identically the well known partial differential equation. Hence we see that

$$\frac{d^2P_n}{dx^2} + \frac{d^2P_n}{dy^2} + \frac{d^2P_n}{dz^2} = 0;$$

when n is greater than 2 this will split up into various equations, because it is identically true. When $n = 2$ it reduces to

$$A + B + C = 0.$$

Thus we may put

$$\iiint P_2\, dx\, dy\, dz = \frac{M}{5}\{B(b^2 - a^2) + C(c^2 - a^2)\}.$$

1011. Lagrange treats the terms which arise from P_4 and P_6 in a similar manner. In both cases he obtains a result of this character: one factor is M, and the other factor is a function of $b^2 - a^2$ and $c^2 - a^2$. This, as he himself observes, confirms, as far as it goes, the important theorem due to Laplace, which we express by saying that the potentials of *confocal* ellipsoids at an external point are as their *masses*.

1012. A memoir by Rumovsky, entitled, *Meditatio de Figura Telluris exactius cognoscenda*, is contained in Vol. XIII. of the *Nova Acta Acad....Petropolitanæ;* the volume is for 1795 and 1796; the date of publication is 1802. The memoir occupies pages 407...417; and there is an account of it in pages 74 and 75 of the historical part of the volume.

1013. Rumovsky says that slightly different results as to the ellipticity of the earth have been deduced from the same data by different writers: he attributes this to the use of approximate formulæ, instead of exact formulæ. Accordingly he undertakes to compare all the degrees of the meridian hitherto measured, both with that of Peru and with that of Lapland, and to determine the value of the ellipticity from every pair. But nevertheless he is really content with an approximation, for he, in fact, assumes that the curvature is constant throughout each separate degree.

Rumovsky obtains by his calculation various values for the ellipticity, lying between $\frac{1}{102}$ and $\frac{1}{666}$. He attributes the discrepancies to the unavoidable errors in determining zenith distances; and as an example of the difficulty of accuracy in such matters, he says that the latitude of the observatory at Paris is still uncertain to the amount of two seconds.

Rumovsky states that if certain corrections are made in the lengths of the measured degrees, they will agree very closely, rejecting the Hungarian arc, in giving $\frac{1}{230}$ as about the value of the ellipticity. The corrections he assigns for the respective degrees in toises are the following: -40 Peru, -105 Cape of Good Hope, $+99$ Pennsylvania, $+75$ Italy, $+60$ North of France,

+ 46 middle of France, + 20 South of France, 0 Piedmont, + 42 middle of Austria, − 90 Lapland.

These numbers accord fairly with those proposed by Frisi for bringing the measures into harmony with the same value of the ellipticity: see page 95 of the work named in Art. 668. The following are some of Frisi's proposed corrections, also in toises: − 50 Peru, − 111 Cape of Good Hope, + 110 Pennsylvania, − 82 Lapland.

But although the corrections proposed by Rumovsky are not extravagant in amount, he agrees with the opinion which he cites from Boscovich:

quaestionem de magnitudine et figura Telluris ex mensura graduum non solum absolutam adhuc non esse, sed vix esse inchoatam.

Finally Rumovsky recommends the measurement of arcs of parallel extending over six or more degrees, the difference of longitudes being determined by the aid of exact chronometers. He maintains, against the opinion of Bouguer, that this is a good practical method for determining the ellipticity of the Earth; but it is almost superfluous to say that the method has never been found really advantageous.

The memoir cannot be considered to be of any importance in the history of our subject.

1014. In the first number of the *Bulletin des Sciences, par la Société Philomatique de Paris*, which was published in April 1797, there is a note on pages 5 and 6, entitled *Formules pour déduire le rapport des axes de la terre, de la longueur de deux arcs du méridien, par le C. R. Prony.*

The object of Prony is to supply a formula more exact than the ordinary approximations; he gives a result without demonstration, but this may be easily supplied.

With the usual notation the length of an arc of the meridian between the latitudes ϕ_1 and ϕ_2 is

$$a(1-e^2)\int_{\phi_1}^{\phi_2}(1-e^2\sin^2\phi)^{-\frac{3}{2}}\,d\phi.$$

Let k denote this length; put β for $\phi_2 - \phi_1$, and γ for $\phi_2 + \phi_1$. Then neglecting powers of e^2 above e^4 we find that

$$\frac{k}{a} = \beta - \frac{e^2}{4}(\beta + 3\sin\beta\cos\gamma) - \frac{3e^4}{16}\left(\frac{\beta}{4} + \sin\beta\cos\gamma - \frac{5}{8}\sin 2\beta\cos 2\gamma\right).$$

Let letters with an accent apply to another arc of the meridian; then by division

$$\frac{k}{k'} = \frac{\beta - \frac{e^2}{4}(\beta + 3\sin\beta\cos\gamma) - \frac{3e^4}{16}\left(\frac{\beta}{4} + \sin\beta\cos\gamma - \frac{5}{8}\sin 2\beta\cos 2\gamma\right)}{\beta' - \frac{e^2}{4}(\beta' + 3\sin\beta'\cos\gamma') - \frac{3e^4}{16}\left(\frac{\beta'}{4} + \sin\beta'\cos\gamma' - \frac{5}{8}\sin 2\beta'\cos 2\gamma'\right)}.$$

If we neglect e^4 we obtain for a first approximation

$$e^2 = \frac{4(k\beta' - k'\beta)}{3(k\sin\beta'\cos\gamma' - k'\sin\beta\cos\gamma)}.$$

This agrees substantially with Prony's formula.

Prony gives the result which is obtained by retaining e^4; it may be easily verified. The subject is fully treated in Puissant's *Traité de Géodésie*; see the third edition of that work, Vol. I. pages 317...320.

1015. In the *Philosophical Transactions* for 1798, published in 1798, there is a memoir by Cavendish, entitled *Experiments to determine the Density of the Earth*. The memoir occupies pages 469...526 of the volume: it was read on June 21, 1798.

1016. This famous memoir, although contributing nothing to the theory with which we are engaged, occupies an important place in the list of experiments and observations connected with the nature of the Earth. The attraction exerted by large balls of lead on adjacent small bodies was observed; and from the result the mean density of the Earth was deduced.

The memoir begins thus:

Many years ago, the late Rev. John Michell, of this Society, contrived a method of determining the density of the earth, by rendering sensible the attraction of small quantities of matter; but, as he was engaged in other pursuits, he did not complete the apparatus till a short time before his death, and did not live to make any experiments with it. After his death, the apparatus came to the Rev. Francis John Hyde Wollaston, Jacksonian Professor at Cambridge, who, not having conveniences for making experiments with it, in the manner he could wish, was so good as to give it to me.

1017. The only part of the memoir with which we are directly concerned is the investigation on pages 523 and 524 of the attraction of a rectangular lamina on a particle which is situated perpendicularly over a corner of the lamina. Cavendish obtains in finite terms the component attraction parallel to an edge of the lamina. But he says that he knows no way of finding the component perpendicular to the lamina except by an infinite series. He gives accordingly two expressions involving infinite series. I have verified the correctness of his result. But this component can be easily expressed in finite terms: see *Statics*, page 317.

Some formulæ are given on page 476 relating to the influence of a resistance which varies as the square of the velocity on the motion of a pendulum: at least they amount to this. I have verified them: but it seems to me that in the last line but one we must read *later* instead of *earlier*.

1018. Cavendish deduces from his experiments that the mean density of the Earth is about 5·48 times that of water. He admits that this differs rather more than he should have expected from the Scheballien experiment, which gave 4·5.

We have sketched the later history of this subject in Art. 733.

1019. A memoir by Trembley, entitled *Observations sur l'attraction et l'équilibre des Sphéroïdes* is contained in the volume for 1799 and 1800 of the Berlin *Mémoires* which was published in 1803. The memoir occupies pages 68...109 of the volume.

1020. In my History of the Theory of Probability I gave an account of several memoirs by Trembley on that subject; the present memoir is of the same character as those. Trembley merely presents in another manner results which are already well known; and his methods in general have no merit to compensate for the want of novelty in the conclusions.

1021. Suppose a mass of rotating fluid in the form of a figure of revolution to be in relative equilibrium. Let r and θ be the usual polar coordinates of a point at the surface; and let ω be the angular velocity. Let V denote the potential of the mass for the assumed point; then we know that for relative equilibrium we must have

$$V + \frac{\omega^2 r^2}{2} \sin^2\theta = \text{constant} \qquad (1).$$

Trembley investigates this equation; see his page 73.

1022. Suppose V_1 to denote the value of V at the pole; then since $\theta = 0$ at the pole, we have by (1)

$$V + \frac{\omega^2 r^2}{2} \sin^2\theta = V_1 \qquad (2).$$

Suppose V_2 to denote the value of V at the equator, and a the equatorial radius of the earth; then from (2) we have

$$V_2 + \frac{\omega^2 a^2}{2} = V_1,$$

so that

$$\frac{\omega^2}{2} = \frac{1}{a^2}(V_1 - V_2).$$

Substitute the value of ω^2 in (2), and we obtain

$$V = \left(1 - \frac{r^2 \sin^2\theta}{a^2}\right) V_1 + \frac{r^2 \sin^2\theta}{a^2} V_2 \qquad (3).$$

Thus if we know the values of V_1 and V_2 we can infer the value of V; and this may be advantageous, because the integrations required to determine the special values V_1 and V_2 may be less complex than the integration required to determine V directly.

This result however is not to be supposed true for every figure, but only for such a figure as is consistent with relative equilibrium. Trembley does not make the assertion explicitly, but we may fairly suspect him of supposing that (3) is an algebraical identity for every figure of revolution.

1023. Trembley determines the value of V_1 for an oblatum on his pages 74...81; his method however is most laborious and repulsive. He keeps the origin of polar coordinates at the centre; if he had imitated the method given by Lagrange in 1773, and put the origin at the pole, the result would have been obtained with simplicity in a page.

The result may be easily verified. Let f, g, h be the coordinates of any point within an ellipsoid or on its surface; then it is known that the resolved attractions parallel to the corresponding axes are respectively Ff, Gg, Hh, where F, G, H are certain constants. Therefore if V be the potential we must have

$$V = \text{constant} - \frac{1}{2}(Ff^2 + Gg^2 + Hh^2).$$

The constant can be determined by actually calculating the value of V for the centre of the ellipsoid.

In the case of an oblatum we shall thus obtain

$$V = \frac{2\pi a^2\sqrt{(1-e^2)}}{e}\sin^{-1}e - \frac{1}{2}(Ff^2 + Gg^2 + Hh^2),$$

where a is the semiaxis major, and e the excentricity of the generating ellipse.

The values of F, G, H are given in elementary books; for the oblatum two of them are equal.

Suppose that H refers to the polar diameter; then F and G are equal: also

$$H = \frac{4\pi}{e^2}\left\{1 - \frac{\sqrt{(1-e^2)}}{e}\sin^{-1}e\right\}.$$

We shall thus find that the value of V at the pole is

$$\frac{2\pi b^2}{\sin^2\phi}\left\{\frac{\phi}{\sin\phi\cos\phi} - 1\right\}, \text{ where } \phi = \sin^{-1}e.$$

Trembley's result is equivalent to this, but in obtaining it he employs a series which is not always convergent.

1024. An algebraical identity occurs in the course of Trembley's investigation, which may deserve to be reproduced. He shews that

$$\Sigma n_r (A + 2r)(A + 2r + 2)......(A + 2r + 2m) = 0,$$

where n_r denotes the rth term in the expansion of $(1 - 1)^n$, and Σ denotes a summation with respect to r from $r = 0$ to $r = n$ inclusive; and m is zero or any positive integer less than $n - 1$.

Suppose $m = 0$; then we have to shew that

$$A - \frac{n}{1}(A + 2) + \frac{n(n-1)}{\underline{|2}}(A + 4) - ... + (-1)^n (A + 2n) = 0;$$

and it is obvious that this is true, for the expression is equivalent to $A(1 - 1)^n - 2n(1 - 1)^{n-1}$.

Then the general proposition can be established by induction. Assume that the expression vanishes when m has a certain value. Change m into $m + 1$; and let U denote the value of the expression thus obtained: so that

$$U = \Sigma n_r (A + 2r)(A + 2r + 2)......(A + 2r + 2m + 2)......(1);$$

while by hypothesis

$$0 = \Sigma n_r (A + 2r)(A + 2r + 2)......(A + 2r + 2m),$$

and therefore by changing A into $A + 2$ we have

$$0 = \Sigma n_r (A + 2r + 2)(A + 2r + 4)......(A + 2r + 2 + 2m)......(2).$$

Multiply (2) by A, and subtract from (1); then we shall find that

$$U = -2n\Sigma(n-1)_r (A + 2r + 4)(A + 2r + 6)......(A + 2r + 4 + 2m),$$

and this is zero, by virtue of our assumption, provided m is less than $n - 1 - 1$, that is, provided m is less than $n - 2$; and this by hypothesis is the case; for we suppose $m + 1$ less than $n - 1$.

1025. Trembley next obtains on his pages 81...83 the value of V_2 for an oblatum; see Art. 1022. With the notation used in Art. 1023, it will be found that this is

$$\frac{\pi a^3 \cos\phi}{\sin^3\phi}\{\phi(2\sin^2\phi - 1) + \sin\phi\cos\phi\}.$$

Trembley's result is equivalent to this; but, as before, in obtaining it he employs a series which is not always convergent.

1026. Let U stand for $\frac{1}{\sqrt{(r^2 - 2rr'\mu + r'^2)}}$; then we know that

$$\frac{d}{d\mu}\left\{(1-\mu^2)\frac{dU}{d\mu}\right\} + r\frac{d^2(rU)}{dr^2} = 0 \text{ .... (1).}$$

Let U be expanded in a series in the form

$$\frac{1}{r'}\left\{1 + P_1\frac{r}{r'} + P_2\frac{r^2}{r'^2} + P_3\frac{r^3}{r'^3} + \dots\right\};$$

then assuming that in (1) the coefficient of each power of r vanishes separately, we have

$$\frac{d}{d\mu}\left\{(1-\mu^2)\frac{dP_n}{d\mu}\right\} + n(n+1)P_n = 0 \text{ .... (2).}$$

This is the way in which (2) is universally obtained. Trembley gives an investigation of (2), in which he does not make the assumption just stated, but starts with the known form of P_n: see Art. 780. I have not verified Trembley's investigation, which occupies his pages 84...89.

1027. Trembley wishes to shew that $P_n = 1$ when $\mu = 1$, and he adopts the following extraordinary method:

In (2) suppose $\mu = 1$; then

$$2\mu\frac{dP_n}{d\mu} = n(n+1)P_n;$$

therefore
$$\frac{2}{P_n}\frac{dP_n}{d\mu} = \frac{n(n+1)}{\mu};$$

therefore
$$(P_n)^2 = H\mu^{n(n+1)}$$

where H is a constant. Thus when $\mu = 1$ we have $P_n = \sqrt{H}$; and as H does not contain n, we see that P_n has the same value when $\mu = 1$, whatever be the value of n. But $P_1 = 1$ obviously when $\mu = 1$; therefore, $P_n = 1$ when $\mu = 1$.

It would be difficult to find worse reasoning. We see that μ is made equal to unity, and yet supposed to be a variable at the same time. And even if the resulting value of $(P_n)'$ had been fairly obtained, the constant H would be merely *constant with respect to* μ; to assume that H is constant with respect to n is to beg the whole question.

1028. On his pages 90...92 Trembley investigates some of the properties of Legendre's coefficients, which Legendre himself gave in his second memoir: see Arts. 825...827. Trembley uses equation (2) of Art. 1026. From this equation by integrating we obtain

$$(1-\mu^2)\frac{dP_n}{d\mu}+n(n+1)\int_0^\mu P_n d\mu = \text{constant};$$

then putting $\mu=0$ to determine the constant we have

$$(1-\mu^2)\frac{dP_n}{d\mu}+n(n+1)\int_0^\mu P_n d\mu = \left(\frac{dP_n}{d\mu}\right)_0,$$

where the suffix 0 indicates the value when $\mu=0$.

Hence putting $\mu=1$ we have

$$n(n+1)\int_0^1 P_n d\mu = \left(\frac{dP_n}{d\mu}\right)_0.$$

If n is even $\frac{dP_n}{d\mu}$ has μ for a factor, and then $\left(\frac{dP_n}{d\mu}\right)_0=0$.

If n is odd $\left(\frac{dP_n}{d\mu}\right)_0=(-1)^{\frac{n-1}{2}}\frac{3.5\ldots n}{2.4\ldots(n-1)}$.

Multiply equation (2) of Art. 1026 by μ^m, and integrate; thus

$$\int_0^\mu \mu^m\frac{d}{d\mu}\left\{(1-\mu^2)\frac{dP_n}{d\mu}\right\}d\mu+n(n+1)\int_0^\mu P_n\mu^m d\mu=0.$$

Integrate the first term by parts, and take unity for the upper limit of integration. Thus when n and m are both even we can arrive at the result given in Art. 825.

When n is even and m is odd we can arrive at the result given in Art. 826. Trembley obtains this result, but he has a superfluous $\pm$ before the right-hand member: nevertheless he says his formula is what Legendre found, which is untrue.

1029. Return to equation (3) of Art. 1022, and suppose that the body is very nearly spherical: put $r = a(1 + \alpha\eta)$ where α is a small quantity, the square of which may be neglected, and η is some function of θ. Thus we obtain approximately

$$V = V_1 \cos^2\theta + V_2 \sin^2\theta + 2\alpha\eta \sin^2\theta (V_2 - V_1),$$

and as we are sure that the difference between V_1 and V_2 must be of the order α, we have by neglecting α^2

$$V = V_1 \cos^2\theta + V_2 \sin^2\theta = V_2 + (V_1 - V_2) \cos^2\theta.$$

See Trembley's pages 93 and 96. He says that as the value of V involves only $\cos^2\theta$ the meridian curve must be an ellipse; but this is mere assertion and not demonstration. He adds:

...On déduit de là le théorème qu'a démontré M. le Gendre, que si l'on suppose qu'une planète en équilibre ait la figure d'un solide de révolution peu différent d'une sphère, et soit partagée en deux parties égales par son équateur, le méridien de cette planète est nécessairement elliptique.

But this understates what Legendre undertook to establish; for Legendre did not limit his figure of revolution to be nearly spherical; see Art. 844: at least he does not confine himself to the first power of the ellipticity.

Trembley suggests that probably the theorem of Legendre will also hold if the figure is nearly spherical, though not necessarily of revolution; he seems ignorant of the fact that Laplace had already established this in his fourth memoir: see Art. 858.

1030. On his page 95 Trembley verifies Laplace's well known equation for the case of an oblatum of small excentricity: see Art. 852. Taking a for the semiaxis major of the oblatum Trembley shews that

$$\frac{1}{2} V + a \frac{dV}{dr} + \frac{2}{3} \pi a^2 = 0.$$

1031. For an oblatum of small excentricity at any point of the surface Trembley finds that approximately

$$V = \frac{M}{r} \left\{ 1 + \frac{1}{10} \frac{e^2 a^2}{r^2} - \frac{3}{10} \cos^2\theta \frac{e^2 a^2}{r^2} \right\},$$

where M is the mass and e the excentricity.

This result is obtained on the assumption which is mentioned in Art. 1022, namely that the oblatum is a form of relative equilibrium for a rotating fluid; so that it is not demonstrated by Trembley. We may accept the result as true because we know that the assumed proposition is true.

Further, Trembley tacitly assumes that this formula is true for *any external point*, when at most all that has been shown by him is that it may be accepted as true for points on the surface.

Of course this assumption may be justified, but Trembley himself says nothing about it. We know that for an external point V will be of the form

$$\frac{M}{r} + \frac{N_2}{r^3} + \frac{N_4}{r^5} + \dots$$

where M denotes the mass; and as the expression may be admitted to hold up to the surface, the values of the constants N_2, N_4, ... may be determined by the aid of the value of V at the surface.

The approximate value of V which Trembley uses may be easily verified; see Art. 1010.

1032. From the approximate value of V given in the preceding Article, Trembley obtains immediately expressions for the attraction resolved in the direction of the radius vector and at right angles to it. These he applies on his pages 99...105 to demonstrate various theorems given by Clairaut; namely, those on Clairaut's pages 203, 236, 245, 220, and 217: see Articles 321, 335, 336, 327, and 323.

1033. Pages 105...109 of Trembley's memoir do not relate to our subject, but to a theorem demonstrated by Laplace respecting the attraction of light by a luminous body, in De Zach's *Ephemerides* for July, 1799. Trembley objects to Laplace's demonstration; I have not examined the point.

1034. In the *Memorie di Matematica.....della Società Italiana*, Vol. VIII., Modena, 1799, we have a memoir entitled *Sopra alcune particolarità concernenti la Gravità terrestre*, by Gregorio Fontana. The memoir occupies pages 124...134 of the volume.

Fontana expresses himself dissatisfied with the demonstrations given of the proposition that the weight of a body resolved along the radius varies inversely as the radius, supposing the Earth to be a homogeneous fluid in relative equilibrium; see Art. 33. Fontana refers specially to a demonstration given by Boscovich in his *De Litteraria Expeditione*, page 443. It does not however seem to me that Boscovich is unsound, though he is brief. I presume that the considerations which Fontana explicitly furnishes were implicitly understood by Boscovich.

Fontana demonstrates the proposition correctly. He also shews that the weight of a given body at any point varies as the normal: see Art. 153. He adds some easy propositions respecting the *angle of the vertical*, that is, the angle between the normal and the radius at any point of the Earth's surface.

The memoir seems to me to have been out of date at its appearance. It might have had interest and value forty years before, but scarcely at the time of publication.

1035. In the *Mémoires de l'Institut*... Vol. II. published in 1799, there is a Report entitled *Rapport sur la mesure de la méridienne de France et les résultats qui en ont été déduits pour déterminer les bases du nouveau système métrique*. This report occupies pages 23...80 of the historical portion of the volume; it was drawn up by Van-Swinden.

This report gives an abstract of the operations for determining the unit of length and the unit of weight in the French metrical system. It contains nothing of importance for our subject, as all the details connected with the measure of the meridian are fully exhibited in the work entitled *Base du Système Métrique*.

1036. We shall now notice the work which gives an account of the Trigonometrical Survey of England and Wales; this consists of three quarto volumes, published respectively in 1799, 1801, and 1811. The work reproduces in substance various papers which were originally published in the *Philosophical Transactions*, with large additions in the third volume: see Art. 984.

The volumes are devoted almost entirely to practical details, and records of observations, and so they do not fall within our prescribed range. Very few pages treat on theory, and these are not of an attractive character.

1037. On page 138 of the first volume we have the formula which is now usually called *General Roy's Rule* for computing the spherical excess in a spherical triangle; the Rule however has been claimed for Mr Dalby: see *Spherical Trigonometry*, Chapter X.

On page 154 a section of some importance is commenced, entitled *Of the horizontal Angles on a Spheroid;* this is probably due to Mr Dalby: compare the corresponding section in the *Philosophical Transactions* for 1790, pages 192...200.

The main design of the section seems to be to establish the following theorem: let there be two points on a surface of revolution, and determine at each point the azimuth of the other point; then the *sum* of the azimuths will be equal to the *sum* for two points on a sphere which have respectively the same latitudes as the points on the surface of revolution, and also the same difference of longitude. But the investigation is obscure and unsatisfactory, as are also other parts of the section. The theorem about the sum of the azimuths is however approximately true if the surface of revolution is nearly spherical: see the *Account... of the Principal Triangulation*, in the Ordnance Survey of Great Britain, 1858, page 236; also the article on the *Figure of the Earth* in the *Encyclopædia Metropolitana*, page 214.

1038. To justify the unfavourable opinion which I have expressed, I will make a remark which may be of service to a reader of the original investigation. It will be seen that page 155 professes to establish some result exactly, that is without approximation; but it is difficult to see precisely which angles are denoted by the letters employed. I believe it will be found that when O is the middle letter, the angle denoted should be the angle between some pair of planes which intersect in the straight line *OS*. Then in the sixteenth line of the page the angle *BOK* means the angle between the planes *BOS* and *KOS*; but in the twentieth line of the page the angle *BOK* is used for the angle between the planes *BOR* and *KOS*: these two meanings of the angle *BOK* are confounded, and the investigation rendered unsound.

On page 171 the following statement is made:

It has also been conjectured, that the degree in Peru is considerably too long, in consequence of the lateral attraction of the high lands where the measurement was performed. (Philos. Trans. 1768.)

I can find no authority in the *Philosophical Transactions* of 1768 for this statement; and I do not think that there is any value whatever in it.

1039. It may be observed that the measure of an arc of the meridian of nearly three degrees presented the same result as the early French operations, namely that the length of a degree appeared to diminish as the latitude increased: see page 109 of the second part of Vol. II. of the work.

For an anomaly as to the latitude of one of the stations see the article on the *Figure of the Earth* in the *Encyclopædia Metropolitana*, page 206.

CHAPTER XXVIII.

LAPLACE, *MÉCANIQUE CÉLESTE*, FIRST AND SECOND VOLUMES.

1040. THE first two volumes of the *Mécanique Céleste* were published in 1799. We shall be principally occupied with the second volume; but a few pages in the first volume are also devoted to our subject. I shall cite the pages of the original edition.

1041. The second Chapter of the Second Book of the *Mécanique Céleste* is entitled *Des équations différentielles du mouvement d'un système de corps soumis à leur attraction mutuelle.* In §§ 11, 12, and 13 of the Chapter Laplace digresses to the subject of attraction. The investigations occupy pages 135...145 of the first volume.

1042. Let V denote what we call the potential of an attracting body on a particle at the point (x, y, z). Laplace gives the well known equation

$$\frac{d^2V}{dx^2} + \frac{d^2V}{dy^2} + \frac{d^2V}{dz^2} = 0 \text{ .... (1);}$$

and then shews how it is to be transformed into polar coordinates by the usual formulæ

$$x = r\cos\theta, \quad y = r\sin\theta\cos\phi, \quad z = r\sin\theta\sin\phi,$$

and putting μ for $\cos\theta$: thus (1) becomes

$$\frac{d}{d\mu}\left\{(1-\mu^2)\frac{dV}{d\mu}\right\} + \frac{1}{1-\mu^2}\frac{d^2V}{d\phi^2} + r\frac{d^2rV}{dr^2} = 0 \text{ .... (2).}$$

We have already recorded the first appearance of these formulæ, and stated that the polar form was that originally given; see Arts. 851 and 866.

1043. If the attracting body be a spherical shell, it is obvious that V will not involve θ or ϕ, so that it will be a function of r only. Thus the partial differential equation for V reduces to

$$\frac{d^2 rV}{dr^2} = 0.$$

Therefore $V = A + \frac{B}{r}$, where A and B are arbitrary constants. And the attraction, being $-\frac{dV}{dr}$ towards the origin, is equal to $\frac{B}{r^2}$.

Now suppose the attracted particle to be at the centre of the spherical shell; then it is obvious that the resultant attraction must be zero: thus $B = 0$, when $r = 0$, and therefore B must always be zero. Hence $\frac{dV}{dr} = 0$ for all points within the shell.

Next suppose the attracted particle to be outside the shell; then it is obvious that when the particle is at an indefinitely great distance the attraction must be the same as if all the attracting mass were collected at its centre. Thus denoting by M the mass of the shell, we must have $\frac{B}{r^2} = \frac{M}{r^2}$ when r is indefinitely great. Hence $B = M$ when r is indefinitely great, and therefore $B = M$ always. Therefore the attraction of a spherical shell on any external particle is the same as if the shell were collected at its centre.

1044. The investigation of the preceding Article is unsatisfactory, because no reason presents itself for the change in the form of V in passing from the hollow part of the shell to the space outside the shell. The fact is that Laplace's fundamental partial differential equation for V is not true when the attracted particle is a constituent particle of the attracting body. It was shewn by Poisson that instead of zero on the right-hand side of (1) we must then have $-4\pi\rho$, where ρ is the density of the attracting body at the point considered. The circumstance that two different determinations of the value of the constant B are required, might have suggested that the fundamental equation for V could not hold *continuously* from the centre of the shell to an infinite distance. For then we should have no means of knowing at what point one form of V should be given up, and the other form taken.

1045. Laplace now proceeds to determine the laws of attraction which make the resultant attraction of a spherical shell on an external particle the same as if the shell were collected at its centre. The problem was first discussed by Laplace in his *Figure des Planetes...*: see Art. 817; the discussion has now passed into the elementary books. In his *Figure des Planetes* Laplace employed the expansion of functions in a series by Taylor's theorem; in the *Mécanique Céleste* he does not employ these expansions: the earlier method has been adopted in our elementary books.

1046. It has been observed that Laplace's solution of the problem involves rather more than it explicitly enunciates; see Schlömilch's *Zeitschrift für Mathematik und Physik*, Vol. [illegible]. page 498. We may put the problem thus: find what must be the law of attraction of the particles in order that the resultant attraction of a spherical shell on an external particle may be the same as if this shell were collected at its centre, and attracted according to *some* law depending on the distance. In this enunciation we do not *assume* that the law of the resultant action is to be the same as the law of the mutual action. We will solve the problem as thus enunciated in the manner of the *Mécanique Céleste*, that is without expansions.

Let r be the radius, δr the thickness of a shell, ρ the density. The attraction of this shell on an external particle at the distance c from the centre may be expressed in the form

$$2\pi\rho r \delta r \frac{d}{dc} \frac{\psi(c+r) - \psi(c-r)}{c}.$$

This is shewn in the *Mécanique Céleste*; see also *Statics*, Chapter XIII.

Let us suppose that the shell is collected at its centre, and that it attracts according to the product of its mass into a certain function of the distance, which we will denote by $\chi(c)$. Then, equating the two expressions of the attraction, we get

$$2\pi\rho r\delta r\,\frac{d}{dc}\,\frac{\psi(c+r)-\psi(c-r)}{c}=4\pi\rho r^2\delta r\,\chi(c)\ \dots\dots\dots\ (3).$$

Integrate with respect to c; thus

$$\psi(c+r)-\psi(c-r)=2cr\int\chi(c)\,dc+Uc\ \dots\dots\dots\dots\ (4),$$

where U is a constant with respect to c, so that it may possibly involve r.

If we represent $\psi(c+r)-\psi(c-r)$ by R we obtain by differentiating (4)

$$\frac{d^2R}{dc^2}=4r\chi(c)+2cr\chi'(c),$$

$$\frac{d^2R}{dr^2}=c\frac{d^2U}{dr^2}.$$

But by the nature of the function R we have

$$\frac{d^2R}{dc^2}=\frac{d^2R}{dr^2};$$

therefore $$4r\chi(c)+2cr\chi'(c)=c\frac{d^2U}{dr^2};$$

therefore $$\frac{2\chi(c)}{c}+\chi'(c)=\frac{1}{2r}\,\frac{d^2U}{dr^2}.$$

Since the first member of this equation is independent of r, and the second member is independent of c, each member must be equal to some constant, which we will denote by $3A$. Hence by integration,

$$\chi(c)=Ac+\frac{B}{c^2},$$

where B is a new constant.

This gives the law of the resultant attraction. We have now to find the law of the mutual attraction. We have from (3)

$$\frac{\psi'(c+r)-\psi'(c-r)}{c}-\frac{\psi(c+r)-\psi(c-r)}{c^2}=2\left(Ac+\frac{B}{c^2}\right)r;$$

therefore

$$c\,[\psi'(c+r)-\psi'(c-r)]-[\psi(c+r)-\psi(c-r)]$$
$$=2(Ac^3+B)\,r \;\ldots\ldots\ldots\ldots\ldots\ldots (5).$$

Differentiate with respect to c; thus

$$c\,[\psi''(c+r)-\psi''(c-r)]=6Ac^2r\,;$$

therefore $$\psi''(c+r)-\psi''(c-r)=6Acr \;\ldots\ldots\ldots\ldots\ldots\ldots (6).$$

Differentiate twice; thus

$$\psi''''(c+r)-\psi''''(c-r)=6Ar \;\ldots\ldots\ldots\ldots\ldots\ldots (7),$$
$$\psi'''''(c+r)-\psi'''''(c-r)=0.$$

This shews that $\psi''''(c)$ must be constant whatever c may be. Denote this by E; then

$$\psi'''(c)=Ec+E_1,$$

where E_1 is another constant. Hence by the aid of (7) we get $E=3A$; and then

$$\psi''(c)=\frac{3Ac^2}{2}+E_1c+E_2,$$

where E_2 is another constant.

By comparing this with (6) we see that $E_1=0$.

Thus $$\psi'(c)=\frac{Ac^3}{2}+E_2c+E_3,$$

and $$\psi(c)=\frac{Ac^4}{8}+\frac{E_2c^2}{2}+E_3c+E_4,$$

where E_3 and E_4 are constants.

Comparing these with (5) we find that

$$E_3=-B.$$

Thus $$\psi'(c)=\frac{Ac^3}{2}+E_2c-B\,;$$

that is $$c\int\phi(c)\,dc=\frac{Ac^3}{2}+E_2c-B,$$

where $\phi(c)$ is the function of the distance which determines the law of mutual action.

Hence $$\int \phi(c)\, dc = \frac{Ac^2}{2} + E_1 - \frac{B}{c},$$

and therefore $$\phi(c) = Ac + \frac{B}{c^2}.$$

Thus the law of mutual action coincides with the law of resultant action.

The same result will follow much more rapidly if we employ expansions. For take (3); expand $\psi(c+r)$ and $\psi(c-r)$ by Taylor's theorem, and equate the coefficients of r^3; thus

$$\frac{d}{dc}\frac{\psi'(c)}{c} = \chi(c),$$

that is $$\phi(c) = \chi(c).$$

And with this value of $\psi'(c)$ all the other powers of r will disappear from (3).

1047. Laplace now proceeds to determine the law of attraction which makes a spherical shell attract an internal particle equally in all directions. This problem occurs here for the first time; and it has since passed into the elementary books. Laplace does not use the expansion of functions. With the notation of the preceding Article we have now

$$\frac{d}{dc}\frac{\psi(r+c)-\psi(r-c)}{c} = 0;$$

therefore $$\psi(r+c) - \psi(r-c) = Uc,$$

where U is a constant with respect to c. Differentiate twice with respect to c; thus

$$\psi''(r+c) - \psi''(r-c) = 0.$$

Since this relation holds for all values of r and c, we must have $\psi''(c)$ constant, whatever c may be; and therefore $\psi'''(c)$ must be zero. But $\psi'(c) = c\int\phi(c)\,dc$; thus

$$2\phi(c) + c\phi'(c) = 0;$$

this gives $\phi(c)=\frac{B}{c^2}$, so that the only law of attraction which satisfies the proposed condition is the law of nature. See Art. 705.

1048. Laplace returns to the general equation (2); he says that the integration is not possible except in certain cases, as for instance that of a sphere. The integration is also possible, he says, when the solid is a cylinder of infinite length on a closed curve as base. In this case if we take the axis of z parallel to the generators of the cylinder we see that V cannot contain z. Hence (1) reduces to

$$\frac{d^2V}{dx^2}+\frac{d^2V}{dy^2}=0 \text{.............................} (8);$$

therefore $V=f_1(x+y\sqrt{-1})+f_2(x-y\sqrt{-1})$, where f_1 and f_2 denote arbitrary functions.

Laplace himself arrives at this result in a less simple way, by using (2) instead of (1).

The condition that the curve is to be closed secures that the differential coefficients of V shall be finite; but V itself becomes infinite for a cylinder of infinite length: Laplace does not notice this.

1049. If we put $x=\rho\cos\phi$ and $y=\rho\sin\phi$, we shall find that (8) transforms to

$$\rho^2\frac{d^2V}{d\rho^2}+\frac{d^2V}{d\phi^2}+\rho\frac{dV}{d\rho}=0 \text{ ....} (9).$$

If the cylinder is a circular cylinder V will be independent of ϕ, and (9) becomes

$$\rho^2\frac{d^2V}{d\rho^2}+\rho\frac{dV}{d\rho}=0.$$

Hence
$$-\frac{dV}{d\rho}=\frac{H}{\rho},$$

where H is an arbitrary constant.

To determine the constant H Laplace supposes that the attracted particle is so remote that the cylinder may be considered to be an infinite rod. Let A denote the base of the cylinder;

then the attraction of the cylinder on a particle at the distance ρ from the axis, when ρ is very great, is thus found to be $\frac{2A}{\rho}$. Thus $H=2A$, when ρ is very great, and therefore $H=2A$ always.

1050. Suppose the attracted particle is within a circular cylindrical shell of constant thickness and infinite length; then also we have $-\frac{dV}{d\rho}=\frac{H}{\rho}$. And as the attraction is zero when the attracted particle is on the axis of the cylinder we must have H zero then; and thus H is zero for all internal points. See however Art. 1044.

1051. The pages from the first volume of the *Mécanique Céleste* which we have been considering, contain valuable matter, which may all be ascribed to Laplace himself.

We have seen in Art. 1048 that Laplace considers that the general integration of (2) is not possible. It must however be remarked, that if we suppose V expanded in powers of r, we obtain from (2) the following equation for determining the coefficient of r^n, which we will denote by u_n,

$$\frac{d}{d\mu}\left\{(1-\mu^2)\frac{du_n}{d\mu}\right\}+\frac{1}{1-\mu^2}\frac{d^2u_n}{d\phi^2}+n(n+1)\,u_n=0.$$

General symbolic forms have been given in recent times for the integral of this equation: see Boole's *Differential Equations*, third edition, pages 433...430.

1052. We now pass to the second volume of the *Mécanique Céleste*.

The Third Book of the *Mécanique Céleste* is entitled *De la figure des corps célestes:* this work is composed of seven Chapters and occupies pages 1...170 of the volume.

1053. The first Chapter of the Third Book is entitled *Des attractions des sphéroïdes homogènes terminés par des surfaces du second ordre.*

The word spheroid is used by Laplace, as in his Second Book, without any definition. It does not mean necessarily a nearly spherical body, for Laplace usually adds this restriction when it is required. In fact *spheroid* with Laplace seems to include every thing which is not exactly a sphere, or at least every thing of which the surface can be determined by one equation between the usual polar variables. So also Lagrange and Poisson use the term spheroid with the like generality.

The Chapter contains a full account of the attraction of a homogeneous ellipsoid on a particle whether external or internal. The Chapter is substantially reproduced from Laplace's fourth memoir. See Chapter XXIII.

1054. Let r, θ, ϕ be the polar coordinates of an element of the attracting mass, the origin being the attracted particle. Let A, B, C denote the resolved attractions parallel to the axes towards the origin. Then, the law of attraction being that of the inverse square of the distance, we have

$$A = \iiint \sin\theta \cos\theta \, dr \, d\theta \, d\phi,$$

$$B = \iiint \sin^2\theta \cos\phi \, dr \, d\theta \, d\phi,$$

$$C = \iiint \sin^2\theta \sin\phi \, dr \, d\theta \, d\phi;$$

the limits of the integrations are to be taken so as to include every element of the attracting mass. These formulæ are now familiar to us from elementary books. Laplace obtains them by transformation from the formulæ referred to rectangular axes; and he also indicates the direct method of obtaining them.

1055. Let a, b, c be the semiaxes of an ellipsoid. Let f, g, h be the coordinates of an attracted particle, parallel respectively to these semiaxes, the centre being the origin. Then if the attracted particle be inside the ellipsoid, or on its surface, the resolved attractions parallel to the semiaxes are determined by the formulæ

$$A = \frac{3fM}{a^3} L, \quad B = \frac{3gM}{a^3} \frac{d\lambda L}{d\lambda}, \quad C = \frac{3hM}{a^3} \frac{d\lambda' L}{d\lambda'},$$

where M denotes the mass of the ellipsoid, and L is put for

$$\int_0^1 \frac{x^2 dx}{\sqrt{\{(1+\lambda^2 x^2)(1+\lambda'^2 x^2)\}}};$$

also

$$\lambda^2 = \frac{b^2-a^2}{a^2}, \quad \lambda'^2 = \frac{c^2-a^2}{a^2}.$$

Laplace puts the attractions in the above form, though he does not use quite the same letters as we do. He uses a, b, c for the coordinates of the attracted particle; and k, $\frac{k}{\sqrt{m}}$, $\frac{k_1}{\sqrt{n}}$ for the semiaxes of the ellipsoid.

1056. The definite integral L involves all the difficulties of indefinite integration. For denote it by $\phi(\lambda^2, \lambda'^2)$. Then if we require the integral between the limits 0 and ξ instead of between the limits 0 and 1, we see that by changing x into $u\xi$ the required result is $\xi^3\phi(\lambda^2\xi^2, \lambda'^2\xi^2)$.

1057. Laplace asserts that the integral L cannot be expressed by means of algebraic, logarithmic, or circular functions. This apparently means that he had demonstrated this result to his own satisfaction; but he never published the demonstration: see Art. 605.

However, the researches of Abel and Liouville in more recent times may be considered to have established the point. See the *Journal de l'École Polytechnique*, Cahier 23, pages 37, 39, 57; also Bertrand's *Calcul Intégral*, pages 89...110.

1058. Laplace gives the well-known expressions for the attraction in the special case in which the ellipsoid becomes an oblatum. We may then put $b = c$, so that $\lambda^2 = \lambda'^2$, and if e be the excentricity of the generating ellipse we have $\lambda^2 = \frac{e^2}{1-e^2}$. The results will be found in the elementary books: see *Statics*, Chapter XIII.

1059. Laplace proceeds to consider the case of the attraction of an ellipsoid on an external particle; see his pages 13...19. It amounts to developing what we now call the potential in a series which will be very convergent when the ellipticities of the principal sections are small, but is not always convergent. The process however is tedious, and requires a reader to perform much work for himself, or to have recourse to Bowditch's notes in the translation of the *Mécanique Céleste*.

1060. Laplace draws from his expansion the remarkable result that the attractions of different ellipsoids which have the same centre, the same position for their axes, and the same foci for their principal sections are as their masses. Laplace himself uses the phrase *the same excentricities*; the word excentricity denotes with him the distance between the centre and a focus, not as in modern books the ratio of this distance to the semiaxis major.

The result just stated we have called *Laplace's theorem*; it is the complete theorem of which Maclaurin gave a special case: see Art. 254. Laplace himself first obtained the theorem in his *Figure des Planetes*: see Art. 806.

I do not reproduce Laplace's method, because it would occupy a great space, and it is now superseded by Ivory's method. As I have already indicated, Bowditch's notes may be consulted with advantage. Also Burckhardt in his German translation of the first two volumes of the *Mécanique Céleste* has commented on Laplace's method. A paper on Laplace's method by Professor Cayley will be found in the *Quarterly Journal of Mathematics*, Vol. I. pages 285...300.

1061. I will place here some remarks which will not be quite intelligible independently of the *Mécanique Céleste*, but may be of interest to the student of that work, or of Professor Cayley's paper.

Laplace gives in Livre III. § 5 a certain partial differential equation which subsists between V and the resolved attractions. Professor Cayley uses a more symmetrical notation than Laplace used; and shews that the partial differential equation resolves itself into two. In his *Figure des Planetes*, where Laplace first gave this process, he started with *three* partial differential equations; and as this book is very scarce, it may be useful to notice here the earlier form. Laplace takes for the equation to the ellipsoid $x^2 + my^2 + nz^2 = k^2$. I follow Professor Cayley in using $lx^2 + my^2 + nz^2 = k$. The coordinates of an attracted external particle are a, b, c; also A, B, C are the resolved attractions parallel to the axes towards the origin; and V is the potential, so that

$$A = -\frac{dV}{da}, \quad B = -\frac{dV}{db}, \quad C = -\frac{dV}{dc}.$$

Then each of the three partial differential equations of Laplace's earlier work may be resolved into two. The two which spring from his first are

$$k\frac{dA}{dk} - l\frac{dA}{dl} - m\frac{dA}{dm} - n\frac{dA}{dn} + a\frac{dA}{da} + b\frac{dA}{db} + c\frac{dA}{dc} - A = 0,$$

$$-(a^2+b^2+c^2)\frac{dA}{dk} + a\frac{dV}{dk} + \frac{1}{2l}\frac{dV}{da} + \frac{dA}{dl} + \frac{dA}{dm} + \frac{dA}{dn}$$
$$- \left(\frac{a}{l}\frac{dA}{da} + \frac{b}{m}\frac{dA}{db} + \frac{c}{n}\frac{dA}{dc}\right) = 0.$$

These equations are true not only for the whole definite integrals which constitute A and V, but also for every element taken separately; should there be found any difficulty in verifying them it will be removed by consulting Bowditch's notes, or Professor Cayley's paper. The other two of Laplace's equations give rise to similar pairs of equations, which involve B and C in the same manner as the first pair involves A.

Now put F for $aA + bB + cC$ as Laplace does. Multiply the first equation of the first pair by a, the first equation of the second pair by b, and the first equation of the third pair by c. Then by addition we shall obtain

$$k\frac{dF}{dk} - l\frac{dF}{dl} - m\frac{dF}{dm} - n\frac{dF}{dn} + a\frac{dF}{da} + b\frac{dF}{db} + c\frac{dF}{dc} - F = 0.$$

This constitutes one of the two parts into which Professor Cayley's first equation may be resolved. The other part of his first equation will consist of

$$-k\frac{dV}{dk} + V - \frac{1}{2}\left(a\frac{dV}{da} + b\frac{dV}{db} + c\frac{dV}{dc}\right) = 0.$$

This equation like the other holds for every element taken separately of the definite integrals.

Again, treat the second equation of each pair in the same manner as we have treated the first; thus by addition we get

$$-(a^2+b^2+c^2)\left(\frac{dF}{dk}-\frac{dV}{dk}\right)-\frac{a}{l}\left(\frac{dF}{da}-\frac{1}{2}\frac{dV}{da}-A\right)$$

$$-\frac{b}{m}\left(\frac{dF}{db}-\frac{1}{2}\frac{dV}{db}-B\right)-\frac{c}{n}\left(\frac{dF}{dc}-\frac{1}{2}\frac{dV}{dc}-C\right)+\frac{dF}{dl}+\frac{dF}{dm}+\frac{dF}{dn}=0.$$

This is Professor Cayley's second equation. It may be abbreviated by putting for A, B, C their values as differential coefficients of V.

1062. Laplace arrives at the following equation on his page 20,

$$a^2+\frac{k'^2}{k'^2+\theta}b^2+\frac{k'^2}{k'^2+\varpi}c^2=k'^2.$$

If θ and ϖ are positive, and all the quantities are given except k', Laplace shews that there is only one real positive value of k'^2. A simpler method than his will be to put the equation in the form

$$\frac{a^2}{k'^2}+\frac{b^2}{k'^2+\theta}+\frac{c^2}{k'^2+\varpi}=1.$$

It is obvious that the left-hand member decreases continually as k'^2 increases, and so cannot have the same assigned value for more than one value of k'^2.

1063. The first Chapter of the Third Book of the *Mécanique Céleste* may be said to contain two very important contributions by Laplace himself to our subject. One of these is the expression by means of a single definite integral of the attraction of an ellipsoid on an internal or superficial point; this, as we have said, actually presented itself to D'Alembert, but was rejected by him: see Art. 805. The other contribution is the theorem which we have called Laplace's, respecting the attraction of an ellipsoid on an external particle. As we have already stated, the Chapter substantially dates from the memoir in the volume of the Paris Academy for 1782.

1064. The second Chapter of the Third Book is entitled *Du développement en série, des attractions des sphéroïdes quelconques.*

This Chapter introduces the functions which we call Laplace's functions; nearly the whole of the Chapter is substantially reproduced from the memoir of 1782: see Chapter XXIII.

1065. Let r, θ, ϕ be the polar coordinates of the attracted point; r', θ', ϕ' the polar coordinates of an element of the attracting body, ρ the density of the attracting body; put μ' for $\cos\theta'$. Then the potential

$$V=\iiint\frac{\rho r'^2\,d\mu'\,d\phi'\,dr'}{\sqrt{(r^2-2rr't+r'^2)}},$$

where t stands for $\cos\theta\cos\theta'+\sin\theta\sin\theta'\cos(\phi-\phi')$. The limits of the integrations are to be so taken as to include the whole attracting body.

We have already given, in Art. 1042, the partial differential equation which V satisfies. Now suppose the attracted particle outside the attracting body, and so far off that r is greater than any value of r'. Let $(r^2-2rr't+r'^2)^{-\frac{1}{2}}$ be expanded in a series

$$\frac{P_0}{r}+P_1\frac{r'}{r^2}+P_2\frac{r'^2}{r^3}+\ldots$$

Substitute in (2) of Art. 1042 and equate the coefficient of each power of r to zero. Thus we obtain equations of which the type is

$$\frac{d}{d\mu}\left\{(1-\mu^2)\frac{dP_i}{d\mu}\right\}+\frac{1}{1-\mu^2}\frac{d^2P_i}{d\phi^2}+i(i+1)P_i=0\ldots\ldots(10).$$

The quantity P_i is called *Laplace's coefficient* of the ith order. Thus Laplace's coefficient of the ith order satisfies the differential equation (10). Any other function of θ and ϕ which satisfies the equation may be called a *Laplace's function* of the ith order. It is of course conceivable that we may have a Laplace's function of the ith order which is more simple or more complex than the coefficient of the ith order.

1066. With respect to the names by which these celebrated functions have been called, a few remarks are necessary. The name *Laplace's coefficients* appears to have been first used by the late Dr Whewell: see *Monthly Notices of the Royal Astronomical Society*, Vol. XXVII. page 211.

The distinction between the *coefficients* and *functions* is given for the first time to my knowledge in Pratt's *Figure of the Earth* 1860, page 21.

When P_i is contemplated as a function of the single variable t it should be more justly called Legendre's coefficient; see Art. 783. It is only when contemplated as a function of the two variables θ and ϕ that Laplace's name is appropriate. The Germans call the functions *Kugelfunctionen*. The name *fonctions sphériques* is used by Resal. Finally the name *spherical harmonics* is used by Sir W. Thomson and Professor Tait.

1067. The first property of Laplace's functions which presents itself to our notice is this: *any function of μ and ϕ can be expanded in a series of Laplace's functions.* Laplace arrives at an indirect demonstration of this theorem in the course of his investigations on attraction: we will explain his method.

Laplace establishes his favourite equation; see Art. 852:

$$-a\frac{dV}{dr} = \frac{2\pi a^2}{3} + \frac{1}{2}V \text{.....................} (11);$$

here the density is denoted by unity. Laplace, without saying so, now begins to restrict himself to the case of homogeneous bodies. And by Art. 1065 we have

$$V = \frac{U_0}{r} + \frac{U_1}{r^2} + \frac{U_2}{r^3} + \dots \text{ ....} (12),$$

where $$U_n = \iiint \rho r'^{n+2} P_n d\mu' d\phi' dr';$$

thus U_n is a Laplace's function of the nth order, for every element of it satisfies (10), and therefore the whole satisfies (10).

From (12) we have

$$-\frac{dV}{dr} = \frac{U_0}{r^2} + \frac{2U_1}{r^3} + \frac{3U_2}{r^4} + \dots$$

Let $a(1 + \alpha y)$ denote the radius vector of the spheroid at the point to which V refers, α being a very small fraction, the square of which may be neglected, and y any function of μ and ϕ. If we neglect quantities of the order α we shall have $V = \dfrac{4\pi a^3}{3r}$. Hence it will follow that U_0 must be equal to $\frac{4\pi a^3}{3}$ increased by a quantity of the order α, which we will denote by U'_0; and also that U_1, U_2, ... are all small quantities of the order α. Substitute $a(1+\alpha y)$ for r in (11) and (12), and neglect the square and higher powers of α. Thus for a point at the surface we have

$$\frac{1}{2}V = \frac{2\pi a^2}{3}(1-\alpha y) + \frac{U'_0}{2a} + \frac{U_1}{2a^2} + \frac{U_2}{2a^3} + \dots$$

$$-a\frac{dV}{dr} = \frac{4\pi a^2}{3}(1-2\alpha y) + \frac{U'_0}{a} + \frac{2U_1}{a^2} + \frac{3U_2}{a^3} + \dots$$

Substitute these values in (11); then we have

$$4\alpha\pi a^2 y = \frac{U'_0}{a} + \frac{3U_1}{a^2} + \frac{5U_2}{a^3} + \frac{7U_3}{a^4} + \dots$$

Thus we have obtained for y, which is any arbitrary function of μ and ϕ, an equivalent series of Laplace's functions.

1068. Laplace gives in his pages 31 and 32 the important proposition that if Y_n and Z_m are two Laplace's functions of different orders, the variables being θ and ϕ,

$$\int_{-1}^{1}\int_{0}^{2\pi} Y_n Z_m \, d\mu \, d\phi = 0.$$

See Arts. 857 and 951.

1069. In order to complete the matter upon which we are engaged we must pass on to Laplace's page 43; we may remark that the Chapter does not seem well arranged by Laplace.

We have in Art. 1067

$$U_n = \iiint \rho r'^{n+2} P_n \, d\mu' \, d\phi' \, dr'.$$

Suppose the spheroid homogeneous, and take $\rho = 1$; and let the spheroid differ but little from a sphere. Let $a(1+\alpha y')$ be the radius vector of the surface corresponding to the coordinates θ' and ϕ'. Then

$$U_n = \frac{a^{n+3}}{n+3}\iint (1+\alpha y')^{n+3} P_n \, d\mu' \, d\phi' \dots\dots\dots\dots\dots (13).$$

Now we have shewn in Art. 1067 that y can be expanded in a series of Laplace's functions; let then

$$y = Y_0 + Y_1 + Y_2 + \dots \quad \dots\dots\dots\dots\dots \text{(14)},$$

and in like manner we shall have

$$y' = Y'_0 + Y'_1 + Y'_2 + \dots$$

where Y'_n is the same function of μ' and ϕ' that Y_n is of μ and ϕ.

Substitute the value of y' in (13); neglect the square of a, and make use of Art. 1068. Thus we obtain

$$U_n = a^{n+3} a \iint Y'_n P_n \, d\mu' \, d\phi'.$$

But by Art. 1067 we have

$$Y_n = \frac{2n+1}{4\pi a a^{n+3}} U_n;$$

therefore
$$\frac{4\pi Y_n}{2n+1} = \iint Y'_n P_n d\mu' \, d\phi'.$$

Hence (14) may be written thus

$$y = \frac{1}{4\pi} \Sigma \iint (2n+1) \; Y'_n P_n \, d\mu' \, d\phi',$$

where Σ refers to n, and implies a summation from $n = 0$ to $n = \infty$.

By Art. 1068 we may if we please put the result thus

$$y = \frac{1}{4\pi} \Sigma \iint (2n+1) \, y' P_n \, d\mu' \, d\phi'.$$

1070. Such then constitutes Laplace's process for expanding any function in a series of Laplace's functions. In Art. 1007 it was shewn that a function could be so expanded, and then in Art. 1069 the last two formulæ give the required expansion explicitly. The demonstration is rather indirect in appearance, and is founded on Laplace's favourite equation, which has been the subject of some controversy. An examination of the value and extent of the demonstration would be more appropriate in a treatise on Laplace's functions than in our history. Other investigations have since been given by Poisson, Dirichlet and Bonnet; also in England we have an investigation by O'Brien and two by Pratt, one founded on O'Brien's; see Pratt's *Figure of the Earth*. These investigations are quite different from Laplace's. Resal, in his *Traité Elémentaire de Mécanique Céleste*, has given a process resembling Laplace's, but more elaborate in two respects: Resal supplies a fuller investigation of Laplace's favourite equation, and also he notices and allows for the circumstance that when a particle is placed near the surface, r may be really less than some of the values of r'.

1071. Laplace shews that a function can only be expanded in one way in a series of Laplace's functions; see his page 32.

1072. Laplace shews that if a be the radius of a sphere of equal volume with the spheroid the term Y_0 will disappear from the value of y. Also if the origin be taken at the centre of gravity the term Y_1 disappears. See his pages 33 and 34.

1073. Hitherto we have treated of the value of V for a particle outside the body, or on the surface. Now we have to find the value of V for an internal particle. This Laplace gives on his pages 35...37 for a homogeneous body.

We have indicated the nature of the formulæ in Art. 925; and we have also remarked that Laplace's investigation is not quite satisfactory: see also Art. 792.

1074. Having thus discussed the attraction of homogeneous spheroids differing but little from spheres, Laplace proceeds to the case in which the density varies, being some function of the parameter a, by which each stratum is particularised: see his pages 37...39. The method is obvious, being, in fact, that which Clairaut had employed: see Art. 323.

1075. Laplace having thus completed his theory of the attraction of spheroids which are nearly spherical, says on his page 39: "Considérons présentement, les sphéroïdes quelconques." This practically means that he intends to develope the value of P_n, which is the Laplace's coefficient of the nth order, and to shew how a rational function of μ, $\sqrt{(1-\mu^2)}\cos\phi$, and $\sqrt{(1-\mu^2)}\sin\phi$ may be most easily transformed into a series of Laplace's functions: see his pages 39...43.

1076. The last section of this Chapter of the *Mécanique Céleste*, which occupies pages 43...49, contains matter which was not in the memoir of 1782.

On his page 44 Laplace gives the remarkable formula

$$\frac{4\pi Y_n}{2n+1} = \int_{-1}^{1}\int_{0}^{2\pi} Y'_n P_n \, d\mu' \, d\phi',$$

where P_n is the nth coefficient, and Y_n is any function of μ and ϕ of the nth order, and Y'_n is the same function of μ' and ϕ': see Art. 657.

On his page 47 Laplace shews that in the case of a solid of revolution if we know the value of V for all the external points which are on the axis of revolution, we know it for all external points. This important theorem was first given by Legendre: see Art. 791.

Laplace extends this theorem and arrives at the following result: if the solid be not of revolution, but be divided into two equal and similar parts by the plane of the equator, then if we know the value of V for all external points which are on the axis, and also for all which are in the plane of the equator, we know the value of V for all external points. See his page 48. We shall see that this result has been generalised by Biot.

It is correctly remarked by Bowditch on page 170 of his translation of the second volume of the *Mécanique Céleste*, that we may omit Laplace's first condition, namely that V is known for all points on the axis; it is sufficient that V should be known for all points in the plane of the equator. Biot's generalisation agrees with this remark.

This Laplace says will hold for the ellipsoid. His own words should be examined. He seems to imply that for an ellipsoid V can be determined with respect to any external point on the axis or in the plane of the equator. One integration can be effected, and so V expressed as a single definite integral. But the second integration could not be effected in finite terms; though it might be in the form of an infinite series. Laplace must mean this, but it seems to me that he has not expressed himself very carefully.

Laplace in this manner obtains another demonstration of the theorem which I call by his name: see Art. 1060.

1077. The second Chapter of the Third Book of the *Mécanique Céleste* may be attributed for the most part to Laplace himself; like the first Chapter it substantially dates from the memoir in the volume for 1782. The Chapter is distinguished by two important features; we have the potential function V extensively used, and we have also the theory of Laplace's functions. The function V was first introduced by Laplace himself, as we have seen in Art. 789. The Laplace's coefficients owe their origin to Legendre, but Laplace's extension of their range justifies the use of his name in connexion with them: see Art. 783.

The Chapter cannot be considered well arranged. The pure analysis and the physical application of it are not kept sufficiently distinct, but this is very characteristic of Laplace, with whom analytical processes seem of little interest apart from the problems in natural philosophy which called them forth.

1078. The third Chapter of the Third Book is entitled *De la figure d'une masse fluide homogène en équilibre, et douée d'un mouvement de rotation.*

The title does not correspond very closely with the subject of the Chapter. Laplace does not profess to investigate what the figure must be in the circumstances proposed; he contents himself with shewing that an oblatum is an admissible figure.

1079. Laplace shews that there cannot be more than two oblata corresponding to a given angular velocity; that there will be only one oblatum if the angular velocity has a certain assigned value; and none at all if the angular velocity exceed this limit: see his pages 56 and 57. An oblongum is not a possible form of relative equilibrium: see his page 59. Finally he shews that corresponding to a given initial moment of rotation there will be one, and only one, oblatum; the phrase *moment of rotation* is not Laplace's, it is used by Resal: see his page 198. Laplace's investigations had all appeared substantially in the *Figure des Planetes*: see Arts. 810...813.

1080. In this Chapter Laplace inherited much from his predecessors. The fact that an oblatum is a possible figure of relative equilibrium was first rigorously established by Maclaurin: see Art. 249. That more than one oblatum might correspond to a given angular velocity was implicitly shewn by Thomas Simpson, and explicitly by D'Alembert: see Art. 580. Laplace himself first shewed that there could not be more than two such oblata: see Art. 585. D'Alembert gave another demonstration which however is not satisfactory: see Art. 657. Cousin also gave a demonstration: see Art. 976. Finally, in the *Mécanique Céleste*, Laplace gave a demonstration different from his first, and rather simpler. Laplace himself also first formally shewed that an oblongum is not a possible form of relative equilibrium, though this result came quite within D'Alembert's reach: see Art. 601.

1081. Laplace gives expressions for determining approximately the excentricities of the two oblata which correspond to the same angular velocity, supposed small. With respect to the oblatum which is nearly spherical, an equivalent to Laplace's expression had already been given by Maclaurin and Thomas Simpson: see Arts. 262 and 283. With respect to the oblatum which deviates much from a sphere, D'Alembert had given the first term of the expression: see Art. 584.

1082. Laplace, as we have said, shews that an oblongum is not a possible figure of relative equilibrium. Plana gives a convenient form to the demonstration; see the *Astronomische Nachrichten*, Vol. XXXVI. page 164.

The fact that an oblongum cannot be a possible form of relative equilibrium may be readily seen by the aid of a diagram.

Let P be any point on an ellipse, PG the normal and PT the tangent at P; let CA be the semiaxis major. Suppose an

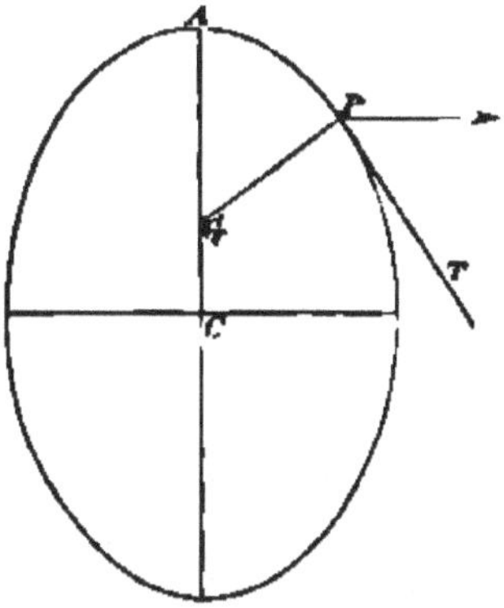

oblongum generated by the revolution of this ellipse around its major-axis.

The attraction of the oblongum at P will be in a direction which is *on the same side of PG as PC is*; this is obvious, for the oblongum may be cut up by planes parallel to the tangent-plane at P into slices, which all have their centres on the diameter through P. The so-called centrifugal force at P will be in a direction at right angles to AC outwards. Hence the attraction, and the so-called centrifugal force, will give rise to a component along PT; and so the fluid cannot be in relative equilibrium.

1083. We may observe that on Laplace's page 58 we have the fraction $\frac{27\lambda + 30\lambda^3 + 7\lambda^5}{(1+\lambda^2)(3+\lambda^2)(9+\lambda^2)}$, which might be reduced to the simpler form $\frac{\lambda(9+7\lambda^2)}{(1+\lambda^2)(9+\lambda^2)}$. Poisson also uses the unreduced form: see his *Traité de Mécanique*, Vol. II. page 542.

1084. If the angular velocity exceeds a certain limit the oblatum is not a possible figure of relative equilibrium. Laplace makes an important remark with respect to this on his page 59: he says that it might have been supposed that this limiting case is that in which the fluid would begin to fly off by reason of the too rapid rotatory motion, but it is easily found that this is not the fact.

Poisson alludes to the matter in the *Connaissance des Tems* for 1829; he says on page 375, after remarking that within a certain limit the ellipsoid of revolution is a solution:

Si l'ellipsoïde était la seule figure qui eût cette propriété, il en résulterait cette conséquence singulière que l'équilibre serait impossible pour une rapidité de la rotation qui n'est pas cependant celle ou le fluide commencerait à se dissiper.

1085. Laplace's theorem that there is only one oblatum corresponding to a given moment of rotation will be found with a different demonstration in Resal's work: see his page 198. An interesting point of analysis is involved.

The problem is reduced to this equation

$$q = (1+\lambda^2)^{\frac{2}{3}} \frac{(3+\lambda^2)\tan^{-1}\lambda - 3\lambda}{\lambda^3},$$

where q is a given positive quantity, and λ has to be found. Then it is demonstrated that there is one, and only one, value of λ between 0 and infinity which satisfies this equation.

It is easily shewn that there is one value of λ which satisfies the equation; for the right-hand member vanishes when λ vanishes, and is infinite when λ is infinite.

Put $\tan\theta$ for λ; then the right-hand member becomes

$$\frac{\theta(1+2\cos^2\theta) - 3\sin\theta\cos\theta}{(\cos\theta)^{\frac{1}{3}}\sin^3\theta};$$

we will denote this by u. It will be found that

$$\frac{du}{d\theta} = \frac{9\sin\theta\cos\theta(1+2\cos^2\theta) + \theta(1-20\cos^2\theta-8\cos^4\theta)}{3(\cos\theta)^{\frac{4}{3}}\sin^4\theta}.$$

Now Resal in effect puts this expression in the following form:

$$\frac{du}{d\theta} = \frac{1}{3(\cos\theta)^{\frac{4}{3}}}\left\{\theta + \frac{9(2+\cos^2\theta)}{\sin^4\theta}\cos^2\theta\left[\frac{1+2\cos^2\theta}{2+\cos^2\theta}\tan\theta - \theta\right]\right\}.$$

This is certainly positive provided $\frac{1+2\cos^2\theta}{2+\cos^2\theta}\tan\theta-\theta$ is positive; and the differential coefficient of the last expression is found to be positive, so that as the expression vanishes with θ it must always be positive as θ changes from 0 to $\frac{\pi}{2}$.

Thus u increases with θ, and so can only once have an assigned value as θ changes from 0 to $\frac{\pi}{2}$.

The point of interest which is involved is the following: if a and b are positive quantities, determine under what conditions $\tan\theta\frac{1+a\tan^2\theta}{1+b\tan^2\theta}-\theta$ is always positive while θ changes from 0 to $\frac{\pi}{2}$.

It will be found that the differential coefficient of the last expression is

$$\frac{(3a-3b+1)\tan^2\theta+(ab+3a-b-b^2)\tan^4\theta+ab\tan^6\theta}{(1+b\tan^2\theta)^2}.$$

Thus the required result is secured if $3a-3b+1$ and $ab+3a-b-b^2$ are both zero or positive. Thus a must not be less than $b-\frac{1}{3}$, and b for perfect security not less than $\frac{3}{5}$. Therefore if a is not less than $\frac{4}{15}$ then $\tan\theta\frac{1+a\tan^2\theta}{1+\left(a+\frac{1}{3}\right)\tan^2\theta}$ is greater than θ, or $\tan\theta$ is greater than $\theta+\frac{1}{3}\frac{\theta\tan^2\theta}{1+a\tan^2\theta}$.

1080. I may remark that Bowditch's notes on this Chapter bring before the reader the peculiar notions which Ivory held as to fluid equilibrium. Like every other person Bowditch objects to these notions; but some of his language in his account of the matter seems to want precision, estimated from our modern notions. Thus on his page 206 he speaks of the "forces which act upon the point I": if by a point he means a small element of the fluid we should require to know the form of that element. Again on page 208 he speaks of "the effort of the fluid......to rise in the branch..." Perhaps he had fallen a little under the influence of Ivory.

A note by Bowditch on his page 222 should be observed. A reader of Laplace might fail to recollect that he uses the revolutionary mode of reckoning hours, minutes, and seconds.

1087. Laplace seems to have attached considerable importance to the proposition that there is only one oblatum corresponding to a given moment of rotation. See the *Mécanique Céleste*, Vol. v. page 10, "mais le véritable problème à résoudre,..." Laplace had been to some extent anticipated by Thomas Simpson: see Art. 286.

1088. The fourth Chapter of the Third Book is entitled *De la figure d'un sphéroïde très-peu différent d'une sphère et recouvert d'une couche de fluide en équilibre.*

The title of the Chapter seems inadequate; for Laplace discusses not only the case in which a solid is covered by a film of fluid, but also the case in which the body is supposed entirely fluid.

The Chapter is mainly composed of matter which Laplace had previously published in memoirs. The §§ 22...28 are from the fourth memoir; the §§ 29 and 30 are from the seventh memoir; the §§ 31 and 32 are from the fifth memoir; § 33 is from the fourth memoir, with the exception of the examination of Bouguer's hypothesis on pages 97...99 which is new; §§ 34 and 35 are new; § 36 is substantially in the *Figure des Planetes*; § 37 is new.

1089. Laplace in his §§ 22...25 treats the case of a homogeneous body which is nearly spherical and fluid, or covered with a film of fluid; when this body rotates with uniform angular velocity, Laplace shews that for relative equilibrium the external surface of the fluid must be that of an oblatum. He does not *assume* that the body is a figure of revolution. The demonstration depends on the use of Laplace's functions. The demonstration is substantially reproduced by Resal: see his pages 200...211.

Laplace gives in his § 26 another demonstration, very curious, and not employing Laplace's functions; it does not seem to have been reproduced in an elementary book. Some remarks on it will be found in Liouville's *Journal de Mathématiques* for June and August 1837, and April 1839. Laplace's method in fact has been shewn to be unsatisfactory; and we shall consider the matter in a later Chapter.

It will be remembered that Legendre first discussed a case of this problem in his second memoir: see Chapter XXII.

1090. Laplace in his §§ 27 and 28 considers the case of a homogeneous fluid which surrounds a spherical nucleus of a density different from that of the fluid. A small part of the investigation is reproduced in Resal's pages 212 and 213.

1091. Laplace in his §§ 29...31 discusses the figure of the Earth considered as a heterogeneous fluid. We have already stated in Art. 908 that Laplace made no substantial addition to the results obtained in Legendre's fourth memoir.

1092. The most important point in these §§ 29...31 is the demonstration that in the expression of the radius vector of any stratum of the body in terms of Laplace's functions, the functions of a higher order than the second must vanish.

I have discussed the various investigations on this important point which have been given by Legendre, Laplace, O'Brien, and Pratt, in a memoir to which I have referred in Art. 933.

1093. Laplace in his § 32 examines the conditions which follow from supposing that the axis of rotation is a principal axis. See Art. 653.

1094. Laplace's § 33 is important. He obtains expressions for the force of gravity, the length of the seconds pendulum, and the length of a degree of the meridian at an assigned latitude. He says on his page 97: "Ces trois expressions ont l'avantage d'être indépendantes de la constitution intérieure de la terre, c'est-à-dire, de la figure et de la densité de ses couches;..." He means that he has only assumed the strata of equal density to be very nearly spherical.

In this section he shews it is impossible to admit Bouguer's hypothesis that the variation in the length of a degree of the meridian is proportional to the fourth power of the sine of the latitude. For this hypothesis see page 298 of Bouguer's *Figure de la Terre*. See also Art. 924.

1095. In his § 34 Laplace considers the particular case in which the body is formed of elliptical strata. This case is that which holds if the body is assumed to be entirely fluid, as appears from the §§ 29...31. Laplace moreover shews that this must be the case if we assume all the strata to be similar and covered with a film of fluid: this case was discussed by Legendre; see Art. 902.

This section reproduces important results given by Clairaut. Thus on Laplace's page 101 we have what we find on Clairaut's page 227: see Art. 329. On Laplace's page 102 we have what we find on Clairaut's page 217: see Art. 323. Also on Laplace's page 102 we have Clairaut's theorem, as on Clairaut's page 250: see Art. 336.

1096. Laplace in his § 35 shews how to calculate the attraction exerted on an external particle by a spheroid, the surface of which is a film of fluid in relative equilibrium. Laplace assumes still that the strata of equal density are nearly spherical.

1097. Laplace in his § 36 finds an expression for the force of gravity, on the supposition that the law of attraction is that of the nth power of the distance, and that the body is nearly spherical and homogeneous, and rotates with uniform angular velocity: see Art. 816.

1098. Laplace in his § 37 shews how to extend the approximation to the square and higher powers of the small quantity a. This matter is more fully discussed by Poisson in a memoir in the *Connaissance des Tems* for 1829.

A misprint at the beginning of this section runs throughout it. The first equation should be

$$\text{constant} = V - \frac{g}{2} r^2 \left(\mu^2 - \frac{1}{3}\right) + \frac{1}{3} g r^2.$$

Laplace omits the term $\frac{1}{3}gr^3$. It should be remarked that on his suppositions the *variable part* of $\frac{1}{3}gr^3$ may be considered as of the second order; but then he is here retaining terms of the second order. The mistake is pointed out by Bowditch.

1099. The fourth Chapter contains, as we see, much that is important. The §§ 22...28 are Laplace's own, and very valuable. With respect to the figure of the Earth, considered as a heterogeneous fluid, we have seen in Art. 801 that Legendre claims the priority.

1100. The fifth Chapter of the Third Book is entitled *Comparaison de la théorie précédente, avec les observations.*

This Chapter is principally from Laplace's seventh memoir; but the following pages are new: 113...125, 141...140, 151...153.

The Chapter consists of two parts; first we have geometrical investigations mainly relating to geodesic lines on a spheroid which differs but little from a sphere; and next we have numerical calculations to determine the figure of the Earth from the measured lengths of degrees at various points of the Earth's surface, and from the observed lengths of the seconds pendulum. Both these subjects have been much developed since Laplace's time. The geometrical investigations would now be studied to most advantage in some work on Geodesy; see for instance the sixth Book of Puissant's *Traité de Géodésie*, third edition, in two quarto volumes, 1842. The practical measurement of degrees on the Earth's surface has been carried on with so much energy in recent times, that the data for numerical computation are now far more extensive than those accessible to Laplace. See for instance the modern works on the English, the Russian, and the Indian surveys.

I may observe that Resal on his pages 244...202 gives geometrical investigations of about the same extent as Laplace's, but by a different method: these would be interesting if they were not so inaccurately printed as to be scarcely intelligible.

1101. Laplace's pages 109...111 consist of generalities about geodesic lines; they might with advantage be put into a more modern shape.

Pages 112...114 contain formulæ suited to the case of a nearly spherical body.

Pages 115...117 treat of the special case in which the geodesic line starts by being parallel to the corresponding plane of the celestial meridian.

Pages 118...122 treat of the special case in which the geodesic line starts by being at right angles to the corresponding plane of the celestial meridian.

Pages 123...126 treat of the radius of curvature of a geodesic line.

1102. We may observe that there is a misprint on Laplace's page 119. He twice puts α before $\frac{ddu'}{d\phi d\psi}$ when it ought not to be there. The misprint was pointed out by Bowditch on his page 394. The misprint is preserved in the national edition of Laplace's works: see the page 139.

Another misprint occurs on Laplace's page 125, and on the corresponding page, namely 146, of the national edition.

Laplace takes for the radius vector of a certain ellipsoid

$$1 - \alpha \sin^2 \psi \, [1 + h \cos 2(\phi + \beta)],$$

when it should be

$$1 - \alpha \sin^2 \psi \, [1 + h \cos 2(\phi + \beta)] + \alpha h \cos 2(\phi + \beta);$$

and in consequence he gives erroneous expressions for the lengths of a degree. For example, he gives for the degree measured perpendicular to the meridian

$$1^\circ + 1^\circ . \alpha [1 + h \cos 2(\phi + \beta)] \sin^2 \psi + 4^\circ \alpha h \tan^2 \psi \cos 2(\phi + \beta),$$

when it should be

$$1^\circ + 1^\circ . \alpha [1 + h \cos 2(\phi + \beta)] \sin^2 \psi - 3^\circ \alpha h \cos 2(\phi + \beta).$$

The corrections were pointed out by Bowditch on his pages 412...416.

The corrections are adopted by Puissant: see Vol. II. pages 393...395 of the work cited in Art. 1100.

1103. We now proceed to the second of the two parts which compose Laplace's fifth Chapter; namely, the numerical calculations as to the figure of the Earth. The various measured lengths of degrees do not agree in giving precisely the same value to the numerical elements of the figure of the Earth; so it is a subject of enquiry to determine the best method of treating the data which are furnished by observation.

Laplace proposes two different methods for treating discordant observations, neither method being that which is known as the method of least squares. The first method is given in his § 39 and the second in his § 40.

1104. Let us first consider the method of § 39. Suppose $a_1, a_2, a_3, \dots$ measured lengths of a degree in different latitudes, and p_1, p_2, p_3, the corresponding squares of the sines of the latitude. If the Earth were accurately an oblatum, and there were no errors of observation, we should have a series of equations of which the type would be, neglecting the square of the ellipticity,

$$a_r - z - yp_r = 0 \text{.........................} (15).$$

But as there will be errors of observation we shall have instead of zero on the right-hand side, an unknown error, which we will denote by e_r. So that the general type of the equations will be

$$a_r - z - yp_r = e_r \text{ ....} (16).$$

Laplace proposes that we should determine y and z by the condition that the numerically greatest of the quantities $e_1, e_2, \dots$ should have the least numerical value. See Arts. 960 and 961.

Laplace sketches a general process of solution which would apply if there were more than *two* quantities to be found like y and z; and then he discusses with greater detail the solution for the actual case.

The problem may be stated verbally thus: to determine the elliptic figure of the Earth so that the greatest deviation from observation may have the least possible value.

From examining equations (15) and (10) we see that the problem which Laplace solves may be put in the following geometrical form: a system of straight lines in a plane is given, required to find the point which has the least possible value for the *relative* distance from the straight line which is most remote from it. By the *relative* distance is here meant the distance measured in a direction which is fixed for each straight line, though in general not the same for any two straight lines.

1105. Laplace's § 40 is devoted to another method of treating the observations. He now proposes to determine the generating ellipse of the Earth's figure by the two conditions that the sum of all the errors is zero, and that the sum of all the errors taken positively is a minimum. Laplace calls this the most probable ellipse. The method is due to Boscovich: see Art. 962.

1106. Bowditch thinks that the method of Boscovich "is not now so much used as it ought to be": see page 434 of the second volume of his translation of the *Mécanique Céleste*. Bowditch objects to the method of least squares as commonly applied to the problem, and proposes a modification of it.

I presume that neither of the two methods which Laplace discusses would now be practically used in such calculations, but the method of least squares.

1107. Laplace's § 41 gives numerical application. He takes seven measures of degrees, and calculates a result by both the methods he has explained: see Art. 961.

Laplace comes to the conclusion that the errors which are thus found in the observations are too large to allow us to adopt the supposition that the figure of the Earth is an oblatum.

Laplace corroborates his opinion that the Earth is not an oblatum, by considering especially the results of operations which had been recently carried on by Delambre and Méchain, for measuring an arc of the meridian between Dunkirk and Barcelona. He applies the method of his § 39; and arrives at an ellipticity of $\frac{1}{150}$, which cannot be reconciled with the phenomena of gravity and of precession and nutation.

Laplace finds the length of a quarter of the terrestrial meridian. He uses the ellipticity $\frac{1}{334}$, which he obtains by combining the French measure of an arc of the meridian with the measure of the arc in Peru. He also settles the length of a metre, defined to be $\frac{1}{10000000}$ of a quarter of the meridian, in terms of the *toise of Peru*.

1108. Laplace in his § 42 discusses the observed lengths of pendulums; he takes fifteen cases: see Art. 965.

1109. Laplace's § 43 is devoted to Jupiter. Assuming that the planet is a homogeneous fluid he determines the ellipticity; he finds that the equatorial diameter would then be to the polar diameter as 1·10067 is to 1.

By a weak analogy from the form of Jupiter Laplace infers that the Earth's ellipticity is less than $\frac{1}{300}$.

1110. There are numerical mistakes on Laplace's pages 139, 142, 148, and 150; the corresponding pages of the national edition are 163, 166, 173, and 175 respectively, where the mistakes are reproduced: the corrections are given by Bowditch on his pages 447, 459, 471 and 477 respectively.

1111. Laplace on his page 140 considers that an error so great as 48·6 double toises cannot have occurred in the arcs measured in Pennsylvania, at the Cape of Good Hope, and in Lapland; and again on his page 141 he considers that an error of 66·26 double toises in the Lapland degree is much too great to be admitted.

If we accept Svanberg's measurement of the arc in Lapland, the error in the original determination of the length of a centesimal degree, which Laplace here uses, is about 200 toises, which exceeds that which Laplace pronounced too great to be admitted: see Art. 197.

As to the arc in Pennsylvania, Bowditch, himself an American, proposes to reject it: see his page 444.

1112. I do not quite follow some remarks made by Laplace on his page 143. He uses four measured arcs of meridian from the recent French operations; and from these by the application of his § 39 he deduces an ellipticity of about $\frac{1}{150}$. Then he shews that this also agrees well with an arc measured perpendicular to the meridian in England. So that on the whole the result may be said to depend on *four* French arcs of meridian, and *one* English arc perpendicular to the meridian.

Now Laplace says:

> Mais il est très-remarquable, que les mesures faites nouvellement en France et en Angleterre, avec une grande précision, dans le sens des méridiens, et dans le sens perpendiculaire aux méridiens, se réunissent à indiquer un ellipsoide osculateur dont l'ellipticité est $\frac{1}{150}$, ...

On this I remark that Laplace's words would seem to suggest that to get this result he had used both French and English arcs of the meridian, and both French and English arcs perpendicular to the meridian; instead of what he really did use. And again he now seems to consider this as the *most probable result* of the observations, whereas he has himself in his § 40 given that name to a different result and obtained on different principles. This may be illustrated by his calculations with respect to the seven selected degrees of § 41. By the method of § 39 Laplace obtains an ellipticity $\frac{1}{277}$, which should have been $\frac{1}{250}$ as Bowditch shews: by the method of § 40 Laplace obtains an ellipticity $\frac{1}{312}$, which is very different from the former.

1113. On his page 147 Laplace notices *fifteen* pendulum observations. Of these he seems to make two divisions, one containing *nine* and the other *eight*: it seems to me that his second division contains only *six*.

1114. On his page 151 Laplace incautiously makes the length of the seconds pendulum vary as the square of the latitude. It should be that the *increment* of the length varies as the square of the sine of the latitude.

1115. The geometrical investigations which constitute § 38 of this Chapter seem to be Laplace's own; at least I have not discovered them in any preceding writer. The method of § 39 seems also his own. The method of § 40 is due to Boscovich, as we have seen in Art. 662. The § 38 is the only part which can be considered now to constitute an essential part of the subject; it consists of the geometrical investigations which we noticed in Art. 1100. The numerical calculations which form the latter part of the Chapter by their nature could only have a temporary value; and they are now superseded by more elaborate work founded on a more extensive supply of measurements and observations.

1116. The sixth Chapter of the Third Book is entitled *De la figure de l'anneau de Saturne.*

The § 44 of this Chapter differs from the corresponding part of Laplace's sixth memoir; but the §§ 45 and 46, which constitute the main part of the Chapter, are substantially the same here as in the memoir.

We may observe that a sketch of the history of the subject, so far as we have gone up to the end of the fifth Chapter, is given by Laplace himself in the pages 1...11 of the fifth volume of the *Mécanique Céleste:* on pages 288...291 he gives a sketch of the labours of Astronomers and Geometers as to the ring of Saturn; we may notice especially the top of page 290. It is stated on page 288 that Herschel saw only *two* rings. This is contrary to what Laplace had anticipated in his sixth memoir.

1117. Laplace says towards the beginning of his § 44 that he will consider a thin stratum of fluid spread over the surface of the rings to be in equilibrium; and he says at the beginning of § 45 that he will consider the ring to be a homogeneous fluid mass. However the two hypotheses come to the same thing; for if we regard the ring as fluid, then, the forces being such as occur in nature, if the condition for the equilibrium of the surface is satisfied, the mass will be in equilibrium throughout.

1118. Laplace does not say distinctly what is the order of approximation which he adopts. The fact is that he replaces a ring by an infinite right cylinder; and he gives no investigation by which we can judge of the amount of error which this involves. The suggestion he makes that we should put

$$V = V' + \frac{1}{a} V'' + \frac{1}{a^2} V''' + \dots,$$

where a is the distance between the centre of Saturn and the centre of the generating curve of the ring, and thus get V in a series, seems of no practical value.

1119. Laplace then really determines the attraction of an infinite cylinder on an external particle. Take the axis of z parallel to the generating lines of the cylinder. Then the potential V must satisfy the equation

$$\frac{d^2V}{dx^2} + \frac{d^2V}{dy^2} = 0;$$

therefore $\qquad V = f(x + y\sqrt{-1}) + F(x - y\sqrt{-1}),$

where f and F denote functions at present arbitrary.

Suppose that from symmetry we know that a change in the sign of y will not change V; then

$$V = f(x - y\sqrt{-1}) + F(x + y\sqrt{-1}).$$

Therefore by addition

$$V = \frac{1}{2}\{f(x + y\sqrt{-1}) + F(x + y\sqrt{-1})\}$$
$$+ \frac{1}{2}\{f(x - y\sqrt{-1}) + F(x - y\sqrt{-1})\}$$
$$= \phi(x + y\sqrt{-1}) + \phi(x - y\sqrt{-1}) \text{ say.}$$

Hence if we find the value of V for the case in which $y = 0$, we can infer the general value of V. Or if we find the value of $\frac{dV}{dx}$ when $y = 0$, we shall in fact determine $\phi'(t)$ when $t = x$; then we can deduce the value of $\phi'(t)$ when $t = x \pm y\sqrt{-1}$.

See the last paragraph of Art. 1048.

1120. We may thus confine ourselves to estimating the attraction of an elliptic cylinder on an external particle, which is in one of the principal planes that contain the axis of the cylinder. Suppose the cylinder decomposed into rods, parallel to the generating lines, of infinitesimal section. The attraction of an infinite straight line we know is represented by $\frac{2}{p}$, where p is the perpendicular from the point on the line.

Let the diagram represent a section of the cylinder by a plane at right angles to the axis, and passing through the attracted

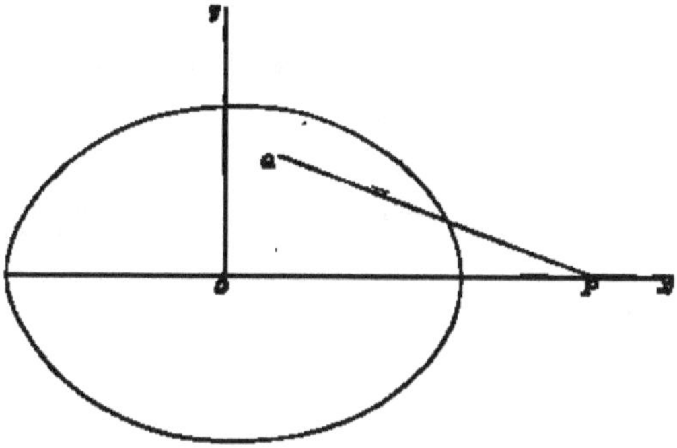

particle P, which is on one of the axes of the elliptic section produced. Let O be the centre of the ellipse, $OP = u$. The attraction of the rod corresponding to Q may be denoted by $\frac{2dxdy}{PQ}$; this is along PQ. Resolve this, and we obtain for the attraction along the axis of x

$$\frac{2dx\, dy\, (u-x)}{y^2 + (u-x)^2}.$$

Hence the resultant attraction of the cylinder is along the axis of x, and its value is

$$2\iint \frac{(u-x)\, dx\, dy}{y^2 + (u-x)^2};$$

the integration is to extend over the whole area of the ellipse, the equation of which may be denoted by $x^2 + \lambda^2 y^2 = k^2$.

Laplace integrates with respect to y, and then states what the integration with respect to x will give.

1121. We may conveniently use polar coordinates in evaluating the definite integral of the preceding Article. Let

$$u - x = r \cos \theta, \qquad y = r \sin \theta.$$

Then the definite integral becomes

$$2 \iint \frac{r \cos \theta \,.\, r dr\, d\theta}{r^2}, \text{ that is } 2 \iint \cos \theta \, dr \, d\theta;$$

and the limits are to be found from the equation

$$(u - r \cos \theta)^2 + \lambda^2 r^2 \sin^2 \theta = k^2,$$

that is $\quad r^2 (\cos^2 \theta + \lambda^2 \sin^2 \theta) - 2ur \cos \theta + u^2 - k^2 = 0 \text{ ...... (17)}.$

Integrate first with respect to r, and use the limits which will be furnished by the last equation, thus we obtain

$$4 \int \frac{\cos \theta \sqrt{\{u^2 \cos^2 \theta - (u^2 - k^2)(\lambda^2 \sin^2 \theta + \cos^2 \theta)\}}\, d\theta}{\lambda^2 \sin^2 \theta + \cos^2 \theta}.$$

Denote this by $4v$. Then

$$\frac{dv}{dk} = \int \frac{k \cos \theta \, d\theta}{\sqrt{\{u^2 \cos^2 \theta - (u^2 - k^2)(\lambda^2 \sin^2 \theta + \cos^2 \theta)\}}}$$

$$= \int \frac{k \cos \theta \, d\theta}{\sqrt{(k^2 - c^2 \sin^2 \theta)}},$$

where $\qquad c^2 = \lambda^2 u^2 + (1 - \lambda^2) k^2.$

The limits of θ are the values of θ for which the two values of r furnished by (17) become equal; it will be found that these are such as make $k^2 - c^2 \sin^2 \theta = 0$.

Therefore $\qquad \dfrac{dv}{dk} = \dfrac{\pi k}{c} = \dfrac{\pi k}{\sqrt{\{\lambda^2 u^2 + (1 - \lambda^2) k^2\}}}.$

Hence we can obtain v; and as v obviously vanishes with k we have

$$v = \frac{\pi}{1 - \lambda^2} [\sqrt{\{\lambda^2 u^2 + (1 - \lambda^2) k^2\}} - \lambda u].$$

And the required attraction is $4v$.

1122. We have in fact in Arts. 1119...1121 a complete account of the attraction of an infinite cylinder on an external particle. As to the action of a cylindrical shell it may be shewn by the aid of Art. 215 of the *Statics* that if the surfaces are similar and similarly situated elliptical cylinders, with a common axis, the attraction on an internal particle is zero.

1123. Laplace however really requires the attraction of an infinite cylinder only for a point at its surface; and this may be found more briefly. Resolve the cylinder as before into rods, parallel to the generating lines, of infinitesimal section. Take the point on the surface as the origin of polar coordinates. Then for the resolved attractions in two directions at right angles to each other in a plane at right angles to the axis of the cylinder we have the expressions

$$X = 2\iint dr\, d\theta \cos\theta, \qquad Y = 2\iint dr\, d\theta \sin\theta.$$

Suppose the cylinder an elliptic cylinder. Let h, k be the coordinates, referred to the centre as the origin, of the point on the surface at which the attraction is required; let $2a$ and $2b$ be the corresponding axes of the ellipse. Then the integration with respect to r is to be taken from $r = 0$ to

$$r = -\frac{\frac{2h\cos\theta}{a^2} + \frac{2k\sin\theta}{b^2}}{\frac{\cos^2\theta}{a^2} + \frac{\sin^2\theta}{b^2}}.$$

The limits with respect to θ are θ_1 and $\theta_1 + \pi$, where θ_1 is such that

$$\frac{h\cos\theta_1}{a^2} + \frac{k\sin\theta_1}{b^2} = 0.$$

Hence we get

$$X = -\int \frac{4hb^2\cos^2\theta\, d\theta}{a^2\sin^2\theta + b^2\cos^2\theta},$$

$$Y = -\int \frac{4ka^2\sin^2\theta\, d\theta}{a^2\sin^2\theta + b^2\cos^2\theta}.$$

Therefore $$\frac{X}{h} + \frac{Y}{k} = -4\int_{\theta_1}^{\theta_1+\pi} d\theta = -4\pi.$$

And $$\frac{Xa^2}{h} + \frac{Yb^2}{k} = -4a^2b^2 \int_{\theta_1}^{\theta_1+\pi} \frac{d\theta}{a^2 \sin^2\theta + b^2 \cos^2\theta}$$
$$= -4ab\pi.$$

Hence, finally,

$$X = -\frac{4\pi bh}{a+b}, \qquad Y = -\frac{4\pi ak}{a+b}.$$

Therefore $$X^2 + Y^2 = \left(\frac{4\pi}{a+b}\right)^2 (b^2h^2 + a^2k^2) = \left(\frac{4\pi ab}{a+b}\right)^2;$$

thus the resultant is constant for all points of the surface of the cylinder.

I cannot find that this simple remark has been made before, though many persons give the formulæ for X and Y; as for instance Laplace in 1787, and Plana in 1819: see also Resal, page 155, and Price's *Infinitesimal Calculus*, Vol. III. page 289.

The *direction* of the resultant attraction at any point of the surface can be readily assigned; the tangent of the angle which this direction makes with the axis of x is equal to $\frac{Y}{X}$, that is to $\frac{ak}{bh}$. Hence the direction is parallel to the corresponding radius of the auxiliary circle.

1124. The substance of Laplace's Chapter on Saturn's ring is reproduced by Resal in his pages 239...243.

1125. The Chapter devoted by Laplace to Saturn's ring is original and interesting; but it does not discuss the subject very fully. The reader who desires to obtain information on this matter will consult the essay by Professor Maxwell, entitled *On the Stability of the Motion of Saturn's Rings*, Cambridge, 1859: for an account of this essay see the *Monthly Notices of the Royal Astronomical Society*, Vol. XIX. page 297.

1126. The seventh Chapter of the Third Book is entitled *De la figure des atmosphères des corps célestes.*

This occupies little more than three pages of the *Mécanique Céleste*. Laplace really adds nothing to what was previously known, and which may be found in the sixth volume of D'Alembert's *Opuscules Mathématiques*: see Art. 639.

It does not seem to me that the Chapter is very clearly written. Laplace for instance says that at the exterior surface $\Pi = 0$; this would be true if the atmosphere were an incompressible fluid, but for an *atmosphere* we cannot have $\Pi = 0$, for as long as there is density there will be pressure. In what follows Laplace gives the equation

$$c = \frac{2}{r} + \alpha r^2 \sin^2 \theta$$

as the equation to the *surface* of the atmosphere; but this is really the equation to any surface of equal pressure.

Laplace afterwards says that the greatest value which the radius vector can have is that where the centrifugal force is equal to the attraction; and this is true, and gives a limit to the extent of the surface.

The subject is treated by Resal in his pages 263...289; he follows the method and principles of E. Roche to whom he refers. The substance of Laplace's Chapter is reproduced in Pratt's *Mechanical Philosophy*, second edition, pages 552...554.

1127. Here we finish our account of the contributions to our subject which are contained in the first two volumes of the *Mécanique Céleste*. They consist of the investigations, collected and improved, which Laplace made during the last quarter of the eighteenth century. Their illustrious author combined the highest mathematical ability with unwearied energy; and he availed himself of the labours of his predecessors, and of his eminent contemporary Legendre. He may be said to have received the theories of Attraction and of the Figure of the Earth immediately from the hands of D'Alembert; and he transmitted them to his successors stamped with the permanent impression of his own genius. Although more than seventy years have elapsed since the publication of the earlier volumes of the *Mécanique Céleste*, they still embody in their pages the standard treatise on those parts of Physical Astronomy of which our history treats.

CHAPTER XXIX.

LAPLACE'S THEOREM.

1128. We shall now proceed to give an account of investigations which have appeared since the publication of the second volume of the *Mécanique Céleste*. We shall consider in separate Chapters the various important points which have been thus discussed. The present Chapter is devoted to Laplace's theorem respecting the attractions of confocal ellipsoids.

1129. We have already noticed Laplace's own demonstration, which first appeared in his treatise of 1784, was improved in his fourth memoir, and finally introduced in the *Mécanique Céleste:* see Arts. 804 and 850. Legendre in his third memoir arrived at the result by a laborious investigation which does not employ infinite series: see Chapter XXIV.

1130. We have first to consider a memoir by Biot, entitled *Recherches sur le calcul aux différences partielles, et sur les attractions des sphéroïdes*; this is contained in the sixth volume of the *Mémoires de l'Institut*...Paris 1806: the memoir occupies pages 201...218 of the volume.

1131. Biot refers to the researches on the subject of the attraction of spheroids by Laplace, Lagrange, and Legendre, before he develops his own method. Let V be the potential of a given body on a particle whose coordinates are a, b, c. Biot starts with the equation

$$\frac{d^2V}{da^2}+\frac{d^2V}{db^2}+\frac{d^2V}{dc^2}=0 \;\;\ldots\ldots\ldots\ldots\ldots\ldots (1).$$

He seems to describe his own method by saying that instead of trying to integrate this partial differential equation he interprets the differential form directly. I should describe it by saying that instead of trying to integrate this differential equation in finite terms he uses an integral in the form of an infinite series.

1132. We may transform (1) by substituting three new variables a', b', c' which are connected with a, b, c by the arbitrary equations

$$a' = F(a, b, c), \quad b' = F_1(a, b, c), \quad c' = F_2(a, b, c) \ldots\ldots\ldots (2).$$

Let V' denote the form which V assumes when for a, b, c we substitute their values in terms of a', b', c' given by (2). Then

$$\frac{dV}{da} = \frac{dV'}{da'}\frac{da'}{da} + \frac{dV'}{db'}\frac{db'}{da} + \frac{dV'}{dc'}\frac{dc'}{da} \ldots\ldots\ldots\ldots (3),$$

and similar expressions hold for $\frac{dV}{db}$ and $\frac{dV}{dc}$.

Similarly we can express $\frac{d^2V}{da^2}$, $\frac{d^2V}{db^2}$, and $\frac{d^2V}{dc^2}$.

Then substitute in (1) and we have a partial differential equation of the second order which we will denote by

$$L = 0 \ldots\ldots\ldots\ldots\ldots\ldots\ldots\ldots\ldots\ldots\ldots (4).$$

The equation (4) like (1) will be linear.

Suppose the value of V' which satisfies (4) to be expanded in powers of a'; say

$$V' = \phi + a'\phi_1 + \frac{a'^2}{\lfloor 2}\phi_2 + \frac{a'^3}{\lfloor 3}\phi_3 + \frac{a'^4}{\lfloor 4}\phi_4 + \ldots\ldots\ldots\ldots\ldots (5),$$

where ϕ, ϕ_1, ϕ_2...denote functions of b' and c' which do not contain a'.

From (5) we must obtain $\frac{dV'}{da'}$, $\frac{dV'}{db'}$, $\frac{dV'}{dc'}$, and the differential coefficients of V' of the second order, and substitute these in (4). Equate to zero the coefficients of the various powers of a'. Thus we shall have equations which will determine ϕ_2, ϕ_3, ϕ_4...in terms of ϕ and ϕ_1, but these will remain quite arbitrary.

This is the main part of Biot's investigation. It is obvious that it is not quite satisfactory. For possibly exceptions might arise when special forms are assigned to the functions denoted by the F, F_1, F_2 of equation (2). Moreover there is nothing to ensure the convergence of (5).

1133. From equation (5) we have

$$\frac{dV'}{da'} = \phi_1 + a'\phi_2 + \frac{a'^2}{\lfloor 2}\phi_3 + \frac{a'^3}{\lfloor 3}\phi_4 + \ldots\ldots$$

$$\frac{dV'}{db'} = \frac{d\phi}{db'} + a'\frac{d\phi_1}{db'} + \frac{a'^2}{\lfloor 2}\frac{d\phi_2}{db'} + \frac{a'^3}{\lfloor 3}\frac{d\phi_3}{db'} + \ldots\ldots$$

$$\frac{dV'}{dc'} = \frac{d\phi}{dc'} + a'\frac{d\phi_1}{dc'} + \frac{a'^2}{\lfloor 2}\frac{d\phi_2}{dc'} + \frac{a'^3}{\lfloor 3}\frac{d\phi_3}{dc'} + \ldots\ldots$$

But in the values of ϕ_2, ϕ_3, ϕ_4,.... we shall find that the function ϕ itself does not occur but only the differential coefficients of ϕ; this arises from the fact that V' itself does not occur in (4) but only the differential coefficients of V'.

Hence the three series just given will be completely determined, when the first terms are known, that is when the values of $\frac{dV'}{da'}$, $\frac{dV'}{db'}$, and $\frac{dV'}{dc'}$ are known corresponding to $a' = 0$.

But these values are connected with the values of $\frac{dV}{da}$, $\frac{dV}{db}$, and $\frac{dV}{dc}$ by (3) and two similar equations; so we shall have the values of $\frac{dV'}{da'}$, $\frac{dV'}{db'}$, and $\frac{dV'}{dc'}$ when $a' = 0$, provided we know the values of $\frac{dV}{da}$, $\frac{dV}{db}$, and $\frac{dV}{dc}$ when $a' = 0$.

And as the particular values of $\frac{dV'}{da'}$, $\frac{dV'}{db'}$, and $\frac{dV'}{dc'}$ when $a' = 0$, suffice to determine the general values of these differential coefficients, it follows also that the general values of $\frac{dV}{da}$, $\frac{dV}{db}$, and $\frac{dV}{dc}$ are determined as soon as we know the particular values which correspond to $a' = 0$.

But the equation $a'=0$ will represent any surface whatever by giving the proper form to $F(a, b, c)$. Hence we obtain the following very general theorem:

In order to know the attractions of a spheroid at any exterior points it will be sufficient to know the attractions of this spheroid at all the points of any exterior surface taken at pleasure.

I have inserted the word *exterior* in Biot's enunciation because we now know that (1) is not true for internal points.

1134. As an example we may take for $a'=0$ the equation to the surface of the attracting spheroid itself.

Biot considers that this includes as a particular case Laplace's theorem respecting ellipsoids. But this becomes more obvious after some developments to which Biot now proceeds. See Art. 1136.

1135. If we wish the arbitrary surface to be a plane we may take $a=0$ for its equation. Then it will not be necessary to transform (1) by the introduction of the new variables a', b', c'. The general value of V derived from (1) will be

$$V=\phi+a\phi_1-\frac{a^2}{\lfloor 2}\left(\frac{d^2\phi}{db^2}+\frac{d^2\phi}{dc^2}\right)-\frac{a^3}{\lfloor 3}\left(\frac{d^2\phi_1}{db^2}+\frac{d^2\phi_1}{dc^2}\right)$$
$$+\frac{a^4}{\lfloor 4}\left(\frac{d^4\phi}{db^4}+2\frac{d^4\phi}{db^2dc^2}+\frac{d^4\phi}{dc^4}\right)+\ldots\ldots\ldots(6);$$

this may be verified by substituting in (1); or it may be obtained as in Boole's *Differential Equations*, third edition, pages 401 and 402.

Biot says that this was first given by Lagrange in the *Mécanique Analytique*, p. 474; this means the first edition of Lagrange's work: see Art. 994.

From the above value of V we can immediately deduce $\frac{dV}{da}$, $\frac{dV}{db}$, and $\frac{dV}{dc}$.

Biot's result includes that of Laplace relative to symmetrical spheroids: see Art. 1070.

1136. The preceding developments apply to all kinds of spheroids; for each particular spheroid the values of ϕ_1 and $\frac{d\phi}{db}$ and $\frac{d\phi}{dc}$ will in general be different.

Suppose we take an ellipsoid; the attraction for any point in the plane of the equator can be obtained, as was shewn in Legendre's third memoir. We have in this case

$$\phi = MU, \quad \phi_1 = 0,$$

where M is the mass of the ellipsoid, and U is a function which involves the coordinates of the external point and the *excentricities* of the ellipsoid; see Art. 1000. The result $\phi_1 = 0$ follows from the fact that the ellipsoid is symmetrical with respect to its equator, and so for any point in that plane the attraction parallel to the axis of a must vanish.

Thus for an ellipsoid we obtain from (6)

$$V = MU - \frac{Ma^2}{\underline{|2}}\left(\frac{d^2U}{db^2} + \frac{d^2U}{dc^2}\right) + \frac{Ma^4}{\underline{|4}}\left(\frac{d^4U}{db^4} + 2\frac{d^4U}{db^2dc^2} + \frac{d^4U}{dc^4}\right) + \dots\dots\dots$$

Hence if there be a second ellipsoid with the same *excentricities*, the value of U will be the same for both; thus if M' be the mass and V' the potential for the second ellipsoid, we have

$$\frac{V'}{M'} = \frac{V}{M}.$$

This constitutes the proof of Laplace's theorem.

1137. Results analogous to those which have been given, but more simple, hold for the case of spheroids of *revolution*.

For a spheroid of revolution we may put

$$b^2 + c^2 = r^2,$$

and V will be a function of r and a.

Thus (1) is transformed into

$$\frac{1}{r}\frac{dV}{dr} + \frac{d^2V}{dr^2} + \frac{d^2V}{da^2} = 0.$$

Hence by the same method as before we arrive at this result: the attraction of a spheroid of revolution will be known for any external point whatever, if it is known for every point of any arbitrary external curve whatever, described in the plane of the meridian. If we suppose this arbitrary curve to be the prolongation of the axis we have the result first given by Legendre: see Art. 791.

Biot himself says: "...et de là résultent, comme cas particulier, les beaux théorèmes démontrés pour la première fois par M. Legendre." I do not know what other theorem Biot has in view besides that to which I have referred.

1138. I must cite another sentence from Biot's memoir; he says on page 208, after introducing the function V,

M. Lagrange a démontré que les coefficiens différentiels

$$\frac{dV}{da}, \frac{dV}{db}, \frac{dV}{dc},$$

pris négativement, expriment les attractions exercées par le sphéroïde sur ce même point, parallèlement aux trois axes rectangulaires. M. Laplace a fait voir ensuite que la fonction V est assujetie à l'équation différentielle partielle

$$\frac{d^2V}{da^2} + \frac{d^2V}{db^2} + \frac{d^2V}{dc^2} = 0.$$

I do not know on what authority the above expressions for component attractions are assigned to Lagrange; to me they appear due to Laplace: see Art. 789, and also pages 70 and 103 of Laplace's *Figure des Planetes*.

1139. Biot's memoir may be said to belong more properly to the subject of partial differential equations than to that of attractions; and its interest for us is rather of a speculative than of a practical character, for it does not really determine the attraction of any spheroid. So far as Laplace's theorem is concerned, we see that the investigation is not quite independent, for it borrows one of the main results of Legendre's abstruse third memoir.

1140. We arrive now at the remarkable simplification effected by Ivory. A memoir by him entitled *On the Attractions of homogeneous Ellipsoids*, was read before the Royal Society on 15th June, 1809; and is printed in the *Philosophical Transactions* for 1809; it occupies pages 345...372.

1141. This memoir is famous for containing the enunciation and demonstration of the theorem which is usually called Ivory's theorem; but which would be more justly called Ivory's demonstration of Laplace's theorem. The memoir is the first communicated by Ivory to the Royal Society; and is I think the best of all his memoirs which relate to our subject. The memoir forms a good treatise on the attractions of homogeneous ellipsoids, and may be read at the present day with interest and profit. There are two improvements which our modern books present to us; Ivory makes frequent use of the process of transformation of the variables in a definite double integral, and this process is now found to be unnecessary; he treats the attraction on an internal particle by the method of series, not always convergent, and we now employ the simple method like that given by Lagrange in 1773, and which is adopted in the *Mécanique Céleste*. It is difficult to see what induced Ivory to use the method of series, when Laplace had solved this part of the problem so much better. The essence of the treatment proposed by Ivory for the attraction of an ellipsoid on an external particle remains in our elementary books; and thus it is unnecessary to enter into particulars respecting it.

1142. Ivory gives a brief sketch of the history of the subject in his introductory pages. He refers to the particular cases discussed by Maclaurin and Legendre, and then passes on to the more general problem which Laplace attacked. He says:

The method of investigation, which La Place has employed for surmounting the difficulties of this last case, although it is entitled to every praise for its ingenuity, and the mathematical skill which it displays, is certainly neither so simple nor so direct, as to leave no room for perfecting the theory of the attractions of ellipsoids in both these respects. It consists in shewing that the expressions for the attractions of an ellipsoid, on any external point, may be resolved into two factors; of which, one is the mass of the ellipsoid, and the other involves only the excentricities of the solid and the co-ordinates of the attracted point: whence it follows, that two ellipsoids, which have the same excentricities, and their principal sections in the same planes, will attract the same external point with forces proportional to the masses of the solids. This theorem includes the extreme case, when the surface of one of the solids passes through the attracted point: and by this means the attraction of an ellipsoid, upon a point placed without it, is made to depend upon the attraction which another ellipsoid, having the same excentricities as the former, exerts upon a point placed in the surface. Le Gendre has given a direct demonstration of the theorem of La Place, by integrating the fluxional expressions of the attractive forces; a work of no small difficulty, and which is not accomplished without complicated calculations.

It will be seen that Ivory speaks of "the theorem of La Place," as I do.

1143. Ivory's own enunciation of his result is contained in the following words:

If two ellipsoids of the same homogeneous matter have the same excentricities, and their principal sections in the same planes; the attractions which one of the ellipsoids exerts upon a point in the surface of the other, perpendicularly to the planes of the principal sections, will be to the attractions which the second ellipsoid exerts upon the corresponding point in the surface of the first, perpendicularly to the same planes, in the direct proportion of the surfaces, or areas, of the principal sections to which the attractions are perpendicular.

The theorem is really the combination of two results both due to Laplace. One result may be thus expressed: the potentials of confocal ellipsoids at a given point external to both are as their masses. The other result is the expression for the attraction of an ellipsoid on a particle at its surface. See Art. 1063.

1144. A peculiarity in Ivory's memoir is his frequent use of the process of transformation of the variables in a definite double integral. Although, as we have seen in Arts. 710 and 677, Lagrange and Legendre had treated of this process, yet I do not think any good account of it had been given at the time of Ivory's memoir. However the cases in which he uses the method would not present any great difficulty.

1145. On the whole we may say that Ivory's memoir goes over the same extent of ground as the first Chapter in the Third Book of the *Mécanique Céleste;* obtaining results equivalent to Laplace's but in a more simple manner.

We may observe that Ivory refers to Laplace's memoir of 1783 by mistake, instead of 1782; see Ivory's page 347: the mistake is the same as Legendre makes in his memoir of 1788. Perhaps Ivory copied it from Legendre: see Art. 876.

1146. The merits of Ivory's process were clearly recognised in France. Thus Legendre says on page 158 of the *Mémoires de l'Institut* for 1810:

> Les difficultés d'analyse que présentait ce problème traité par tant de moyens différens, disparaissent ainsi tout d'un coup, par le procédé de M. Yvory, et une théorie qui appartenait à l'analyse la plus abstruse, peut maintenant être exposée dans toute sa généralité, d'une manière presque entièrement élémentaire.

This seems the passage which Dr Thomas Young has in view though I do not understand his reference involved in the words: "...say Legendre and Delambre (*M. Inst.* 1812)." See Young's *Works*, Vol. II. page 581.

1147. A memoir by Plana entitled *Sulla teoria dell' attrazione degli sferoidi elittici* is contained in the *Memorie di Matematica ...della Società Italiana*, Vol. XV. Modena, 1811. The memoir occupies pages 370...390 of the first part of the volume. It was communicated on the 24th November, 1810.

I have already stated that Laplace's proof of his theorem was published by him in the fourth section of his treatise *De la Figure... des Planetes;* and afterwards given in an improved form in the first Chapter of the Third Book of the *Mécanique Céleste;* see Arts. 804 and 1060. Plana's memoir is simply a reproduction of the section from the treatise *De la Figure...des Planetes*, a little expanded by giving the steps of the work in some cases where

Laplace only records the result. Plana probably regarded Laplace's earlier form of the argument as the more natural, though the later form is briefer.

Plana's memoir adds nothing to the knowledge of the subject; though it might save some trouble to young mathematicians in their study of Laplace's method.

I may draw attention to the following words which occur on Plana's page 376:

... bellissimo teorema del Sig. *Legendre* mediante il quale, l'attrazione di un elissoide sopra un punto esteriore alla sua superficie dipende in ogni caso da quella dei punti situati sulla superficie.

I cannot admit the propriety of calling this theorem by Legendre's name; Legendre really established only a part of this: see Art. 782. The extension of the theorem to the generality thus ascribed to it by Plana is really due to Laplace, being in fact involved in the theorem which I call by his name.

1149. A note by Biot entitled *Sur l'attraction des Sphéroïdes* is given in the *Nouveau Bulletin...la Société Philomatique* for March 1812, pages 44...48. This note may be considered as an appendix to the memoir of 1800.

In the memoir Biot had shewn that the attraction of an ellipsoid at any external point might be deduced by simple differentiations from a particular expression, which is theoretically known when the attraction is known for all points situated in the plane of one of the principal sections.

Now he says that the demonstration would cease to be applicable in the case in which the projection of the external point on the assigned plane falls within the principal section of the ellipsoid. For the expressions which give the values of the attractions are different according as the point is within or without the ellipsoid, and so the results cannot be comprehended in the same formulæ.

Biot proposes to surmount the difficulty by a transformation of the coordinates.

Let x, y, z be the old coordinates of the external point, where x and y refer to the assigned principal plane. Let x', y', z' be the new coordinates of the point, connected with the old coordinates by the equations

$$x' = x + z \tan\alpha \cos\beta, \quad y' = y + z \tan\alpha \sin\beta, \quad z' = z \sec\alpha.$$

Then the equation (1) of Art. 1131 transforms into

$$(1 + \tan^2\alpha \cos^2\beta) \frac{d^2V}{dx'^2} + (1 + \tan^2\alpha \sin^2\beta) \frac{d^2V}{dy'^2}$$

$$+ \sec^2\alpha \frac{d^2V}{dz'^2} + 2 \tan^2\alpha \sin\beta \cos\beta \frac{d^2V}{dx'dy'}$$

$$+ 2 \tan\alpha \sec\alpha \left(\cos\beta \frac{d^2V}{dx'dz'} + \sin\beta \frac{d^2V}{dy'dz'}\right) = 0.$$

This is still a linear equation. Hence as in Art. 1132 we find that

$$V = A_0 + A_1 z' + A_2 \frac{z'^2}{\underline{|2}} + A_3 \frac{z'^3}{\underline{|3}} + \ldots,$$

where $A_0, A_1, A_2, \ldots$ are functions of x' and y'. Then the inference is of the same kind as in Art. 1133.

We may take α and β so that (x', y') falls without the ellipsoid, for the case of any assigned external point. Thus the difficulty is surmounted.

Although Biot's method in his memoir and in this note is interesting, yet the use of infinite series of which the convergence is not secured, cannot be accepted as rigorous.

1149. We have next to consider a memoir by Legendre entitled *Mémoire sur l'attraction des ellipsoïdes homogènes.* This is published in the *Mémoires de l'Institut* for 1810, second part: the date of publication is 1811. The memoir occupies pages 155...183 of the volume. The memoir was read on the 5th October, 1812.

1150. Legendre begins by an historical sketch of the subject. He says that the problem of the attraction of an ellipsoid on an internal particle had been completely solved with much elegance by Maclaurin in his prize essay on the Tides. Maclaurin however, as we have seen, explicitly considered only the case of an ellipsoid of revolution, though his methods admitted of obvious extension to the case of the general ellipsoid.

With regard to the attraction of an ellipsoid on an external particle Legendre adverts to the theorem of Maclaurin; to the extension which he himself gave of it in the memoir which I have called his first; to the further extension given by Laplace; and to his own investigations in the memoir which I have called his third. He says of the second part of this memoir:

J'avoue néanmoins que cette partie de mon Mémoire n'a que le mérite d'être directe, et de montrer, dès l'abord, la possibilité de la solution, mais que d'ailleurs l'analyse en est d'une extrême complication. Il était donc à desirer qu'on découvrit une route plus facile pour parvenir au même résultat.

Legendre then refers to Biot for his happy idea of applying to the equation of the attraction the integral which Lagrange had given for another object. Legendre says that Biot's result joined to the first part of his own memoir, the third, completed in a satisfactory manner the theory of the attraction of homogeneous ellipsoids; and so there was little hope of acquiring any new degree of perfection.

But Ivory, whom Legendre calls Yvory, had thrown a fresh light on the subject by an ingenious transformation. Accordingly Legendre proposes to avail himself of Ivory's discovery in order to present the whole theory of the attraction of ellipsoids in its simplest form. This he effects by combining Ivory's demonstration with the mode of solution for an internal particle given by Lagrange, this mode being simpler than that adopted by Ivory, which depends on development in a series.

1151. Legendre's memoir forms a very good account of the attraction of homogeneous ellipsoids; it is clear, simple, and comprehensive, and might be reprinted at the present time as an elementary treatise on the subject.

Although there is nothing really new in the methods employed, yet there are some subordinate results of interest which appear for the first time, and these we will indicate.

1152. What we call Ivory's theorem is given by Legendre substantially in Ivory's manner, though slightly improved. We need not delay on this, but pass to the case in which the attracted particle is within the ellipsoid or on its surface.

Let a, b, c be the semiaxes of the ellipsoid; let f, g, h be the corresponding coordinates of an attracted particle; let A, B, C be the corresponding resolved parts of the attraction. Then Legendre shews that

$$A = 2f \iint \frac{\sin\theta \cos^2\theta \, d\theta \, d\phi}{\cos^2\theta + \frac{a^2}{b^2}\sin^2\theta\cos^2\phi + \frac{a^2}{c^2}\sin^2\theta\sin^2\phi},$$

$$B = \frac{2a^2 g}{b^2} \iint \frac{\sin^3\theta \cos^2\phi \, d\theta \, d\phi}{\cos^2\theta + \frac{a^2}{b^2}\sin^2\theta\cos^2\phi + \frac{a^2}{c^2}\sin^2\theta\sin^2\phi},$$

$$C = \frac{2a^2 h}{c^2} \iint \frac{\sin^3\theta \sin^2\phi \, d\theta \, d\phi}{\cos^2\theta + \frac{a^2}{b^2}\sin^2\theta\cos^2\phi + \frac{a^2}{c^2}\sin^2\theta\sin^2\phi}.$$

The limits for both θ and ϕ are 0 and π.

Consider the expression for A; the integration with respect to ϕ can be effected: thus

$$A = 2f\pi \int \frac{\sin\theta \cos^2\theta \, d\theta}{\sqrt{\left(\cos^2\theta + \frac{a^2}{b^2}\sin^2\theta\right)}\sqrt{\left(\cos^2\theta + \frac{a^2}{c^2}\sin^2\theta\right)}}.$$

Put M for the volume of the ellipsoid, that is for $\frac{4\pi abc}{3}$. Also put x for $\cos\theta$. Thus we get

$$A = \frac{3Mf}{a} \int_0^1 \frac{x^2 dx}{\sqrt{\{a^2 + (b^2 - a^2)x^2\}}\sqrt{\{a^2 + (c^2 - a^2)x^2\}}}.$$

1153. Now instead of making use of the formulæ given above for B and C, we may if we please dedüce values of B and C from the value of A, by appropriate changes of the letters. Thus we have

$$B = \frac{3Mg}{b} \int_0^1 \frac{x^2 dx}{\sqrt{\{b^2 + (c^2 - b^2)x^2\}}\sqrt{\{b^2 + (a^2 - b^2)x^2\}}},$$

$$C = \frac{3Mh}{c}\int_0^1 \frac{x^2 dx}{\sqrt{[c^2 + (a^2 - c^2)x^2]}\,\sqrt{[c^2 + (b^2 - c^2)x^2]}}.$$

All this of course was well known.

Legendre however observes that if we do make use of the formulæ given in Art. 1152 we shall obtain

$$B = \frac{3Mg}{a(c^2 - b^2)}\left[\frac{c}{b} - \int_0^1 \frac{\sqrt{[a^2 + (c^2 - a^2)x^2]}\,dx}{\sqrt{[a^2 + (b^2 - a^2)x^2]}}\right],$$

$$C = \frac{3Mh}{a(c^2 - b^2)}\left[-\frac{b}{c} + \int_0^1 \frac{\sqrt{[a^2 + (b^2 - a^2)x^2]}\,dx}{\sqrt{[a^2 + (c^2 - a^2)x^2]}}\right].$$

Legendre adds that it is easy to convince ourselves that the different formulæ agree in value. I will supply the process.

1154. We have in fact by integration by parts

$$\int \frac{\sqrt{[a^2 + (c^2 - a^2)x^2]}}{\sqrt{[a^2 + (b^2 - a^2)x^2]}}\,dx = \frac{x\sqrt{[a^2 + (c^2 - a^2)x^2]}}{\sqrt{[a^2 + (b^2 - a^2)x^2]}}$$

$$- \int \frac{x^2(c^2 - a^2)\,dx}{\sqrt{[a^2 + (c^2 - a^2)x^2]}\,\sqrt{[a^2 + (b^2 - a^2)x^2]}}$$

$$+ \int \frac{x^2(b^2 - a^2)\sqrt{[a^2 + (c^2 - a^2)x^2]}}{[a^2 + (b^2 - a^2)x^2]^{\frac{3}{2}}}\,dx$$

$$= \frac{x\sqrt{[a^2 + (c^2 - a^2)x^2]}}{\sqrt{[a^2 + (b^2 - a^2)x^2]}} + \int \frac{a^2(b^2 - c^2)x^2\,dx}{[a^2 + (c^2 - a^2)x^2]^{\frac{1}{2}}\,[a^2 + (b^2 - a^2)x^2]^{\frac{3}{2}}}.$$

Take the integrals between the limits 0 and 1; thus we get from Art. 1153

$$B = 3Mga\int_0^1 \frac{x^2\,dx}{[a^2 + (c^2 - a^2)x^2]^{\frac{1}{2}}\,[a^2 + (b^2 - a^2)x^2]^{\frac{3}{2}}},$$

this is another form for B.

Now put $\dfrac{x}{\sqrt{[a^2 + (b^2 - a^2)x^2]}} = \dfrac{y}{b}$; thus

$$B = \frac{3Mga}{b}\int_0^1 \frac{x^2}{[a^2 + (c^2 - a^2)x^2]^{\frac{1}{2}}}\,\frac{dy}{a^2};$$

also $x^2 = \frac{a^2y^2}{b^2-(b^2-a^2)y^2}$, and $a^2+(c^2-a^2)x^2 = \frac{a^2b^2+a^2(c^2-b^2)y^2}{b^2-(b^2-a^2)y^2}$;

so that $B = \frac{3Mg}{b}\int_0^1 \frac{y^2dy}{\sqrt{\{b^2+(c^2-b^2)y^2\}}\sqrt{\{b^2+(a^2-b^2)y^2\}}}$,

which was to be shewn.

1155. If the ellipsoid differs but little from a sphere we may obtain an approximation by a series. Take the expression for A in Art. 1152 as an example.

Let $\mu = \frac{a^2-b^2}{a^2}$, and $\nu = \frac{a^2-c^2}{a^2}$. Then

$$A = \frac{3Mf}{a^2}\int_0^1 \frac{x^2dx}{\sqrt{(1-\mu x^2)}\sqrt{(1-\nu x^2)}}.$$

We should expand *before integration*. Let

$$(1-\mu x^2)^{-\frac{1}{2}}(1-\nu x^2)^{-\frac{1}{2}} = 1 + P_1x^2 + P_2x^4 + P_3x^6 + \ldots\ldots$$

Then it is easy to shew that

$$P_1 = \frac{1}{2}(\mu+\nu),$$

$$P_2 = \frac{1.3}{2.4}(\mu^2+\nu^2) + \frac{1}{2}\cdot\frac{1}{2}\mu\nu,$$

$$P_3 = \frac{1.3.5}{2.4.6}(\mu^3+\nu^3) + \frac{1.3}{2.4}\cdot\frac{1}{2}\mu\nu(\mu+\nu),$$

$$P_4 = \frac{1.3.5.7}{2.4.6.8}(\mu^4+\nu^4) + \frac{1.3.5}{2.4.6}\cdot\frac{1}{2}\mu\nu(\mu^2+\nu^2) + \frac{1.3}{2.4}\cdot\frac{1.3}{2.4}\mu^2\nu^2.$$

....

The law is obvious.

Thus $$A = \frac{3Mf}{a^2}\left\{\frac{1}{3} + \frac{1}{5}P_1 + \frac{1}{7}P_2 + \frac{1}{9}P_3 + \ldots\right\}.$$

1156. If however we wish to avoid series, or if the series are not convergent, it is convenient to have recourse to elliptic functions.

Let $m^2 = b^2 - a^2, \quad n^2 = c^2 - a^2, \quad k^2 = 1 - \frac{m^2}{n^2}.$

In the expression for A given at the end of Art. 1152, put $x = \frac{a}{n} \tan \phi$; then we get

$$A = \frac{3Mf}{n^3} \int_0^{\phi_1} \frac{\tan^2 \phi \, d\phi}{\sqrt{(1 - k^2 \sin^2 \phi)}},$$

where ϕ_1 is such that $\sin \phi_1 = \frac{n}{c}$, $\cos \phi_1 = \frac{a}{c}$, $\tan \phi_1 = \frac{n}{a}$.

Again, in the expression for B given at the beginning of Art. 1153 put $x = \frac{b \sin \phi}{n \sqrt{(1 - k^2 \sin^2 \phi)}}$; then wo get

$$B = \frac{3Mg}{n^3} \int_0^{\phi_1} \frac{\sin^2 \phi \, d\phi}{(1 - k^2 \sin^2 \phi)^{\frac{3}{2}}}.$$

Lastly, in the expression for C given at the beginning of Art. 1153 put $x = \frac{c \sin \phi}{n}$; then we get

$$C = \frac{3Mh}{n^3} \int_0^{\phi_1} \frac{\sin^2 \phi \, d\phi}{\sqrt{(1 - k^2 \sin^2 \phi)}}.$$

Legendre states the results which are obtained by expressing the integrals now left in A, B, C by elliptic integrals: he refers for the formulæ required to his *Exercices de Calcul Intégral*, 1re partie, No. 138. Poisson works out the transformations in his memoir of 1835, to which I have referred in Art. 887.

Thus, to take the simplest of the three expressions for example, we have $$C = \frac{3Mh}{k^2 n^3} \left\{ \int_0^{\phi_1} \frac{d\phi}{\sqrt{(1 - k^2 \sin^2 \phi)}} - \int_0^{\phi_1} \sqrt{(1 - k^2 \sin^2 \phi)} \, d\phi \right\}.$$

1157. Legendre gives an algebraical relation between A, B, and C, which he deduces in three ways from his formulæ: we will take the simplest way. From the formulæ of Art. 1152 we see that

$$\frac{A}{f} + \frac{B}{g} + \frac{C}{h} = 2 \int_0^{\pi} \int_0^{\pi} \sin \theta \, d\theta \, d\phi = 2\pi \int_0^{\pi} \sin \theta \, d\theta = 4\pi.$$

Legendre speaks of this result as an "équation qui ne paraît pas avoir été remarquée jusqu'à présent, et qui doit être regardée comme un théorème nouveau."

1158. Legendre gives another result somewhat like that of the preceding Article; namely

$$\frac{Aa^2}{f} + \frac{Bb^2}{g} + \frac{Cc^2}{h} = \frac{3M}{n} \int_0^{\phi_1} \frac{d\phi}{\sqrt{(1-k^2 \sin^2 \phi)}}:$$

this he obtains from the values which he has found for A, B, and C in terms of elliptic integrals.

It is easy to verify this result; for the formulæ of Art. 1152 give

$$\frac{Aa^2}{f} + \frac{Bb^2}{g} + \frac{Cc^2}{h} = 2a^2 \int_0^{\pi} \int_0^{\pi} \frac{\sin\theta \, d\theta \, d\phi}{\cos^2\theta + \frac{a^2}{b^2}\sin^2\theta\cos^2\phi + \frac{a^2}{c^2}\sin^2\theta\sin^2\phi}$$

$$= 2\pi a^2 \int_0^{\pi} \frac{\sin\theta \, d\theta}{\sqrt{\left(\cos^2\theta + \frac{a^2}{b^2}\sin^2\theta\right)}\sqrt{\left(\cos^2\theta + \frac{a^2}{c^2}\sin^2\theta\right)}}$$

$$= 3Ma \int_0^1 \frac{dx}{\sqrt{\{a^2 + (b^2-a^2)x^2\}}\sqrt{\{a^2 + (c^2-a^2)x^2\}}}.$$

Then, as in Art. 1156, put $x = \frac{a}{n}\tan\phi$, and the integral becomes $\frac{3M}{n} \int_0^{\phi_1} \frac{d\phi}{\sqrt{(1-k^2\sin^2\phi)}}$.

1159. Legendre's memoir is reproduced in his *Exercices de Calcul Intégral*, Vol. II. 1817, pages 512...531, and also in his *Traité des Fonctions Elliptiques*, Vol. I. 1825, pages 539...556. The historical sketch is omitted, and some slight changes occur in the notation, but the memoir remains substantially as it was originally. All that is added in these later editions of the memoir consists of two remarks, of which we will give the substance.

(1) The expressions for A, B, and C in Art. 1156 have an inconvenience when the ellipsoid differs but little from a sphere; n^2 is then very small, and as this occurs in the denominators, the numerators must be calculated with great accuracy. This also holds with respect to the form of these expressions when two of the axes of the ellipsoid are equal.

(2) The definite integrals in the expressions for A, B, and C depend on only *two* quantities, namely k and ϕ_1, although in general there are six elements involved in the problem, namely the three semiaxes and the three coordinates of the attracted particle.

1160. An account of Ivory's memoir by Poisson is given in the *Nouveau Bulletin......la Société Philomatique* for November, 1812, pages 176...180, and for January, 1813, page 210.

This is a clear and satisfactory exposition of the essence of the memoir of Yvory, as Poisson here spells the name. The following incidental points may be noticed.

Poisson says that Maclaurin demonstrated his proposition for points situated on the prolongations of the axes in the case of solids of revolution. Poisson, like other eminent French mathematicians, here underestimates what Maclaurin really effected. See Art. 260.

Poisson observes that Ivory treated the attraction on an internal particle by the method of series; but that it would be better to adopt the method of direct integration after the manner of Lagrange. A few days previously Legendre's memoir had been presented to the French *Institut*, in which Poisson's suggestion was anticipated.

Poisson extends the range of Ivory's theorem, by shewing that it is true whatever may be the function of the distance which expresses the law of attraction; this is now familiar to us, for it has passed into the elementary books.

Poisson observes that Gauss had recently sent to the *Institut* an extract from a memoir on the subject; and that he had made use of the same transformation as Ivory for expressing the coordinates of a point at the surface of an ellipsoid in terms of two independent coordinates; see Art. 1141. Poisson says: "Cette transformation est le point principal de l'analyse de M. Yvory, et c'est aussi celui de l'analyse de M. Gauss, qui ne paraît pas avoir eu connaissance du Mémoire du géomètre anglais."

The transformation to which Poisson refers is, that the equation $\frac{x^2}{a^2}+\frac{y^2}{b^2}+\frac{z^2}{c^2}=1$ is satisfied by putting

$$x=a\cos\theta,\quad y=b\sin\theta\sin\phi,\quad z=c\sin\theta\cos\phi.$$

It may be doubted, however, if we can call this transformation the principal point in the analysis either of Ivory or of Gauss. The modern exhibitions of Ivory's theorem do not employ this transformation.

1161. Poisson, as we have just seen, was the first to point out that Ivory's theorem holds when the law of attraction is expressed by any function of the distance, as well as when the attraction varies inversely as the square of the distance. The theorem is now demonstrated with Poisson's extension in elementary works. Moreover, as I have shewn in my *Statics*, in treating on Ivory's theorem, the demonstration establishes rather more than the enunciation states.

The extension which Poisson gave to Ivory's theorem does not apply to what we call Laplace's theorem; that is to say, it is not true for any law of attraction that the potentials of confocal ellipsoids at the same external point are as their masses.

Let a_1, b_1, and c_1 be the semiaxes of an ellipsoid; let a_2, b_2, and c_2 be the semiaxes of a second ellipsoid confocal with the former. Let P denote a point external to both. Let a third ellipsoid confocal to the former two have the point P on its surface. Let P_1 denote the point on the first ellipsoid which corresponds to P, and let P_2 denote the point on the second ellipsoid which corresponds to P.

Let E_1, E_2, and E denote the ellipsoids.

Then for any law of attraction we have, estimating the attractions parallel to the third axes,

$$\frac{\text{Attraction of } E_1 \text{ at } P}{\text{Attraction of } E \text{ at } P_1}=\frac{a_1b_1}{ab},$$

$$\frac{\text{Attraction of } E_2 \text{ at } P}{\text{Attraction of } E \text{ at } P_2}=\frac{a_2b_2}{ab}.$$

Therefore

$$\frac{\text{Attraction of } E_1 \text{ at } P}{\text{Attraction of } E_2 \text{ at } P} \times \lambda = \frac{a_1 b_1}{a_2 b_2},$$

where λ denotes the ratio of the attraction of E at P_2 to its attraction at P_1. According to the ordinary law of attraction we find that $\lambda = \frac{c_2}{c_1}$; this depends on the fact that the attraction of an ellipsoid on a particle at the surface, estimated perpendicular to a principal plane, varies as the distance from that plane: this fact however does not hold for all laws of attraction. It does hold as we know for the law of the inverse square; and it also holds for the law of the direct distance.

1162. We now arrive at a memoir by Gauss entitled *Theoria attractionis corporum sphaeroidicorum ellipticorum homogeneorum methodo nova tractata.*

This memoir was communicated to the Royal Society of Göttingen on the 18th of March, 1813, and published in the *Comm. Societat. Reg.... Gott.* Vol. II. 1813. The memoir occupies pages 1...22 of Vol. V. of the collected works of Gauss; and I have studied it in this reprint. A notice of the memoir by Gauss himself occupies pages 279...286 of the volume, being reprinted from the *Göttingische gelehrte Anzeigen* of April, 1813.

This notice is also reprinted in De Zach's *Monatliche Correspondenz*, Vol. XXVII.; and there is a German translation of Gauss's memoir in Vol. XXVIII. of the same series.

1163. An account of Gauss's method for the attraction of ellipsoids by Professor Cayley will be found in the *Quarterly Journal of Mathematics*, Vol. I. pages 162...166.

1164. Gauss's writings are distinguished for the combination of mathematical ability with power of expression: in his hands Latin and German rival French itself for clearness and precision.

1165. Gauss gives a short sketch of the history of the problem of the attraction of ellipsoids. He begins with *ipse summus*

Newton, and then passes to *sagax Maclaurin;* he does not make the mistake of D'Alembert and others: see Art. 260. He refers to Lagrange's memoirs of 1773 and 1793, to the first and the third of Legendre's memoirs, to the writings of Laplace, and to the memoirs by Biot and by Plana, which we have already noticed in this Chapter. He does not here refer to Ivory, nor to the memoir by Legendre of 1812: but, as we shall see, he became acquainted with Ivory's memoir after his own was finished.

1166. Laplace's proof of his theorem was in the opinion of Gauss an elegant specimen of analytical skill; but left with geometers a desire for a more simple and direct method. The efforts made by Biot and Plana to simplify the discussion must also be considered very intricate applications of analysis.

1167. Then with respect to his own solution Gauss says:

Gratam itaque analystis atque astronomis fore speramus solutionem novam problematis celebratissimi per viam plane diversam procedentem, et ni fallimur ea simplicitate gaudentem, ut nihil amplius desiderandum linquat.

Certainly he succeeds completely in his design: his solution is both simple and elegant.

1168. Gauss, before he proceeds to the actual problem, gives various theorems which may be useful on other occasions and which he therefore develops more fully than was absolutely necessary for the purposes of immediate application. We will reproduce some of these theorems, without retaining his order.

1169. This is his fifth theorem: let ds be an element of the surface of a body, r the distance of the element from a fixed point, ϕ the angle between r and the normal to the surface measured outwards: then the volume of the body is equal to $-\frac{1}{3}\int r \cos \phi ds$, the integration being extended over the whole surface of the body.

It is obvious that this theorem holds for such a body as the ellipsoid, where there are no singularities, like folds, in the surface; for we can cut the body up into infinitesimal cones having their vertices at the fixed point: and $-\frac{1}{3} r \cos\phi\, ds$ expresses the volume of an element. Gauss however gives a very careful investigation of this and other theorems which he enunciates, so as to shew that they hold even when the surface has folds.

1170. This is Gauss's fourth theorem: with the notation of the preceding Article the integral $\int \frac{\cos\phi}{r^2}\, ds$, extended over the entire surface, is equal to 0, -2π, or -4π, according as the fixed point is outside the body, on the surface, or within.

The demonstration is the same as that which is now well known of a similar proposition in the theory of Potentials: see *Statics*, Arts. 243 and 244. It depends on the fact that if we describe a sphere of radius unity round the fixed point as centre, the element $\frac{\cos\phi}{r^2}\, ds$ is numerically equal to the corresponding element for the surface of this sphere.

1171. Now we will give Gauss's third theorem; this consists of a general expression for the resolved attraction in a given direction of a given body at a given point.

Suppose the law of attraction to be denoted by $f(r)$, where r denotes the distance. Let a, b, c be the coordinates of the given point; then the attraction of the element $dx\, dy\, dz$ at the point (x, y, z) resolved parallel to the axis of x will be $\frac{x-a}{r} f(r)\, dx\, dy\, dz$. Integrate with respect to x; thus we obtain $dy\, dz\, [F(r_2) - F(r_1)]$, where $F(r)$ is the integral of $f(r)$, and r_1 and r_2 are the limiting values of r. Thus we have, in fact, an expression for the resolved attraction parallel to the axis of x, produced by a strip of the body parallel to the same direction. Hence the attraction in this direction of the whole body will be found by integrating this expression with respect to y and z.

Let N denote the angle made with the axis of x by the normal to the surface drawn outwards at the point x, y, z. Then instead of integrating $dy\, dz\, [F(r_2) - F(r_1)]$ we may integrate $ds\, F(r) \cos N$

over the whole surface of the body. In fact if N_1 and N_2 denote the values of N corresponding to the points to which r_1 and r_2 respectively belong, we get

$$ds \cos N_1 = -dx\,dy$$

and
$$ds \cos N_2 = dx\,dy.$$

As we have said in Art. 1169, it is easy to see at once that the theorem is true for such a body as an ellipsoid; but Gauss shows also that it is true for other bodies, for example, for a body in which a straight line parallel to the axis of x meets the surface *four* times instead of twice.

The theorem then is that the resolved attraction parallel to the axis of x is equivalent to $\int ds\,F(r)\cos N$, where the integration is to extend over the whole surface of the body.

Similar expressions may be found for the resolved attractions parallel to the other axes.

1172. We will now give Gauss's sixth theorem, for the particular case in which the law of attraction is that of nature. The object of the theorem is to find a new expression for the resolved attraction.

Let the attracted particle be the vertex of an indefinitely thin cone which cuts the body. Let $d\sigma$ be the element of the surface of a sphere of radius unity, having its centre at the particle, which this cone intercepts. We may take for an element of the body $r^2\,dr\,d\sigma$, so that the resolved attraction of the element parallel to the axis of x will be $dr\,d\sigma\cos\chi$, where χ is the angle between the direction of r and the axis of x. Suppose for facility of conception that the attracted particle is outside the body. Integrate with respect to r; thus we get $(r_2 - r_1)\,d\sigma\cos\chi$, where r_1 and r_2 are the limiting values of r. The resolved attraction of the whole body will be found by integrating this expression with respect to σ over limits corresponding to the part of the surface of the sphere which we have to consider.

This integral we may transform to $\int \frac{r\,ds}{r^2} \cos Q \cos \chi$, that is to $\int \frac{ds}{r} \cos Q \cos \chi$, where Q is the angle between the direction of r and that of the normal to ds measured outwards: the integration is to extend over the whole surface of the body.

If the attracted particle is inside the body we shall arrive at the same result. As before, Gauss shews that the theorem is true for bodies of every form.

Similar expressions may be found for the resolved attractions parallel to the other axes.

1173. We can now apply these general formulæ to the case of the attraction of an ellipsoid.

Let the equation to the ellipsoid be

$$\frac{x^2}{A^2} + \frac{y^2}{B^2} + \frac{z^2}{C^2} = 1.$$

Let the coordinates of the attracted point be a, b, c. Gauss now introduces two new variables, p and q, which are given by the following relations:

$$x = A \cos p, \quad y = B \sin p \cos q, \quad z = C \sin p \sin q.$$

We need not follow Gauss in his transformation of ds into an expression in terms of the new variables, because by a process now familiar to students of the Integral Calculus it is easy to shew that

$$ds = ABC\psi \sin p\,dp\,dq,$$

where ψ stands for $\left(\frac{x^2}{A^4} + \frac{y^2}{B^4} + \frac{z^2}{C^4}\right)^{\frac{1}{2}}$.

Let X denote the attraction parallel to the axis of x towards the origin. Then, by Art. 1171,

$$X = \iint \frac{BCx}{rA} \sin p\,dp\,dq = BC \iint \frac{\sin p \cos p\,dp\,dq}{r}.$$

Put $X = ABC\xi$; thus

$$\xi = \frac{1}{A} \iint \frac{\sin p \cos p\,dp\,dq}{r} \quad\dots\dots\dots\dots(1).$$

Again, by Art. 1172, we have

$$\xi = -\iint \frac{a-x}{r^3}\left\{\frac{(a-x)\,x}{A^2} + \frac{(b-y)\,y}{B^2} + \frac{(c-z)\,z}{C^2}\right\}\sin p\,dp\,dq \ldots (2).$$

And from Art. 1170 we have, supposing the attracted particle external to the body,

$$0 = \iint \frac{1}{r^3}\left\{\frac{(a-x)\,x}{A^2} + \frac{(b-y)\,y}{B^2} + \frac{(c-z)\,z}{C^2}\right\}\sin p\,dp\,dq \ldots (3).$$

Now suppose we pass from the ellipsoid considered to another, having its principal axes coincident in direction with those of the former, but infinitesimally different in magnitude. Let δA, δB, δC denote the changes then made in A, B, C respectively. Moreover let these changes be consistent with the conditions

$$A^2 - B^2 = \text{constant}, \qquad A^2 - C^2 = \text{constant};$$

so that the ellipsoids will have their principal sections *homofocal*. From (1)

$$A\xi = \iint \frac{dp\,dq \sin p \cos p}{r};$$

therefore $\qquad \xi\delta A + A\delta\xi = -\iint \frac{1}{r^2}\,\delta r \sin p \cos p\,dp\,dq.$

Now $\qquad r^2 = (a-x)^2 + (b-y)^2 + (c-z)^2;$

so that $\qquad r\delta r = (x-a)\,\delta x + (y-b)\,\delta y + (z-c)\,\delta z$

$$= (x-a)\cos p\,\delta A + (y-b)\sin p\cos q\,\delta B + (z-c)\sin p\sin q\,\delta C$$

$$= (x-a)\,\frac{x}{A}\,\delta A + (y-b)\,\frac{y}{B}\,\delta B + (z-c)\,\frac{z}{C}\,\delta C$$

$$= A\delta A\left\{\frac{x(x-a)}{A^2} + \frac{y(y-b)}{B^2} + \frac{z(z-c)}{C^2}\right\},$$

since $A\delta A = B\delta B = C\delta C$.

Thus $\xi\delta A + A\delta\xi$

$$= -A\delta A\int\frac{1}{r^3}\left\{\frac{x(x-a)}{A^2} + \frac{y(y-b)}{B^2} + \frac{z(z-c)}{C^2}\right\}\sin p\cos p\,dp\,dq.$$

But from (2)

$$\xi\delta A = \delta A \iint \frac{a-x}{r^3}\left\{\frac{x(x-a)}{A^2} + \frac{y(y-b)}{B^2} + \frac{z(z-c)}{C^2}\right\}\sin p\, dp\, dq.$$

Subtract this equation from that which immediately precedes it; thus

$$A\delta\xi = -\delta A \iint \frac{a}{r^3}\left\{\frac{x(x-a)}{A^2} + \frac{y(y-b)}{B^2} + \frac{z(z-c)}{C^2}\right\}\sin p\, dp\, dq.$$

Hence by (3) we have

$$\delta\xi = 0.$$

Therefore ξ is constant. Thus the attraction of confocal ellipsoids at a given external point parallel to the axis of x varies as the mass. This is Laplace's theorem.

1174. Gauss also uses the result which his fourth theorem gives for the case of an internal particle, and thus obtains formulæ for the calculation of the attraction of an ellipsoid on an internal particle.

1175. Gauss finishes with the following paragraph in which he refers to Ivory's researches:

Additamentum.

Postquam haecce jam perscripta essent, innotuit, indicante ill. Laplace, commentatio egregia cl. Ivory in *Philosophical Transactions* ad A. 1809; ubi idem argumentum per methodum ab iis, quibus usi erant ill. Laplace et Legendre, prorsus diversam tractatur. Summa elegantia ille geometra attractionem puncti externi ad attractionem puncti interni reducere docuit, i.e. problematis partem, quae semper pro difficiliori habita est, ad faciliorem. Methodus autem, per quam hanc alteram partem tractavit, longe magis complicata est, partimque perinde ut methodus, qua ill. Laplace pro punctis externis usus erat, considerationi seriei rum infinitarum non semper convergentium innititur, quam utique evitare licuisset. Ceterum haec solutio clar. Ivory, quae obiter spectata quandam similitudinis speciem cum nostra prae se ferre videri posset, proprius examinata principiis omnino *diversis* inniti invenietur, nec fere quidquam utrique solutioni commune est, nisi usus indeterminatarum a nobis per p, q denotatarum.

It will be seen that the criticism of Gauss on Ivory's memoir resembles that expressed by Legendre: see Art. 1150.

1176. We have next to consider a memoir entitled *Mémoire sur l'attraction des sphéroïdes, par M. Rodrigues, Docteur ès-sciences.*

This memoir is published in the *Correspondance sur l'Ecole Royale Polytechnique,*...Vol. III. 1816. The memoir occupies pages 361...385 of the volume. The memoir is stated to have been the subject of a thesis for the degree of Doctor, which was maintained on the 28th of June, 1815.

1177. The memoir is divided into two parts. The first part which occupies pages 361...374 gives the general formulæ for the attraction of any body, and applies the formulæ to the sphere and the ellipsoid.

There are no new results in this part; but there are two matters which are treated in rather a novel manner. One of these is the partial differential equation for V with respect to an internal particle; and this will be conveniently discussed in the next Chapter. The other matter is a demonstration of Laplace's theorem, and the investigation of the attraction of an ellipsoid on an external particle.

1178. The method of Rodrigues would seem to have been suggested by that of Gauss; but no reference is given to Gauss. An analysis of the method of Rodrigues is given by Professor Cayley in the *Quarterly Journal of Mathematics,* Vol. II. pages 333...337, where it is observed that "the method is very similar to that given two years before by Gauss." So also Poisson in the *Comptes Rendus,*...Vol. VII. page 3, remarks:

Au reste, la démonstration que M. Rodrigues a rapportée dans sa thèse, est celle que M. Gauss a donnée en 1813, et qui est fondée sur la transformation des variables employées par M. Ivory, et sur une propriété générale des surfaces fermées.

1179. The memoir is not difficult when it is carefully studied; but some attention is necessary in order to follow the processes.

Consider the ratio of the potential of a homogeneous body to the mass of the body; this ratio might be called the *relative* potential: we will denote it by W.

Let h, k, l be the coordinates of a point. Let a, b, c be the semiaxes of an ellipsoid. Let the symbol δ be used to denote an infinitesimal variation in a, b, c, or any function of them, such that

$$a\delta a = b\delta b = c\delta c = \tau \text{ say.}$$

By this variation in fact we pass from the ellipsoid whose semiaxes are a, b, c to an adjacent *confocal* ellipsoid.

Let W refer to the first ellipsoid, and $W + \delta W$ to the second; then Rodrigues investigates the value of δW.

He shews that for an external particle

$$\delta W = 0;$$

and that for an internal particle

$$\delta W = \frac{3\tau}{2abc}\left(\frac{h^2}{a^2} + \frac{k^2}{b^2} + \frac{l^2}{c^2} - 1\right).$$

We will now explain how he arrives at these results.

1180. Let the density be denoted by unity; let V denote the potential at (h, k, l); then

$$V = \iiint \frac{dx\,dy\,dz}{R},$$

where $$R = \{(x-h)^2 + (y-k)^2 + (z-l)^2\}^{\frac{1}{2}}.$$

The integration is to extend throughout the ellipsoid.

Assume

$$x = ar\cos\theta, \quad y = br\sin\theta\cos\phi, \quad z = cr\sin\theta\sin\phi;$$

then $$V = \int_0^{2\pi}\int_0^{\pi}\int_0^1 \frac{abc\,r^2\sin\theta\,d\phi\,d\theta\,dr}{R}.$$

And $W = \dfrac{V}{\frac{4}{3}\pi abc}$, so that

$$\frac{4\pi}{3}W = \int_0^{2\pi}\int_0^{\pi}\int_0^1 \frac{r^2\sin\theta\,d\phi\,d\theta\,dr}{R}.$$

Therefore $\frac{4\pi}{3}\delta W = \int_0^{2\pi}\int_0^{\pi}\int_0^1 \delta\left(\frac{1}{R}\right) r^2 \sin\theta \, d\phi \, d\theta \, dr.$

Now $\delta x = r\cos\theta\delta a = \frac{x}{a}\delta a = \frac{xa\delta a}{a^2} = \frac{x\tau}{a^2};$

similarly $\delta y = \frac{y\tau}{b^2}, \quad$ and $\delta z = \frac{z\tau}{c^2}.$

Thus $\delta\frac{1}{R} = -\frac{\tau}{R^3}\left\{(x-h)\frac{x}{a^2} + (y-k)\frac{y}{b^2} + (z-l)\frac{z}{c^2}\right\},$

so that $\frac{4\pi}{3}\delta W =$

$$-\tau\int_0^{2\pi}\int_0^{\pi}\int_0^1 \frac{1}{R^3}\left\{(x-h)\frac{x}{a^2} + (y-k)\frac{y}{b^2} + (z-l)\frac{z}{c^2}\right\} r^2 \sin\theta \, d\phi \, d\theta \, dr.$$

1181. We have thus a certain triple integral, say

$$\int_0^{2\pi}\int_0^{\pi}\int_0^1 N r^2 \sin\theta \, d\phi \, d\theta \, dr;$$

the integral extends throughout the ellipsoid. Rodrigues transforms it into a single integral. Consider the shell which is bounded by the ellipsoidal surface whose semiaxes are ra, rb, rc, and that whose semiaxes are $(r+dr)a$, $(r+dr)b$, $(r+dr)c$. Let dS denote an element of one of the surfaces of the shell, and ϵ the corresponding thickness of the shell; then the element of volume $abc\,r^2 \sin\theta \, d\phi \, d\theta \, dr$ may be replaced by ϵdS.

Let (x, y, z) denote the point on the inner surface of the shell; λ, μ, ν the direction cosines of the normal there. Thus

$$\frac{x^2}{a^2} + \frac{y^2}{b^2} + \frac{z^2}{c^2} = r^2,$$

$$\frac{(x+\lambda\epsilon)^2}{a^2} + \frac{(y+\mu\epsilon)^2}{b^2} + \frac{(z+\nu\epsilon)^2}{c^2} = (r+dr)^2;$$

therefore $\left(\frac{\lambda x}{a^2} + \frac{\mu y}{b^2} + \frac{\nu z}{c^2}\right)\epsilon = r\,dr.$

And putting for λ, μ, ν their values we get

$$\epsilon = \frac{rdr}{\left(\frac{x^2}{a^4}+\frac{y^2}{b^4}+\frac{z^2}{c^4}\right)^{\frac{1}{2}}} = pr\,dr \text{ say.}$$

Thus we have

$$\frac{1}{R^3}\left\{(x-h)\frac{x}{a^2}+(y-k)\frac{y}{b^2}+(z-l)\frac{z}{c^2}\right\} r^2 \sin\theta\, d\phi\, d\theta\, dr$$

$$=\frac{1}{R^3}\left\{(x-h)\frac{\lambda}{p}+(y-k)\frac{\mu}{p}+(z-l)\frac{\nu}{p}\right\}\frac{\epsilon dS}{abc}$$

$$=\frac{rdr}{R^2}\left\{\frac{x-h}{R}\lambda+\frac{y-k}{R}\mu+\frac{z-l}{R}\nu\right\}\frac{dS}{abc}$$

$$=-\frac{rdr}{R^2}\cos\psi\,\frac{dS}{abc},$$

where ψ is the angle between the straight line which joins (h, k, l) to (x, y, z) and the normal to the ellipsoid at (x, y, z) drawn outwards.

Hence $$\frac{4\pi}{3}\,\delta W = \frac{\tau}{abc}\int\left(\int\frac{\cos\psi\, dS}{R^2}\right) rdr.$$

The integral relative to S extends over the whole surface of the ellipsoid with semiaxes ra, rb, and rc.

Now as we have seen in Art. 1170

$$\int\frac{\cos\psi}{R^2}\,dS = 0 \quad \text{or} \quad -4\pi,$$

according as the fixed point (h, k, l) is outside or inside the surface.

Hence if (h, k, l) be outside the surface

$$\delta W = 0.$$

If (h, k, l) be inside the surface

$$\delta W = -\frac{3\tau}{abc}\int rdr.$$

1182. In the preceding Article we transformed the triple integral by cutting up the ellipsoid into shells; these shells are bounded by *homothetical* ellipsoids: this is in fact the mode of decomposition adopted by Rodrigues himself.

The investigation is given in another form in the memoir by Professor Cayley which I have cited in Art. 1178.

But Rodrigues does not determine any thing respecting the *attraction* of one of these shells just spoken of. This was first considered by Poisson in his memoir of 1835, which is cited in Art. 887.

1183. Let us return to the results obtained in Art. 1181.

For an external point we have $\delta W = 0$; so that $\frac{V}{M}$ does not depend on the absolute lengths of the semiaxes, but on the excentricities of the ellipsoid. This is in fact Laplace's theorem.

For an internal particle

$$\delta W = -\frac{3\tau}{abc}\int r\,dr;$$

the integration is to be taken so as to correspond to all the shells *outside* the particle, say from $r = r'$ to $r = 1$; and r' is determined by the equation

$$\frac{h^2}{a^2} + \frac{k^2}{b^2} + \frac{l^2}{c^2} = r'^2;$$

so that

$$\delta W = \frac{3\tau}{2abc}\left(\frac{h^2}{a^2} + \frac{k^2}{b^2} + \frac{l^2}{c^2} - 1\right).$$

1184. The remainder of the process given by Rodrigues consists in obtaining expressions for the attraction from the above formula for δW. We shall not reproduce it, but briefly deduce a symmetrical expression for V.

Let $\qquad a^2 = \alpha^2 + t, \quad b^2 = \beta^2 + t, \quad c^2 = \gamma^2 + t;$

then τ becomes equivalent to $\frac{1}{2}dt$; thus in the ordinary language of the Differential Calculus,

$$\frac{dW}{dt} = \frac{3}{4\sqrt{\{(\alpha^2+t)(\beta^2+t)(\gamma^2+t)\}}}\left(\frac{h^2}{\alpha^2+t} + \frac{k^2}{\beta^2+t} + \frac{l^2}{\gamma^2+t} - 1\right).$$

where $$W = \frac{V}{M} = \frac{3V}{4\pi\sqrt{[(\alpha^2+t)(\beta^2+t)(\gamma^2+t)]}}.$$

Integrate from $t=0$ to $t=\infty$, observing that W vanishes when t is infinite; thus

$$-\frac{3V}{4\pi\alpha\beta\gamma} = \frac{3}{4}\int_0^\infty \frac{dt}{\sqrt{[(\alpha^2+t)(\beta^2+t)(\gamma^2+t)]}}\left(\frac{h^2}{\alpha^2+t}+\frac{k^2}{\beta^2+t}+\frac{l^2}{\gamma^2+t}-1\right).$$

Here V is the potential of the ellipsoid whose semiaxes are α, β, γ at the internal point (h, k, l); and we have

$$V = -\pi\alpha\beta\gamma\int_0^\infty \frac{dt}{\sqrt{[(\alpha^2+t)(\beta^2+t)(\gamma^2+t)]}}\left(\frac{h^2}{\alpha^2+t}+\frac{k^2}{\beta^2+t}+\frac{l^2}{\gamma^2+t}-1\right).$$

Then for an external point, by Laplace's theorem, the value of W is the same as it would be for an ellipsoid having the semi-axes $\alpha_1, \beta_1, \gamma_1$, where

$$\alpha_1^2 - \alpha^2 = \beta_1^2 - \beta^2 = \gamma_1^2 - \gamma^2 = t_1 \text{ say,}$$

and $$\frac{h^2}{\alpha^2+t_1}+\frac{k^2}{\beta^2+t_1}+\frac{l^2}{\gamma^2+t_1} = 1.$$

Thus for the external point

$$\frac{V}{\alpha\beta\gamma} = -\pi\int_0^\infty \frac{L\,dt}{\sqrt{[(\alpha^2+t_1+t)(\beta^2+t_1+t)(\gamma^2+t_1+t)]}},$$

where $$L = \frac{h^2}{\alpha^2+t_1+t}+\frac{k^2}{\beta^2+t_1+t}+\frac{l^2}{\gamma^2+t_1+t}-1;$$

that is, putting $t_1+t=t'$, $V=$

$$-\pi\alpha\beta\gamma\int_{t_1}^\infty \frac{dt'}{\sqrt{[(\alpha^2+t')(\beta^2+t')(\gamma^2+t')]}}\left(\frac{h^2}{\alpha^2+t'}+\frac{k^2}{\beta^2+t'}+\frac{l^2}{\gamma^2+t'}-1\right).$$

1185. We pass to the second part of the memoir, which occupies pages 374...385. This treats on the attraction of spheroids, which differ very slightly from a sphere, and on the general development of the potential function.

1186. The most remarkable matter in this part consists in

the treatment of Laplace's coefficients. Rodrigues says that his mode of analysis had been employed to a great extent by Ivory in the *Philosophical Transactions* for 1812, and by Legendre in his *Exercices de Calcul Intégral;* but he had not been acquainted with these works when he composed his thesis.

Nevertheless Rodrigues went in some respects beyond Ivory and Legendre; Heine does not seem to have been acquainted with the memoir which we are now examining: see Art. 784.

1167. Laplace's m^{th} coefficient may be expressed in the form

$$M + \frac{2 \sin\theta \sin\theta' \cos(\varpi - \varpi')}{m(m+1)} \frac{d^2 M}{d\mu\, d\mu'}$$

$$+ \frac{2 \sin^2\theta \sin^2\theta' \cos 2(\varpi - \varpi')}{(m-1)\, m\, (m+1)(m+2)} \frac{d^4 M}{d\mu^2\, d\mu'^2}$$

$$+ \ldots$$

$$+ \frac{2 \sin^m\theta \sin^m\theta' \cos m(\varpi - \varpi')}{|\underline{2m}} \frac{d^{2m} M}{d\mu^m\, d\mu'^m}.$$

Legendre had obtained this result: see Art. 950. Afterwards Ivory gave it, see page 60 of his memoir of 1812. Where Rodrigues has the advantage is that he finds for M a certain compact form which had not been previously obtained, namely

$$M = \frac{1}{2^{2m} |\underline{m}\, |\underline{m}} \frac{d^{2m}}{d\mu^m\, d\mu'^m} (1-\mu^2)^m (1-\mu'^2)^m.$$

This includes a very simple formula for Legendre's m^{th} coefficient, namely

$$P_m = \frac{1}{2^m |\underline{m}} \frac{d^m (\mu^2 - 1)^m}{d\mu^m}.$$

This is given by Rodrigues. It presents itself as $J \dfrac{d^m (1-\mu^2)^m}{d\mu^m}$, where J is a constant which he does not explicitly determine; but this constant comes at once from his value given above for M.

The result was given by Ivory in the *Philosophical Transactions* for 1822; and so is ascribed to him by Heine, on his page 9.

1188. A result, which is ascribed by Heine to himself and Bertram, on his page 89, seems to me to have been also anticipated by Rodrigues: see pages 370...378 of the memoir.

1189. Another result is given by Rodrigues which has been claimed for a later writer. This result expressed in the most symmetrical form is

$$\frac{(x^2-1)^{\frac{n}{2}}}{\underline{|m+n}}\,\frac{d^{m+n}(x^2-1)^m}{dx^{m+n}}=\frac{(x^2-1)^{-\frac{n}{2}}}{\underline{|m-n}}\,\frac{d^{m-n}(x^2-1)^m}{dx^{m-n}},$$

m and n being positive integers, and n not greater than m.

Heine on his page 117 says: "Diese schöne, von Jacobi zuerst gegebene Formel..."; and in a note he refers to Crelle's *Journal für Mathematik*, Vol. II. page 225: the date of this volume of Crelle's Journal is 1827.

The mode in which the result presents itself to Rodrigues may be noticed.

Suppose T_m to denote Laplace's mth coefficient, then we know that

$$\frac{d}{d\mu}\left\{(1-\mu^2)\frac{dT_m}{d\mu}\right\}+\frac{1}{1-\mu^2}\frac{d^2T_m}{d\varpi^2}+m(m+1)\,T_m=0.$$

Suppose T_m expressed in the form

$$y_0+y_1(A_1\sin\varpi+B_1\cos\varpi)+y_2(A_2\sin 2\varpi+B_2\cos 2\varpi)+\dots;$$

then the general coefficient y_n is determined by

$$\frac{d}{d\mu}\left\{(1-\mu^2)\frac{dy_n}{d\mu}\right\}+\left\{m(m+1)-\frac{n^2}{1-\mu^2}\right\}y_n=0.$$

If we assume $y_n=(1-\mu^2)^{\frac{n}{2}}x_n$ we obtain

$$(m-n)(m+n+1)\,x_n-2(n+1)\,\mu\frac{dx_n}{d\mu}+(1-\mu^2)\frac{d^2x_n}{d\mu^2}=0.$$

If we assume $y_n=(1-\mu^2)^{-\frac{n}{2}}x_{-n}$ we obtain

$$(m+n)(m-n+1)\,x_{-n}+2(n-1)\,\mu\frac{dx_{-n}}{d\mu}+(1-\mu^2)\frac{d^2x_{-n}}{d\mu^2}=0.$$

Rodrigues integrates these two equations; and thus obtains

$$x_n = \frac{1}{(1-\mu^2)^n}\frac{d^{m-n}}{d\mu^{m-n}}\left\{C + D\int\frac{d\mu}{(1-\mu^2)^{m+1}}\right\}(1-\mu^2)^m.$$

$$x_{-n} = (1-\mu^2)^n\frac{d^{m+n}}{d\mu^{m+n}}\left\{F + K\int\frac{d\mu}{(1-\mu^2)^{m+1}}\right\}(1-\mu^2)^m.$$

Then as we have $x_{-n} = (1-\mu^2)^n x_n$, by virtue of our original assumptions, we obtain

$$x_n = \frac{d^{m+n}}{d\mu^{m+n}}\left\{F + K\int\frac{d\mu}{(1-\mu^2)^{m+1}}\right\}(1-\mu^2)^m.$$

Here C, D, F and K are arbitrary constants.

Thus we have two different general forms of x_n; and we have also the two following particular forms

$$\frac{C}{(1-\mu^2)^n}\frac{d^{m-n}}{d\mu^{m-n}}(1-\mu^2)^m, \text{ and } F\frac{d^{m+n}(1-\mu^2)^m}{d\mu^{m+n}}.$$

These are both *rational* and *integral* functions of μ; hence as the equation which x_n satisfies is a *linear* equation, we are certain that by properly determining the ratio of the constant C to the constant F these particular expressions will be identical. By determining the ratio, Rodrigues arrives at the result stated above.

1190. Let $R_s = \dfrac{d^s(1-\mu^2)^s}{d\mu^s}$; and suppose that

$$Z_m = B_0 R_m + (1-\mu^2)^{\frac{1}{2}}\frac{dR_m}{d\mu}(A_1 \sin \varpi + B_1 \cos \varpi)$$
$$+ \ldots (1-\mu^2)^{\frac{r}{2}}\frac{d^r R_m}{d\mu^r}(A_r \sin r\varpi + B_r \cos r\varpi) + \ldots$$

$$W_n = b_0 R_n + (1-\mu^2)^{\frac{1}{2}}\frac{dR_n}{d\mu}(a_1 \sin \varpi + b_1 \cos \varpi)$$
$$+ \ldots + (1-\mu^2)^{\frac{r}{2}}\frac{d^r R_n}{d\mu^r}(a_r \sin r\varpi + b_r \cos r\varpi) + \ldots$$

so that Z_m and W_n express Laplace's functions of the order m and n

respectively. Then Rodrigues shews that $\int_{-1}^{1}\int_{0}^{2\pi} Z_m W_n \, d\mu d\varpi = 0$, when m and n are unequal; and he gives the form of the result when m and n are equal. Legendre had already done this; see Art. 951; but Rodrigues expresses the result when m and n are equal more compactly as he was in possession of the formula of Art. 1187, namely

$$\int_{-1}^{1}\int_{0}^{2\pi} Z_n W_n \, d\mu d\varpi = \frac{2^{2n+1}\,\underline{|n}\,\underline{|n}\,\pi}{2n+1} \Sigma \left(A_r a_r + B_r b_r\right) \frac{\underline{|n+r}}{\underline{|n-r}},$$

where Σ relates to r, and applies a summation from $r=0$ to $r=n$ inclusive; observing that $A_0 a_0 + B_0 b_0$ must be replaced by $2B_0 b_0$. The method of Rodrigues is good.

1191. Rodrigues gives the formula for the attraction of spheroids which differ very little from spheres; there is nothing important in this part of his memoir: he briefly investigates Laplace's equation which we shall discuss in the next Chapter, and expresses no doubts respecting it.

1192. Rodrigues insists on the necessity of having the series convergent; see his pages 375 and 385. Nevertheless he seems unaware of the difficulty which Poisson subsequently discussed, and to which I allude in Art. 843.

1193. From what we have said on this memoir by Rodrigues, it is obvious that the following general remarks may be made. So far as relates to the attraction of an ellipsoid the memoir contains a simple solution of the problem, but adds nothing to what had been previously established. But so far as relates to Laplace's functions the memoir is very important, and deserves a prominent place in the history of this branch of analysis.

1194. Thus in the present Chapter we have noticed four complete discussions of the problem of the attraction of an ellipsoid; namely those by Ivory, Legendre, Gauss and Rodrigues: omitting that part of Ivory's which relates to the internal particle, all are eminent for simplicity, rigour, and completeness. We have also noticed two other memoirs; that by Biot, which may be regarded as a commentary on the writings of Legendre and Laplace; and that by Plana, which is a commentary on the investigation in Laplace's earlier treatise. A memoir by Poisson on this subject will come before us in a subsequent Chapter. Practically speaking, the method of Ivory has superseded all the others; although those of Gauss and Rodrigues are very striking. It is remarked by Chasles, *Mémoires...par divers Savants...*, Vol. IX. page 636:

Mais l'élégant théorème de M. Ivory, qui, joint à l'analyse de Lagrange pour le cas des points intérieurs, complétait une solution *facile* et briève de la question, fixa tellement l'attention des géomètres, que le beau mémoire de M. Gauss, et la solution remarquable aussi de M. Rodrigues, où se trouvait, implicitement, la considération d'une couche infiniment mince comprise entre deux ellipsoïdes semblables, restèrent, pour ainsi dire, inaperçus.

CHAPTER XXX.

LAPLACE'S EQUATION.

1195. It has been shewn that Laplace gave repeatedly a certain equation relative to the potential of a nearly spherical homogeneous body at a point on its surface: see Art. 1067.

The equation was finally introduced in the *Mécanique Céleste*, Livre III. § 10; and in two forms. There is the general form in which the attraction is supposed to vary as the nth power of the distance, and the particular form in which n is taken to be -2.

1196. The particular form of the equation is

$$\frac{1}{2} V + a \frac{dV}{dr} + \frac{2\pi a^2}{3} = 0 \text{ ....} (1).$$

Here V is the potential at any point of the surface, r is the distance of that point from a fixed origin which is very near the centre of gravity of the spheroid, a is the radius of a sphere which differs very little from the spheroid in volume, and $\frac{dV}{dr}$ is the differential coefficient of V with respect to r, supposing the direction of r unchanged; so that $V + \frac{dV}{dr} dr$ would ultimately be the value of the potential at a second point, normally over the point to which V refers, and distant dr from it.

1197. A memoir by Lagrange on this subject is contained in the *Journal de l'Ecole Polytechnique*, Cahier XV. Volume VIII. The memoir is entitled *Eclaircissement d'une difficulté singulière qui se rencontre dans le Calcul de l'Attraction des Sphéroïdes très-peu différens de la Sphère.* The memoir occupies pages 57...67 of the volume, which was published in December, 1809.

1198. Lagrange remarks that D'Alembert was the first who calculated the attraction of spheroids which differ but little from spheres; and that Laplace treated the matter in a new and more general manner. Speaking of Laplace's theory Lagrange says:

Sa théorie est fondée sur un beau théorème très-remarquable par sa simplicité autant que par sa généralité; mais ce théorème donne lieu à une difficulté singulière, qui paraît n'avoir encore été remarquée par personne, et qui mérite d'être examinée.

Lagrange then investigates equation (1), and establishes its truth; but shews that by an error, which might naturally occur, a different result would present itself.

We will give briefly the substance of Lagrange's process.

1199. Resolve the spheroid into two parts, a sphere of radius a nearly coinciding with the spheroid, and an additional part contained between the surface of the sphere and the surface of the spheroid. Let the radius-vector of the spheroid be $a(1+y')$, where y' is so small that its square may be neglected; let y be the value of y' at the point considered. Let V_1 denote the part of V which arises from the sphere, and v the part which arises from the additional matter; so that $V = V_1 + v$.

Take the centre of the sphere for the origin of r. Then the value of V_1 is $\frac{4\pi a^3}{3r}$; and that of $\frac{dV_1}{dr}$ is $-\frac{4\pi a^3}{3r^2}$. Hence

$$\frac{1}{2}V_1 + a\frac{dV_1}{dr} = \frac{2\pi a^3}{3r} - \frac{4\pi a^4}{3r^2}.$$

This is exact. Put for r on the right-hand side its value $a(1+y)$, and neglect y^2; thus we get

$$\frac{1}{2}V_1 + a\frac{dV_1}{dr} = -\frac{2\pi a^2}{3} + 2\pi a^2 y.$$

Hence to establish (1) it will be necessary to shew that

$$\frac{1}{2}v + a\frac{dv}{dr} = -2\pi a^2 y \ldots\ldots\ldots\ldots\ldots\ldots\ldots\ldots (2).$$

1200. The value of v is obtained by approximation. It is assumed that the additional matter may be supposed condensed on the surface of the sphere of radius a; this assumption we have had to make on former occasions: see Arts. 424 and 852. Thus

$$v = \int \frac{ay'd\sigma}{\sqrt{(r^2 - 2ar\mu + a^2)}},$$

where $d\sigma$ represents an element of the spherical surface, and μ is the cosine of the angle between the direction of r and the radius drawn to this element.

As v is obviously of the order of y, it is assumed that ultimately in $\frac{1}{2}v + a\frac{dv}{dr}$ we may put $r = a$, and still have our result true to the order of y.

Now we easily see that

$$\frac{1}{2}v + a\frac{dv}{dr} = -\frac{r^2 - a^2}{2}\int \frac{ay'd\sigma}{(r^2 - 2ar\mu + a^2)^{\frac{3}{2}}} \quad \ldots\ldots\ldots (3);$$

then if we put $r = a$ we might at first suppose that the right-hand member would be zero; and thus we should have a result different from (2). This constitutes substantially the difficulty which Lagrange undertakes to explain.

The fact is that if we make $r = a$ the expression under the integral sign becomes infinite in the course of integration, namely when $\mu = 1$. Thus although the first factor on the right-hand side vanishes when $r = a$, the second factor may become infinite; and we must determine the exact value of the expression.

Refer the surface of the sphere to the usual polar coordinates, taking the radius through the point under consideration as the straight line from which θ is measured. Then we may put $a^2 \sin\theta\, d\theta\, d\phi$ for $d\sigma$, and $\cos\theta$ for μ; therefore the integral becomes

$$-\frac{r^2 - a^2}{2}a^3 \iint \frac{y' \sin\theta\, d\theta\, d\phi}{(r^2 - 2ar\cos\theta + a^2)^{\frac{3}{2}}}.$$

Put $y' - y + y$ for y'; thus we obtain

$$-\frac{y(r^2 - a^2)a^3}{2}\iint \frac{\sin\theta\, d\theta\, d\phi}{(r^2 - 2ar\cos\theta + a^2)^{\frac{3}{2}}}$$

$$-\frac{(r^2 - a^2)a^3}{2}\iint \frac{(y' - y)\sin\theta\, d\theta\, d\phi}{(r^2 - 2ar\cos\theta + a^2)^{\frac{3}{2}}}.$$

Assuming then for the present that the second integral does not become infinite when $r=a$, the second term will vanish when $r=a$ in consequence of the factor r^2-a^2.

The integral

$$\iint \frac{\sin\theta\, d\theta\, d\phi}{(r^2-2ar\cos\theta+a^2)^{\frac{3}{2}}} = 2\pi \int \frac{\sin\theta\, d\theta}{(r^2-2ar\cos\theta+a^2)^{\frac{3}{2}}}.$$

This single integral can be immediately found; the limits for θ are 0 and π, and thus we shall obtain finally

$$\frac{2\pi}{ar}\left(\frac{1}{r-a}-\frac{1}{r+a}\right), \text{ that is } \frac{4\pi}{r(r^2-a^2)}.$$

Hence the right-hand member of (3) becomes

$$-\frac{y(r^2-a^2)a^2}{2} \times \frac{4\pi}{r(r^2-a^2)}, \text{ that is } -2\pi a^2 y \text{ ultimately.}$$

Thus (2) is demonstrated.

1201. We must now examine if the assumption we have made is satisfactory. We have assumed that

$$\iint \frac{(y'-y)\sin\theta\, d\theta\, d\phi}{(r^2-2ar\cos\theta+a^2)^{\frac{3}{2}}}$$

remains finite when $r=a$.

Lagrange's own treatment consists in integration by parts. Consider the integration with respect to θ, and put μ for $\cos\theta$. Then

$$-\int \frac{(y'-y)\,d\mu}{(r^2-2ar\mu+a^2)^{\frac{3}{2}}} = -\frac{y'-y}{ar(r^2-2ar\mu+a^2)^{\frac{1}{2}}}$$
$$+\frac{1}{ar}\int \frac{1}{(r^2-2ar\mu+a^2)^{\frac{1}{2}}}\frac{dy'}{d\mu}\,d\mu.$$

The limits for μ are 1 and -1. When $\mu=1$ we have $y'-y=0$. Suppose that Y is the value of y' when $\mu=-1$. Thus we obtain

$$-\frac{Y-y}{r+a}+\frac{1}{ar}\int_1^{-1} \frac{1}{(r^2-2ar\mu+a^2)^{\frac{1}{2}}}\frac{dy'}{d\mu}\,d\mu.$$

Therefore if the latter integral is finite when $r=a$, so also is the original integral. Lagrange in fact assumes that the latter integral is finite. This cannot be safely admitted however if $\frac{dy'}{d\mu}$ should become infinite within the range of the integration.

But in the original integral it is only when μ is very nearly unity that the function to be integrated can possibly become very large. So the process of integration by parts need only be employed with respect to that part of the integral for which μ is nearly equal to unity. Hence practically the limitation to which Lagrange's process must be subjected is this: $\frac{dy'}{d\mu}$ must be zero or finite when $\mu=1$.

1202. Another method might be used. The expression

$$\iint \frac{(r^2-a^2)(y'-y)\,d\mu\,d\phi}{(r^2-2ar\mu+a^2)^{\frac{3}{2}}}$$

vanishes when $r=a$, provided that the quantity under the integral sign is zero or finite when $\mu=1$. When $\mu=1$ this quantity takes the indeterminate form $\frac{0}{0}$; by evaluating it in the usual way we obtain

$$-\frac{(r^2-a^2)\frac{dy'}{d\mu}}{3ar(r^2-2ar\mu+a^2)^{\frac{1}{2}}}.$$

This is finite when $\mu=1$ and $r=a$, provided $\frac{dy'}{d\mu}$ is not infinite.

Thus on the whole we may admit the truth of (1) provided $\frac{dy'}{d\mu}$ is not infinite when $\mu=1$.

1203. We have next to notice a memoir by Ivory entitled *On the Grounds of the Method which Laplace has given in the second Chapter of the third Book of his Mécanique Céleste for computing the Attractions of Spheroids of every Description.* This memoir is published in the *Philosophical Transactions* for 1812;

it occupies pages 1...45 of the volume. The memoir is composed of two parts. The first part consists of pages 1...33; this was written before Ivory had seen Lagrange's memoir in the *Journal de l'Ecole Polytechnique*, and was read to the Royal Society on July 4th, 1811. The second part consists of pages 34...45; this was written after Ivory had seen Lagrange's memoir, and was read to the Royal Society on November 7th, 1811.

1204. Let us consider the first part of Ivory's memoir. This is intended to shew that Laplace's demonstration of equation (1) is defective and erroneous. With respect to this equation Ivory says on his pages 7 and 8:

...The theorem, it may be remarked, is merely laid down by the author, and the truth of it confirmed by a demonstration; it does not naturally arise in the course of the analysis; and the reader of the *Mécanique Céleste* is at a loss to conjecture by what train of thought it may have been originally suggested. It may be doubted whether the theorem was introduced for the sake of demonstrating a method of investigation previously known to be just from other principles; or whether it preceded in the order of invention, and led to the method of investigation.

The history of Laplace's writings, which we have traced, settles the point thus raised by Ivory; the theorem was given by Laplace at a very early period, and did precede the application which he afterwards made of it to the expansion of a function in what we call Laplace's functions.

1205. Ivory remarks on his page 9:

It is also to be observed that the *Mécanique Céleste* has now been many years before the public: and although the problem of attractions is the foundation of many important researches, and is more particularly recommended to the notice of mathematicians by the novelty and uncommon turn of the analysis; on which account it may be supposed to have been scrutinised with more than an ordinary degree of curiosity; yet nobody has hitherto called in question the accuracy of the investigation.

Ivory might have increased the force of his remark by adverting to the long time which elapsed between the first publication of the theorem and the reproduction of it in the *Mécanique Céleste*.

Ivory concludes the paragraph which contains the preceding sentence by saying that

The writings of no author on any subject deserve to have more respect and deference paid to them, than the writings of Laplace on the subject of physical astronomy; with this no one can be more deeply impressed than the author of this discourse; and it was not till after much meditation that, yielding to the force of the proofs which are now to be detailed, he has ventured to advance anything in opposition to the highest authority, in regard to mathematical and physical subjects, that is to be found in the present times.

1206. I shall not reproduce Ivory's process. It seems to me longer and more elaborate than was really necessary. I think his objections to Laplace's equation may be fairly epitomized by saying, as in Art. 1202, that we have no ground for asserting the truth of the equation at any point unless we assume that $\frac{dy'}{d\mu}$ is finite for that point; and he would allow the truth of the equation if y' is such that $\frac{dy'}{d\mu}$ is always finite.

1207. Let us consider where the difficulty really lies in Laplace's process. We take the suppositions of Art. 1199, except that we now let the sphere *touch* the spheroid at the point considered; so that the y of that Article in fact is zero.

Let there be a particle of mass λ situated on the surface of the sphere, and at an angular distance γ from the direction to which r belongs. Let

$$f = \sqrt{\{2a^2(1 - \cos\gamma)\}},$$

and

$$f' = \sqrt{\{(a + dr)^2 - 2a(a + dr)\cos\gamma + a^2\}}.$$

Then $\frac{\lambda}{f}$ represents the part of V which arises from this particle, and $\frac{\lambda}{f'}$ represents the part of $V + \frac{dV}{dr}dr + \ldots$ which so arises.

If f be finite we have when dr is small enough

$$f' = f\left(1 + \frac{dr}{2a}\right) \dots\dots\dots\dots\dots\dots (4),$$

so that
$$\frac{\lambda}{f'} = \frac{\lambda}{f} - \frac{\lambda}{f}\frac{dr}{2a}.$$

Hence *so far as this particle is concerned* we get $\dfrac{dV}{dr} = -\dfrac{\lambda}{f}\cdot\dfrac{1}{2a}$; so that $2a\dfrac{dV}{dr} = -V$, that is $2a\dfrac{dV}{dr} + V = 0$.

Therefore we may say that considering only those particles which are at a *finite* distance from the point considered, Laplace's equation holds. But at the same time when f itself is infinitesimal, the assertion in (4) cannot be accepted as satisfactory.

1208. Ivory proceeds thus. We have

$$f'^2 - (dr)^2 = \left(1 + \frac{dr}{a}\right) f^2;$$

therefore
$$\frac{1}{f} = \frac{1}{f'}\left(1 + \frac{dr}{a}\right)^{\frac{1}{2}} \left\{1 - \frac{(dr)^2}{f'^2}\right\}^{-\frac{1}{2}};$$

and to obtain $\dfrac{1}{f'} - \dfrac{1}{f}$ Ivory expands the factor $\left\{1 - \dfrac{(dr)^2}{f'^2}\right\}^{-\frac{1}{2}}$ in ascending powers of $\dfrac{dr}{f'}$. But this expansion is very unsatisfactory when f' becomes infinitesimal as it does ultimately.

1209. Ivory himself has no doubt of the soundness of his investigation. He says that "It completely overturns the demonstration of Laplace...".

1210. Laplace, as we have seen in Art. 1067, employed the equation we are considering to shew that any function of the usual angular polar coordinates could be expressed in a series of Laplace's functions. Of course any limitation which may be found to apply to the theorem we are considering, applies also to the deduction from it. Accordingly, Ivory does not allow that *any* function can be expanded in a series of Laplace's functions; but allows that any rational integral function of μ, $\sqrt{(1-\mu^2)}\cos\phi$, and $\sqrt{(1-\mu^2)}\sin\phi$ can be so expanded, where μ stands for $\cos\theta$.

1211. With respect to Laplace's demonstration of his theorem it must be observed that he expressly supposes the sphere to *touch* the spheroid. Ivory does not advert to this supposition. It is obvious that if this supposition be made the condition required in Lagrange's investigation will be satisfied; and so that investigation will furnish an adequate proof of Laplace's theorem: see Art. 1200. And by proper precautions we may also fortify the weak part of Laplace's own investigation.

Laplace's supposition amounts to the condition, in modern language, that there are to be no *singular points* on the surface of the spheroid: so that applied to the Earth it would exclude such irregularities as chasms or craters, and ridges or peaks, and mountains or valleys with vertical faces.

No doubt this condition limits to a corresponding degree the range of the theorem about the expansibility of a function in a series of Laplace's functions; or rather limits Laplace's own demonstration of it.

1212. We have shewn in Arts. 1200 and 1201 that if $\frac{dy'}{d\mu}$ is finite when $\mu=1$ the limit when $r=a$ of

$$(r-a)\int_0^{\pi}\int_0^{2\pi}\frac{a^2y'\sin\theta\,d\theta\,d\phi}{(r^2-2ar\cos\theta+a^2)^{\frac{3}{2}}}\text{ is }2\pi y.$$

Under the same condition it may be shewn that the limit when $r=a$ of

$$(r-a)^{i-1}\int_0^{\pi}\int_0^{2\pi}\frac{a^2y'\sin\theta\,d\theta\,d\phi}{(r^2-2ar\cos\theta+a^2)^{\frac{i+1}{2}}}\text{ is }\frac{2\pi y}{i-1}.$$

Ivory establishes this result: see his page 22.

1213. We may notice the way in which Ivory shews that a rational integral function of the usual variables can be expanded in a series of Laplace's functions.

Let $d\sigma$ denote an element of the surface of a sphere of radius unity; then with the notation which has been explained in Art. 785 we have

$$\int \frac{y' d\sigma}{\sqrt{(r^2 - 2ar\mu + a^2)}} = \int \frac{y'}{r} \left\{1 + \frac{a}{r} P_1 + \frac{a^2}{r^2} P_2 + \frac{a^3}{r^3} P_3 + \ldots\right\} d\sigma.$$

Differentiate with respect to r; thus

$$\int \frac{(a\mu - r) y' d\sigma}{(r^2 - 2ar\mu + a^2)^{\frac{3}{2}}} = -\int \frac{y'}{r^2} \left\{1 + \frac{2a}{r} P_1 + \frac{3a^2}{r^2} P_2 + \ldots\right\} d\sigma.$$

Multiply the second result by $2r$ and add to the former; thus

$$\frac{r+a}{a} \int \frac{(r-a)\, y' a^2 d\sigma}{(r^2 - 2ar\mu + a^2)^{\frac{3}{2}}} = \int \frac{ay'}{r} \left\{1 + \frac{3a}{r} P_1 + \frac{5a^2}{r^2} P_2 + \ldots\right\} d\sigma.$$

Now suppose $r = a$; then the limit of the left-hand is $4\pi y$ by Art. 1212. Therefore

$$4\pi y = \int y' \{1 + 3P_1 + 5P_2 + 7P_3 + \ldots\}\, d\sigma.$$

This is equivalent to Laplace's result: see Art. 1069.

The method here given is substantially that which Poisson adopted in various places of his writings for establishing the possibility of the expansion.

1214. Ivory makes some remarks on what we have called in Art. 1195 the general form in which Laplace's equation presents itself, when the attraction is supposed to vary as the nth power of the distance. Then

$$\frac{dV}{dr} - \frac{n+1}{2a} V = \frac{dV_1}{dr} - \frac{n+1}{2a} V_1,$$

where V belongs to the whole spheroid, and V_1 to the sphere of radius a: see Art. 814. It is easy to anticipate what will be the nature of Ivory's conclusion. He admits that if n be positive the equation is true, and also if n be negative and numerically less than 2. In fact if we proceed after Lagrange's method, as given in this Chapter, we shall see that our integrals remain finite as long as n is algebraically greater than -2.

It may be taken I think as universally admitted that the equation cannot be considered established if n is negative and numerically greater than 2.

1215. Some incidental matters which occur in the memoir may be noticed.

If θ and ϕ are the usual polar coordinates of a point on the surface of a sphere of radius unity, we know that the expression for the element of surface will be $\sin\theta\, d\theta\, d\phi$. If we take another pole and transform our expression, and so introduce corresponding angles θ_1 and ϕ_1 the expression for the element of surface will be $\sin\theta_1\, d\theta_1\, d\phi_1$. This is a simple but important transformation: Ivory uses it on his page 18: by the aid of it in Art. 424 some simplification was effected of an investigation by D'Alembert.

On his page 32 Ivory says, with reference of course to Laplace:

... On this account the analysis in No. 25, Liv. 3e, cannot be admitted as satisfactory: and indeed from the words at the beginning of No. 26, we may infer that the author himself was not perfectly satisfied with the strictness and universality of his investigation.

Laplace's words however at the beginning of his § 26 seem to me to shew that the investigation which he was about to bring forward was intended to remove the doubts of other people rather than his own, as to the soundness of his antecedent method: the words will be quoted in Art. 1256.

On his page 33 Ivory says:

... Although the analysis which Laplace has traced out for the attractions of spheroids must be allowed to be very ingenious and masterly, yet still there are some considerations which cannot but lead us to think, that it falls short of that degree of perfection which it is laudable to aim at. And in particular the coefficients of the several terms of the expansion are, in his procedure, formed one after another, beginning with the last term: so that the first terms of the series cannot be found without previously computing all the rest. This is no doubt an imperfection of some moment:...

Ivory I presume refers to the process which occurs towards the end of Laplace's Livre III. § 16. The objection does not seem to me of much weight: it might be said that we could give the name of *first* terms to those which he calls the *last*, for if we have to find a set of terms it is not of much importance at which end we begin.

1216. The pages 34...45 of Ivory's memoir form an appendix to what had preceded; in these pages Ivory gives an account of the memoir by Lagrange which we have already noticed in Arts. 1197...1202.

Ivory begins thus:

Some time before the end of May last, a paper of mine was presented to the Royal Society, in which I entered on an examination of a fundamental proposition in the second chapter of the third book of the *Mécanique Céleste*. About three months after that paper was in the possession of the Society, towards the middle of August, a large collection of foreign books, imported from the Continent, was received in London; among which were several *Cahiers* of the *Journal de l'École Polytechnique*. In the 15th *Cahier*, which had been published at Paris in December 1809, although it did not find its way into this country prior to the above date, there is a short memoir by Lagrange on the same subject treated of in my paper: and in this Appendix I shall lay before the Society a short account of Lagrange's memoir, pointing out what are the views of that celebrated mathematician in regard to the conclusions obtained in my paper.

1217. Ivory illustrates on his pages 38 and 39 an important point to which we are now accustomed to pay due attention; we should describe it by saying that if a function to be integrated becomes infinite within the range of integration it will be necessary to determine by careful examination what the real value of the integral is.

1218. Lagrange himself in his memoir expressed no doubt as to the accuracy of Laplace's equation; Ivory however seeks to draw a confirmation of his own opinion from Lagrange's memoir. But it does not appear to me that in this Appendix Ivory adds any matter of consequence to what he had already given.

1219. A second memoir by Ivory occurs in the same volume of the *Philosophical Transactions* entitled, *On the Attractions of an extensive Class of Spheroids.* We shall give a notice of this memoir hereafter; at present it is sufficient to say that Ivory just alludes to the equation which we are considering, and reasserts that there is an error in Laplace's process: see page 73 of the memoir.

1220. Laplace himself returned to the equation in a memoir on the Figure of the Earth which was published in the *Mémoires... de l'Institut* for 1817, and was reproduced in the fifth volume of the *Mécanique Céleste*: see pages 24...27 of the volume. The following is the substance of the addition which Laplace here makes to his original demonstration.

Take the expression which forms the right-hand member of (3) in Art. 1200; and suppose the sphere and the spheroid to touch so that y' vanishes at the common point. Since r is to be made equal to a ultimately, we are sure that the expression will vanish if we except for a moment that part of the integral which arises from elements close to the point considered; this part requires special examination. But since we suppose the sphere and the spheroid *to touch* we have *y' varying as $r^2 - 2ar\mu + a^2$ ultimately* in the neighbourhood of the common point; and this ensures that the corresponding part of the integral vanishes.

1221. Laplace puts his investigation in a form resembling that which we have adopted in Art. 852. The potential is separated into two parts which we have denoted by V and V_1; and the equation is obtained which we express thus:

$$-a\frac{dV_1}{dr} = \frac{1}{2}V_1.$$

This agrees with Laplace's equation (a) on his page 24; he uses V'' for our V_1, and he supposes a to be unity.

1222. Laplace says on his page 27:

Telle est la démonstration que j'ai donnée de cette équation, dans l'endroit cité de la Mécanique céleste. Quelques géomètres ne l'ayant pas bien saisie, l'ont jugée inexacte. Lagrange, dans le tome VIII. du Journal de l'Ecole Polytechnique, a démontré cette équation, par une analyse à peu près semblable à celle qui me l'avait fait découvrir (Mémoires de l'Académie des Sciences, année 1775, page 83). C'est pour simplifier cette matière, que j'ai préféré de donner dans la Mécanique céleste, la démonstration précédente.

I do not know who are meant by the words *quelques géomètres*; the words seem to imply that more than one person had attacked the demonstration, but I have found no other besides Ivory, up to the date of publication of Laplace's fifth volume.

Laplace does not offer to defend the general form of his equation: see Art. 1214.

1223. We find something bearing on the point under discussion in a memoir by Poisson on the distribution of heat in solid bodies, which was published in the *Journal de l'Ecole Polytechnique*, Vol. XII.: see page 159 of the volume. Although Poisson puts his remarks in a more elaborate form, yet the following simple statement constitutes their sum.

Take the expression which forms the right-hand member of (3) in Art. 1200. When we make $r - a$ infinitesimal the only part of the integral which can be sensible is that which arises from elements close to the point considered. Thus we may regard y' as constant; so that the expression becomes

$$-\frac{r^2-a^2}{2}\,ay\int\frac{d\sigma}{(r^2-2ar\mu+a^2)^{\frac{3}{2}}}.$$

Then continuing as in that Article we arrive at the result $-2\pi a^2 y$.

Although no condition is stated by Poisson, it is obvious that it will not be safe to say that y' may be considered constant unless we are sure that $\frac{dy'}{d\mu}$ is not infinite when $\mu = 1$.

1224. Ivory returned to the equation in a memoir entitled, *On the expansion in a series of the attraction of a Spheroid*, published in the *Philosophical Transactions* for 1822: we shall here notice only so much of the memoir as relates to the equation under discussion.

Ivory says on pages 106 and 107 of this memoir:

We come next to consider the differential equation that takes place at the surface of a spheroid. Of this equation, three demonstrations have been published; one, in the second chapter of the third book of the *Mécanique Céleste*; another by the same author, not precisely the same with the former, but similar to it, in a memoir read to the Academy of Sciences in 1818; and a third by M. Poisson, in an interesting and profound memoir on the distribution of heat in solid bodies. The two last demonstrations are fundamentally the same; but as M. Poisson has stated the reasoning more fully, and fixed the sense of the proof more precisely, I wish to refer to his memoir.

I do not understand what Ivory means by speaking of the last two demonstrations as *fundamentally the same*: it appears to me that there is an appreciable difference between them. The memoir by Laplace to which Ivory here refers is that noticed in Art. 1220.

1225. In this memoir Ivory seems to object not so much to the theorem itself as to the inference which was drawn from it, namely that any function could be expanded in a series of Laplace's functions.

1226. Laplace himself in a conversation with Sir Humphry Davy bore testimony to the talents and labours of Ivory: see the *Abstracts of the Papers printed in the Philosophical Transactions...* Vol. IV. page 409. The end which Ivory aimed at in his criticisms on Laplace's equation may be considered now substantially attained; for the equation seems to have been relinquished by most mathematical writers. The theory of Laplace's functions, and the theory of attractions, are now usually exhibited quite independently of the equation.

1227. A paper on this matter by G. B. Airy, now Astronomer Royal, is contained in the *Cambridge Philosophical Transactions*, Vol. II. 1827. The paper occupies pages 379...390 of the volume; it was read May 8th, 1826.

Mr Airy commences thus:

In two papers printed in the Philosophical Transactions for 1812, and in a third in the Transactions for 1822, Mr Ivory has objected to some parts of Laplace's investigation of the attraction of spheroids differing little from a sphere. That there are difficulties in that theory cannot be denied, but that Mr Ivory has pointed out correctly the errors from which the obscurities arise appears to me quite doubtful. After considering the subject attentively, I have come to the conclusion, that in the part to which Mr Ivory has most strongly objected, Laplace's investigation may, by a slight alteration, be made free from error; but that an assertion of Laplace which Mr Ivory has admitted without scruple, is absolutely unsupported by any demonstrative evidence whatever......

1228. The manner in which Mr Airy establishes Laplace's equation is very interesting, and deserves to be studied. Instead of reproducing it I will give a process which is founded on the same principles.

Let there be a sphere nearly equal in volume to the spheroid and having a common point and nearly coinciding with the spheroid; but the two need not necessarily touch at the common point.

Laplace's equation holds for the sphere. It is also admitted to hold for that part of the excess of the spheroid over the sphere which is not close to the common point; so that we need only consider the part close to the common point. Let v be the value of the potential for the portion of this excess which is near the common point; and which we may suppose to be bounded by a sphere of radius f described round the common point as centre. Then for this small portion of the excess we shall shew that v and $\frac{dv}{dr}$ may both be considered zero ultimately.

First let us examine the value of v. Resolve the small portion we have to consider into infinitesimal cones or pyramids which have their vertex at the common point. Suppose that ω represents the surface which one of these cones or pyramids would cut from a sphere of radius unity; then $\frac{1}{2}f^2\omega$ will be the part of v arising from this cone or pyramid. Hence the value of v will be ultimately zero. For f is ultimately indefinitely small. If we suppose with Laplace that the sphere and the spheroid touch, the sum of all the values of ω will be infinitesimal. And if we suppose with Mr Airy that the sphere and the spheroid do not touch, still the sum of all the values of v will be very small. Moreover in this case there is a species of compensation; for as the sphere will be partly within and partly without the spheroid the elementary cones will contribute partly positive and partly negative values to v. Thus we may safely admit v to be ultimately zero to the order which Laplace's equation involves.

Next let us examine the value of $\frac{dv}{dr}$. The part of $\frac{dv}{dr}$ which arises from one of the infinitesimal cones or pyramids expresses the attraction of that cone or pyramid on a particle at the vertex estimated in the direction which is almost at right angles to the axis of the cone or pyramid. It is therefore obvious that this must be infinitesimal. Hence too the aggregate arising from all the cones or pyramids will be ultimately zero to the order we have to regard.

1229. Mr Airy then touches on a matter of less importance. He says:

The accuracy of the investigation in Liv. III. No. 10. being supposed to be established, I proceed to No. 11. In this article as it stands, there is certainly an obscurity (attended however with no erroneous results) which a small change in the notation will entirely remove. In the preceding article Laplace has taken $r=a$ at the attracted point, he now supposes $r=a(1+\alpha y)$: and whether this be an inaccuracy, or the origin of co-ordinates be supposed to be changed, it is equally incomprehensible to the reader, and equally likely to lead him into error.

It does not seem to me that there is any serious difficulty here; in his § 10 Laplace makes r *nearly* equal to a, but he does not require that r should be *exactly* equal to a.

1230. A more important point is then introduced. Laplace undertook to shew that a function y cannot be expanded into two different series of Laplace's functions. For suppose if possible there are two different expansions, say

$$y = Y_0 + Y_1 + Y_2 + Y_3 + \dots$$

and

$$y = X_0 + X_1 + X_2 + X_3 + \dots$$

Hence by subtraction

$$0 = Y_0 - X_0 + Y_1 - X_1 + Y_2 - X_2 + \dots$$

Let Z_i be a Laplace's function of the ith order; multiply the last equation by Z_i and integrate between the usual limits; then by a well known property of the functions we have

$$\int_{-1}^{1}\int_{0}^{2\pi} (Y_i - X_i)\, Z_i \, d\mu \, d\phi = 0 \dots\dots\dots\dots (5).$$

Laplace says that if we take for Z_i the most general function of its kind this equation cannot be true except $Y_i - X_i = 0$.

Mr Airy considers that there is not the slightest evidence for this assertion; and so that the theorem which Laplace wished to establish "rests at present entirely on Laplace's unsupported assertion."

Mr Airy does not attempt to complete the demonstration of the result $Y_i - X_i = 0$, but gives another investigation of the proposition that a rational function of μ, $\sqrt{(1-\mu^2)}\cos\phi$, and $\sqrt{(1-\mu^2)}\sin\phi$ can be expanded in only one series of Laplace's functions.

I do not however consider that Laplace's assertion is destitute of evidence: here, as in many cases, Laplace leaves his readers to do much for themselves, but I think there is little doubt as to how we should here proceed.

For an example suppose $i = 2$. Then he would say that $Y_2 - X_2$ must be of the form

$$C_0 f_0(\mu) + C_1 f_1(\mu) \cos\phi + C_2 f_2(\mu) \cos 2\phi$$
$$+ E_1 f_1(\mu) \sin\phi + E_2 f_2(\mu) \sin 2\phi,$$

where C_0, C_1, C_2, E_1, E_2 are constants, and $f_0(\mu)$, $f_1(\mu)$, $f_2(\mu)$ are certain functions of μ.

Now take for Z_2 an expression of the form

$$H_2 f_2(\mu) \sin 2\phi,$$

where H_2 is a constant.

Substitute in (5); then we see that it reduces to

$$E_2 H_2 \iint [f_2(\mu) \sin 2\phi]^2 \, d\mu \, d\phi;$$

and this is impossible unless $E_2 = 0$.

In like manner we can shew that the other constants which occur in $Y_2 - X_2$ must all be zero by reason of (5).

We may observe that the inference that $Y_i = X_i$ is drawn from (5) in modern books in a simpler manner by the aid of an important equation which Laplace himself gave; namely the equation (1) of the *Mécanique Céleste*, Vol. II. page 44: see Arts. 857 and 1069. Poisson uses this manner in his *Théorie...de la Chaleur*, page 225.

With respect to this paper by Professor Airy, the reader may consult an article also by him in the *Philosophical Magazine* for June 1827.

1231. The equation is investigated by Pontécoulant in his *Théorie Analytique du Système du Monde*, Vol. II. 1829, pages 374...380. Pontécoulant's process is substantially the same as Lagrange's, which we have noticed in Arts. 1197...1202.

1232. The matter discussed in the present Chapter is also considered by Bowditch in his notes to the translation of the *Mécanique Céleste;* see pages 88 and 92 of his second volume. But he adds nothing to what had been previously given.

1233. A very interesting paper on the subject by the late James Mc Cullagh is published in the *Transactions of the Royal Irish Academy*, Vol. XVII. 1837, pages 237...239; the paper was read on May 28, 1832. The object of the paper is to demonstrate a certain exact theorem of which Laplace's equation may be considered an approximate form. We will give a sketch of the process.

If ϕ be the solid angle of an infinitesimal cone or pyramid of length r it is easily seen that the potential at the vertex is $\frac{1}{2}\phi r^2$, and the attraction there is ϕr; the density being taken as unity.

Now consider a solid of any shape, regular or irregular, terminated at one end by a plane; in this plane take any point P, and from P draw a straight line at right angles to the plane meeting the solid again at Q. Let there be a sphere of any magnitude whose diameter $P'Q'$ is parallel to PQ. Let P'' be another fixed point; and from the points P, P', P'' draw three parallel straight lines $Pp, P'p', P''p''$, the first two terminated respectively by the solid and the sphere, and the third equal to the difference of the other two, without regarding which of them is the greater. Suppose all the points p'' taken according to this law to trace out a third solid.

Let $Pp, P'p', P''p''$, be edges of three infinitesimal pyramids, with their other edges proceeding from P, P', P'' all parallel; they will have the same solid angle, which we will denote by ϕ. Let r, r', r'' denote the respective lengths, and V, V', V'' the potentials at the vertices. From p draw pR perpendicular to PQ; the attraction of the pyramid corresponding to Pp in the direction of PQ will be $\phi \times PR$; call this A. Let a be the radius of the sphere.

Since r'' is the difference of r and r' we have

$$r^2 + r'^2 - r''^2 = 2rr' = 2PR \times P'Q';$$

therefore $$\frac{1}{2}\phi r^2 + \frac{1}{2}\phi r'^2 - \frac{1}{2}\phi r''^2 = 2a\phi \times PR,$$

that is $$V + V' - V'' = 2aA.$$

This result then holds for every other three pyramids similarly related to each other throughout the whole extent of the three solids. Thus if we now denote by V, V', V'' the *whole* potentials for the three solids, and by A the *whole* attraction of the first solid parallel to PQ on a particle at P, we shall have

$$V + V' - V'' = 2aA.$$

This is M^c^Cullagh's exact equation. To express it in Laplace's notation he observes that the attraction A is synonymous with $-\frac{dV}{dr}$, and that V' for the sphere is equal to $\frac{4}{3}\pi a^3$. Substituting these values we obtain

$$V + 2a\frac{dV}{dr} = -\frac{4}{3}\pi a^3 + V''.$$

M'Cullagh then shows that if the original solid differs but slightly from a sphere, and we choose a suitably, then V'' will be a small quantity of the second order.

1234. A few remarks may be made on this ingenious paper.

The plane which terminates the first solid may if we please be a tangent plane, and then PQ becomes the normal at P.

The normal attraction at P, in Laplace's use of his equation, is not necessarily *exactly* $-\frac{dV}{dr}$; but it will not differ from this expression by more than a quantity of the second order at most. See the *Mécanique Céleste*, Vol. v. page 26.

M'Cullagh refers to a certain case, as he says, "because both Lagrange and Ivory have used this case to show that the reasonings of Laplace are incorrect." But I do not think Lagrange professed to shew that the reasonings of Laplace are incorrect. Lagrange shewed that an error might very naturally be made; but he did not assert or imply that Laplace had made any error.

1235. The matter is considered by Plana; see the *Astronomische Nachrichten*, Vol. XXXVIII., page 226: but the article by Plana may be most conveniently studied in connexion with Laplace's Livre XI. Chapitre II. §§ 2...5, to which it relates.

Resal investigates Laplace's equation; see his pages 164...166: the method may be considered to resemble that of Art. 1228.

CHAPTER XXXI.

PARTIAL DIFFERENTIAL EQUATION FOR V.

1236. LET V denote the potential of a given mass at a point whose coordinates are x, y, z. Then, as we have seen in Art. 866, Laplace obtained for V the partial differential equation

$$\frac{d^2 V}{dx^2} + \frac{d^2 V}{dy^2} + \frac{d^2 V}{dz^2} = 0;$$

for abbreviation we shall denote this by

$$\nabla V = 0.$$

About thirty years elapsed before it was discovered that the equation is not universally true; it is not true if the point (x, y, z) is a point of the body: it is only true if the point (x, y, z) is outside the body or within some cavity of the body.

1237. The correction was furnished by Poisson in a note published in the *Nouveau Bulletin......Société Philomatique*, Dec. 1813. The note occupies pages 388.....392 of Volume III. of the *Nouveau Bulletin*; it is entitled *Remarques sur une équation qui se présente dans la théorie des attractions des sphéroides.*

Poisson's method is now familiar to us, for it has passed into the elementary books. He divides the body into two parts, a sphere which includes the point (x, y, z), and the rest of the body. Let V be separated into two corresponding parts, V_1 and V_2, of which V_1 belongs to the sphere and V_2 to the rest of the body; then we have

$$\nabla V_1 = -4\pi\rho,$$

$$\nabla V_2 = 0,$$

where ρ is the density at the point (x, y, z).

1238. Poisson says, and justly, that Laplace's form is true when the point (x, y, z) is without the body, or within a cavity formed by the body. He adds "ces deux cas sont, à la vérité, les seuls pour lesquels on ait fait usage de l'équation...". I do not consider this quite correct so far as it implies that no error had been made hitherto; for I have shewn that some of Laplace's processes are rendered unsatisfactory by the tacit assumption that ∇V is always zero: see Arts. 1044 and 1030.

1239. Poisson gives two applications of his formulæ. One is to determine the attraction at an external or internal point exerted by a sphere in which the density is a function of the distance from the centre. The other is to establish a result first obtained by Legendre; see Art. 1157.

1240. The matter is considered by Rodrigues in a memoir which has been noticed in Chapter XXIX.: see Art. 1177. I will give his process.

Let V denote the potential of a given body at the point (a, b, c). Let A, B, C denote the corresponding resolved attractions; so that

$$A = -\frac{dV}{da}, \quad B = -\frac{dV}{db}, \quad C = -\frac{dV}{dc}.$$

Take (a, b, c) as the origin of the usual polar coordinates; let ρ denote the density: then

$$A = -\iiint \rho \cos\theta \sin\theta \, d\theta \, d\phi \, dr,$$

$$B = -\iiint \rho \sin^2\theta \cos\phi \, d\theta \, d\phi \, dr,$$

$$C = -\iiint \rho \sin^2\theta \sin\phi \, d\theta \, d\phi \, dr.$$

We will suppose (a, b, c) within the body. Then the limits of integration for r are 0 and r_1, where r_1 denotes some function of θ and ϕ, which is known from the equation to the surface; the limits for θ are 0 and π; and the limits for ϕ are 0 and 2π.

Put $x' = r\cos\theta$, $y' = r\sin\theta\cos\phi$, $z' = r\sin\theta\sin\phi$; then ρ will be a function of the variables $a + x'$, $b + y'$, and $c + z'$; thus

$$\frac{d\rho}{da} = \frac{d\rho}{dx'}, \quad \frac{d\rho}{db} = \frac{d\rho}{dy'}, \quad \frac{d\rho}{dc} = \frac{d\rho}{dz'}.$$

Now $\frac{dA}{da}$ will consist of two terms; one arising from the fact that ρ involves a, and the other from the fact that the limit r_1 also involves a.

Hence
$$\frac{dA}{da} = -\iiint \frac{d\rho}{da} \cos\theta \sin\theta \, d\theta \, d\phi \, dr$$
$$- \iint \rho_1 \frac{dr_1}{da} \cos\theta \sin\theta \, d\theta \, d\phi,$$

where ρ_1 denotes the density at the point of the surface which corresponds to r_1.

Thus we get
$$\frac{d^2V}{da^2} + \frac{d^2V}{db^2} + \frac{d^2V}{dc^2} = \iiint \frac{\sin\theta \, d\theta \, d\phi \, dr}{r} \left(x' \frac{d\rho}{dx'} + y' \frac{d\rho}{dy'} + z' \frac{d\rho}{dz'}\right)$$
$$+ \iint \frac{\rho_1 \sin\theta \, d\theta \, d\phi}{r_1} \left(x_1' \frac{dr_1}{da} + y_1' \frac{dr_1}{db} + z_1' \frac{dr_1}{dc}\right),$$

where as before the suffix 1 is used to denote a value at the point of the surface.

Now
$$x' \frac{d\rho}{dx'} + y' \frac{d\rho}{dy'} + z' \frac{d\rho}{dz'} = r \frac{d\rho}{dr}.$$

Let the equation from which r_1 is to be found be denoted by

$$F(a + r_1 \cos\theta, \; b + r_1 \sin\theta \cos\phi, \; c + r_1 \sin\theta \sin\phi) = 0;$$

thus
$$\frac{dF}{da} + \left(\frac{dF}{da} \cos\theta + \frac{dF}{db} \sin\theta \cos\phi + \frac{dF}{dc} \sin\theta \sin\phi\right) \frac{dr_1}{da} = 0,$$

and we have two similar equations; the three give

$$x_1' \frac{dr_1}{da} + y_1' \frac{dr_1}{db} + z_1' \frac{dr_1}{dc} = -r_1.$$

Hence
$$\frac{d^2V}{da^2} + \frac{d^2V}{db^2} + \frac{d^2V}{dc^2} = \iiint \sin\theta \, d\theta \, d\phi \, dr \, . \, \frac{d\rho}{dr} - \iint \sin\theta \, d\theta \, d\phi \rho_1.$$

But $\int \frac{d\rho}{dr} dr = \rho_1 - \rho_0$, where ρ_0 is the density at the origin of polar coordinates, that is the point (a, b, c). Thus

$$\frac{d^2 V}{da^2} + \frac{d^2 V}{db^2} + \frac{d^2 V}{dc^2} = -\rho_0 \iint \sin\theta \, d\theta \, d\phi = -4\pi\rho_0.$$

1241. Poisson took up the matter again in his *Mémoire sur la Théorie du Magnétisme en Mouvement;* this memoir is contained in the *Mémoires de l'Académie*, Vol. VI. which is for 1823, and was published in 1827. The memoir was read on the 10th of July, 1826. We are now concerned with pages 455...463 of the volume containing the memoir; and we will indicate the method which Poisson adopts.

1242. We suppose the body homogeneous, and take unity for its density. Let $dx' \, dy' \, dz'$ denote an element of volume of the body; and put r for $\sqrt{\{(x-x')^2 + (y-y')^2 + (z-z')^2\}}$, so that

$$V = \iiint \frac{dx' \, dy' \, dz'}{r}.$$

Therefore
$$\frac{dV}{dx} = -\iiint \frac{x-x'}{r^3} \, dx' \, dy' \, dz',$$

and
$$\frac{d^2 V}{dx^2} = -\iiint \frac{d}{dx}\left(\frac{x-x'}{r^3}\right) dx' \, dy' \, dz'$$

$$= \iiint \frac{d}{dx'}\left(\frac{x-x'}{r^3}\right) dx' \, dy' \, dz'.$$

Then by applying the same kind of transformation that Gauss had used, but without referring to him, Poisson puts this result in the form

$$\frac{d^2 V}{dx^2} = \iint \frac{x-x'}{r^3} \cos l \, d\omega,$$

where $d\omega$ is an element of the surface of the body, and l is the angle between the positive direction of the axis of x and the normal to the surface at the element $d\omega$ drawn outwards. The integration is to extend over the whole surface of the body. Compare Art. 1171.

Similar transformations hold for $\frac{d^2 V}{dy^2}$ and $\frac{d^2 V}{dz^2}$; thus finally

$$\nabla V = \iint \left(\frac{x-x'}{r} \cos l + \frac{y-y'}{r} \cos m + \frac{z-z'}{r} \cos n \right) \frac{d\omega}{r^2} \ldots\ldots (1).$$

Poisson, like Gauss, is careful to consider the cases in which a straight line parallel to an axis of coordinates meets the surface more than twice, so that (1) may be established with adequate generality.

Let i denote the angle between the straight line drawn from (x, y, z) to $d\omega$ and produced, and the normal to the surface at $d\omega$ drawn outwards; then

$$\cos i = \frac{x'-x}{r} \cos l + \frac{y'-y}{r} \cos m + \frac{z'-z}{r} \cos n.$$

Let $d\theta$ be the element of a spherical surface of radius unity, which is cut out by the cone having its vertex at (x, y, z) and circumscribed round $d\omega$. Then

$$\cos i \, d\omega = \pm r^2 \, d\theta,$$

where the upper or lower sign is to be taken according as i is acute or obtuse. Thus we have from (1)

$$\nabla V = \iint (\mp) \, d\theta \ldots\ldots\ldots\ldots\ldots\ldots\ldots\ldots (2).$$

Now consider separately the various positions of the point (x, y, z).

I. Let the point (x, y, z) be outside the body. Then if a cone be drawn from this point as vertex to circumscribe the body, the surface of the body will be divided into two parts: in one part i is acute, and in the other obtuse. The double integral which forms the right-hand member of (2) has for these parts values which are numerically equal but of opposite signs. In this case then

$$\nabla V = 0.$$

II. Let the point (x, y, z) be on the surface of the body.

Then the upper sign will have to be taken in (2), and the double integral will extend over half the surface of the sphere.

In this case then

$$\nabla V = -2\pi.$$

III. Let the point (x, y, z) be inside the body.

Then the upper sign will have to be taken in (2), and the double integral will extend over the whole surface of the sphere.

In this case then

$$\nabla V = -4\pi.$$

1243. Hitherto we have taken the body to be homogeneous with the density unity. Now let ρ' which is any function of x', y', and z' denote the density at the point (x', y', z'). Then

$$\nabla V = \iiint \rho' \nabla \left(\frac{1}{r}\right) dx'\, dy'\, dz'.$$

When (x, y, z) is not a point of the body $\nabla \left(\frac{1}{r}\right)$ vanishes throughout the triple integral.

When (x, y, z) is a point of the body we divide the body into two parts, namely one which does not contain the point and for which $\nabla \left(\frac{1}{r}\right)$ always vanishes; and the other which contains the point; we may take this part so small that ρ' may be considered constant throughout it, and may therefore be put equal to ρ, where ρ denotes the density at the point (x, y, z).

Thus $$\nabla V = \rho \iiint \nabla \left(\frac{1}{r}\right) dx'\, dy'\, dz',$$

where the triple integral extends over that part of the body which contains (x, y, z). Hence by what has been shewn in the preceding Article we obtain $\nabla V = -2\pi\rho$ for a point on the surface of the body, and $\nabla V = -4\pi\rho$ for a point within the body.

Hence finally

$$\nabla V = 0, \quad \text{or} \ -2\pi\rho, \quad \text{or} \ -4\pi\rho,$$

according as the point (x, y, z) is without the body, or on its surface, or within the body.

1244. Poisson says on his page 463 respecting the three cases just considered:

Les géomètres ont remarqué le premier cas depuis long-temps; j'ai été conduit à la troisième valeur, il y a plusieurs années, par une analyse moins directe que la précédente; j'y joins maintenant la seconde; ce qui ne laissera plus rien à désirer touchant cette équation, dont on connaît l'importance dans un grand nombre de questions, et qui nous sera bientôt utile.

1245. The process of Art. 1242 is doubtless perfectly satisfactory for the case of an external point; it does not carry conviction to my mind for the other two cases which Poisson considers. As to the internal particle Poisson's original treatment seems to me conclusive; the result is now universally accepted as one of the standard theorems in the subject of attraction. As to the particle at the surface however the case seems different; I do not think that Poisson's result has been ever generally accepted or used. I shall hereafter return to this point: see Art. 1253.

1246. Poisson gives another investigation of the formula $\nabla V = -4\pi\rho$ in his memoir in the *Connaissance des Tems* for 1829. We shall return to this in Chapter XXXV.

If we take the usual polar coordinates the equation for an internal particle becomes

$$r\frac{d^2(rV)}{dr^2} + \frac{1}{\sin\theta}\frac{d}{d\theta}\left(\sin\theta\frac{dV}{d\theta}\right) + \frac{1}{\sin^2\theta}\frac{d^2V}{d\phi^2} = -4\pi r^2\rho.$$

In this memoir Poisson does not give his result for a particle on the surface.

1247. We have next to notice a paper by Ostrogradsky entitled *Note sur une intégrale qui se rencontre dans le calcul de l'attraction des Sphéroïdes.* This is published in the *Mémoires de l'Académie...St Pétersbourg*, sixth series, Vol. I. 1831. The note was read on the 2nd July, 1828.

Ostrogradsky after some preliminary remarks cites Poisson's equations for an internal particle, and also that for a particle at the surface; he numbers the former (3) and the latter (4). Then he says:

C'est M. Poisson qui a trouvé les équations (3) et (4), de mon côté j'ai trouvé l'équation (3) sans connaître la remarque de M. Poisson que j'ai vue depuis dans le Bulletin des sciences; quant à l'équation (4) j'ignore, encore maintenant comment l'illustre géomètre que je viens de citer y est parvenu.

1248. Ostrogradsky states that his object is to indicate how we ought to replace the equation given by Poisson for a point at the surface when that equation does not hold. We may express the object by saying that Ostrogradsky investigates what the equation ought to be for singular points of the surface.

Thus for example at the corner of a rectangular parallelepiped Ostrogradsky considers that instead of Poisson's $-2\pi\rho$ we ought to have $-\frac{\pi}{2}\rho$.

1249. I do not regard the paper by Ostrogradsky as of any interest or value in the theory of attractions; though it may deserve a little attention from a writer on the Integral Calculus. The point involved is the treatment of a definite integral when the function to be integrated has an infinite value; Ostrogradsky's process resembles the well-known one of Cauchy. But I cannot say that I have any confidence in the method which Ostrogradsky pursues; and as to his results the remark made at the end of Art. 1245 applies.

1250. One important statement Ostrogradsky makes without demonstration; but says merely "*Nous avons fait voir ailleurs...*" I have not been able to find the place to which he thus vaguely alludes.

1251. In Ferussac's *Bulletin......Sciences Mathématiques...* Vol. XIV. 1830, there is a notice of the researches of Poisson and Ostrogradsky on the point which we are considering. The notice is on pages 81...88 of the volume, signed S., which I presume stands for Sturm.

This notice repeats the historical statement to which I have objected in Art. 1138. With respect to the point to which I have alluded in Art. 1250, the notice says: "Ostrogradsky has proved...." If S. knew where Ostrogradsky had given the proof he should have supplied the reference; if he did not know he should have said not, "Ostrogradsky has proved...", but, "Ostrogradsky states that he has proved...."

The notice asserts that Laplace's form of the equation is true in some sense even for an internal point: but I do not understand in what sense.

It says that Ostrogradsky's investigation is far less simple than Poisson's; and with this I entirely agree.

And finally it throws doubt on the truth of Poisson's equation for the case of a point at the surface; for instance, supposing the body to be a sphere, it shews that we may very naturally obtain $-4\pi\rho$ instead of Poisson's $-2\pi\rho$.

1252. Bowditch notices the correction for the case of an internal particle in the second volume of his translation of the *Mécanique Céleste*, published in 1832. Bowditch says on his page 67:

It is somewhat remarkable, that this defect in the formula, as it was first published by La Place, should have remained unnoticed, nearly half a century; particularly as he had expressly called the attention of mathematicians to the necessity of having the limits of the integrals independent of the co-ordinates of the attracted point;...and had also conformed to this restriction, in the calculations of the first volume.

It is a great fault in Bowditch's work that he gives scarcely any references. Thus in the present case, although he attributes the correction to Poisson, he does not say where Poisson first published the correction. But as Bowditch speaks of an interval of nearly half a century, it would appear that he was not acquainted with Poisson's paper of Dec. 1813. But yet Bowditch himself, on his page 64, uses a method like that of Poisson's paper; namely, he divides the body into two parts, one part being a sphere which contains the point considered.

I do not agree with what is implied by Bowditch, that the first volume of the *Mécanique Céleste* is perfectly free from error as to the equation for V; see Arts. 1044 and 1050.

1253. It remains only to shew that Poisson's equation for a point at the surface is unsatisfactory. This is in fact considered by Gauss in his celebrated memoir entitled *Allgemeine Lehrsätze... Anziehungs- und Abstossungs-Kräfte*, Leipsic, 1840; the memoir is reprinted in Vol. v. of the edition of the collected works of Gauss: see pages 204...206 of the volume.

Consider a sphere of radius a, and density ρ; take the centre as origin of co-ordinates. Let x, y, z be the coordinates of any point. Then we know that for an internal point

$$V = 2\pi\rho a^2 - \frac{2}{3}\pi\rho(x^2 + y^2 + z^2),$$

and for an external point

$$V = \frac{4\pi\rho a^3}{3(x^2 + y^2 + z^2)^{\frac{1}{2}}}.$$

Hence we find that for an internal point

$$\frac{d^2V}{dx^2} = -\frac{4\pi\rho}{3},$$

and for an external point

$$\frac{d^2V}{dx^2} = \frac{4\pi\rho a^3(3x^2 - r^2)}{3r^5},$$

where r stands for $(x^2 + y^2 + z^2)^{\frac{1}{2}}$.

The two values of $\frac{d^2V}{dx^2}$ do not agree at the surface; so that we must say that $\frac{d^2V}{dx^2}$ is not determinate *at the surface*, but has two distinct values.

In like manner ∇V has not a determinate value at the surface. In fact ∇V is an aggregate of three terms, each of which has two values; so that there are in all eight combinations, of which one gives the value of ∇V agreeing with that found for an internal particle, and the other gives the value of ∇V agreeing with that found for an external particle; the other six remain without meaning.

Gauss says he cannot admit the reasoning by which some mathematicians have deduced the value $-2\pi\rho$ for a point at the surface.

CHAPTER XXXII.

LAPLACE'S SECOND METHOD OF TREATING LEGENDRE'S PROBLEM.

1254. D'ALEMBERT attempted to demonstrate that among figures of revolution an oblatum is the only form of relative equilibrium for homogeneous fluid rotating with uniform angular velocity; but his process is a failure, as we have seen in Art. 570. Laplace contributed a little to the investigation in his two earliest memoirs, but Legendre was the first who discussed the problem with tolerable success; and I have therefore called it by his name: see Arts. 744 and 763. Laplace subsequently gave two demonstrations, without assuming that the fluid is a figure of equilibrium, but only that it is nearly spherical; one of these demonstrations depends on the expansion of the radius vector in a series of Laplace's functions, while the other does not employ any expansion. The demonstrations date from 1782, and are reproduced in the fourth Chapter of the Third Book of the *Mécanique Céleste*.

1255. The second demonstration remained for more than fifty years unchallenged. At last it was shewn to be unsatisfactory by Liouville in a note published in his *Journal de Mathématiques*, Vol. II. 1837. The note is entitled *Sur un passage de la Mécanique céleste, relatif à la Théorie de la Figure des Planètes*. The note occupies pages 200...219 of the volume. We shall devote the present Chapter to the matter.

1256. After completing his first demonstration Laplace proceeds thus in his § 20:

L'analyse précédente nous a conduits à la figure d'une masse fluide homogène en équilibre, sans employer d'autres hypothèses que celle d'une figure très-peu différente de la sphère: elle fait voir que la figure elliptique qui, par le Chapitre précédent, satisfait à cet équilibre, est la seule alors qui lui convienne. Mais comme la réduction du rayon du sphéroïde, dans une série de la forme $a(1 + \alpha Y_0 + \alpha Y_1 + \ldots)$, peut faire naître quelques difficultés, nous allons démontrer directement et indépendamment de cette réduction, que la figure elliptique est la seule figure d'équilibre d'une masse fluide homogène, douée d'un mouvement de rotation; ce qui, en confirmant les résultats de l'analyse précédente, servira en même temps, à dissiper les doutes que l'on pourroit élever contre la généralité de cette analyse.

1237. We have now to explain Laplace's process, and also the objection to which it is exposed. Let a be the radius of a sphere nearly coinciding with the spheroid. Let the radius vector of the spheroid be denoted by $a(1 + \alpha Y')$, where α is very small, and Y' is a function of the usual angular coordinates μ' and ϖ'. Let Y be the corresponding value of Y' when we put μ and ϖ for μ' and ϖ' respectively.

Let V denote the *potential* at the point (μ, ϖ) on the surface of the spheroid. Then we know that to the order of the first power of α

$$V = \frac{4\pi a^3}{3r} + \alpha a^2 \int_{-1}^{1} \int_0^{2\pi} \frac{Y' \, d\mu' \, d\varpi'}{\sqrt{(2 - 2\lambda)}} \quad \ldots\ldots\ldots\ldots\ldots (1),$$

where r stands for $a(1 + \alpha Y)$, and λ for the cosine of the angle between the radius vector to (μ, ϖ) and the radius vector to (μ', ϖ'): see Art. 832.

Now Laplace in fact transforms the second term in the above value of V, by changing the variables; and thus arrives substantially at the result

$$\int_{-1}^{1} \int_0^{2\pi} \frac{Y' \, d\mu' \, d\varpi'}{\sqrt{(2 - 2\lambda)}} = \int_0^{\pi} \int_0^{2\pi} Y' \sin p \, dp \, dq' \quad \ldots\ldots\ldots\ldots (2).$$

This result constitutes a large part of Laplace's process; we shall now examine it, though not quite in Laplace's way. He is very brief, and introduces his usual phrases, *il est facile de voir* and *on trouvera facilement*. The connection between the old variables and the new variables will be made manifest in the progress of our investigation.

1258. In the double integral which we propose to transform we shall suppose that μ' and ϖ' are the polar coordinates of a point on the surface of a sphere of radius unity.

Let O be the centre of the sphere, OP the fixed radius from which θ and θ' are measured, where as usual $\mu = \cos\theta$, and $\mu' = \cos\theta'$.

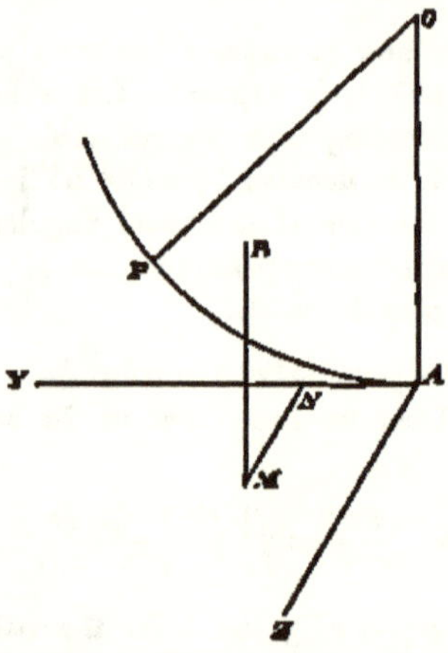

Let A be the point (μ, ϖ), so that $POA = \theta$.

Take A as the origin of a new set of coordinates; let AO be the axis of x, AY the tangent to AP the axis of y, and AZ which is perpendicular to the plane of the paper the axis of z. Then the equation to the surface of the sphere will be

$$2x = x^2 + y^2 + z^2 \text{ .... } (3).$$

Transform to polar coordinates by the usual relations

$$z = \rho \cos p, \quad x = \rho \sin p \cos q, \quad y = \rho \sin p \sin q.$$

Then from (3) we obtain for the radius vector ρ the value

$$\rho = 2 \sin p \cos q.$$

Now we know by the Integral Calculus that the polar expression for an element of a surface is

$$\sqrt{\left\{\rho^2 \sin^2 p + \sin^2 p \left(\frac{d\rho}{dp}\right)^2 + \left(\frac{d\rho}{dq}\right)^2\right\}}\, \rho\, dp\, dq.$$

In the present case this becomes $2\rho \sin p\, dp\, dq$.

And we know that $\sqrt{(2-2\lambda)}$ expresses the same thing as ρ, that is the distance between (μ, ϖ) and (μ', ϖ'), that is the distance between the new origin and (ρ, p, q).

Hence, finally, $\iint \frac{Y' d\mu' d\varpi'}{\sqrt{(2-2\lambda)}}$ transforms to $\iint \frac{Y' 2\rho \sin p\, dp\, dq}{\rho}$, that is to $2\iint Y' \sin p\, dp\, dq$.

And from the diagram we see that the limits of the integrations are $-\frac{1}{2}\pi$ and $\frac{1}{2}\pi$ for q, and 0 and π for p.

1259. We have now to connect the old variables with the new.

Let B denote the point (μ', ϖ') on the surface of the sphere, that is the point (ρ, p, q). Let M be the projection of B on the plane of (y, z), and N the projection of M on the axis of y. The straight lines AB and OB may be supposed to be drawn.

We shall project AB on OP in two ways, and equate the results.

First consider AB as made up of the components AO and OB. Thus we get as the projection

$$\cos\theta' - \cos\theta.$$

Next consider AB as made up of the components AN, NM, and MB. Thus we get as the projection

$$y \sin\theta - x\cos\theta.$$

Therefore $\qquad \cos\theta' - \cos\theta = y\sin\theta - x\cos\theta,$

so that $\qquad \cos\theta' = (1-x)\cos\theta + y\sin\theta.$

Put for x and y their values in terms of p and q; hence we get

$$\mu' = \mu \cos^2 p - \sin^2 p \cos(2q + \theta) \quad\ldots\ldots\ldots\ldots\ldots\ldots(4).$$

Again $\varpi' - \varpi$ is the angle between the planes POA and POB. Hence the perpendicular from B on the plane POA is $\sin\theta' \sin(\varpi' - \varpi)$; and this perpendicular is equal to MN, so that

$$\sin\theta' \sin(\varpi' - \varpi) = z,$$

that is $$\sin\theta' \sin(\varpi' - \varpi) = 2\sin p \cos p \cos q \quad\ldots\ldots\ldots\ldots(5).$$

The equations (4) and (5) are theoretically sufficient to connect the old variables with the new; but another equation will also be useful in some cases, namely,

$$\sin\theta' \cos(\varpi' - \varpi) = \sin\theta - y\cos\theta - x\sin\theta.$$

This may be obtained thus: suppose B projected on the plane POA, let L denote the projection; so that L is in fact the point $(x, y, 0)$. Then the equation just written may be obtained by two ways of projecting OL on a straight line at right angles to OP; in one way OL is projected immediately, and in the other way it is made up of the components OA, AN, and NL.

The equation becomes by putting for x and y their values

$$\sin\theta' \cos(\varpi' - \varpi) = \sin\theta - 2\sin^2 p \cos q \sin(q + \theta) \quad\ldots\ldots(6).$$

The equation (6) is not independent of (4) and (5); it will be found that if we square and add (5) and (6) so as to eliminate $\varpi' - \varpi$ we obtain a result which is equivalent to (4).

1260. Laplace himself does not give equation (5) nor equation (6); because he does not require them. He begins by assuming that the figure of the solid required is to be one of revolution; and afterwards gives a supplementary investigation for the case in which the figure is not assumed to be one of revolution. Thus in this part of his process Y' is a function of μ' only; but it is convenient for us here to take the most general supposition, namely, that Y' is a function of ϖ' as well as μ'.

1261. In (4) put $2q + \theta = q'$; thus

$$\mu' = \mu \cos^2 p - \sin^2 p \cos q' \quad\ldots\ldots\ldots\ldots\ldots\ldots\ldots(7).$$

Hence from the result of Art. 1258, we have

$$\int_{-1}^{1}\int_{0}^{2\pi}\frac{Y'\,d\mu'\,d\varpi'}{\sqrt{(2-2\lambda)}}=\int_{0}^{\pi}\int_{-\pi+\theta}^{\pi+\theta}Y'\sin p\,dp\,dq'\text{.........(8).}$$

Thus it will be seen that (8) does not quite agree with (2) because the limits of q' are not the same in the two formulæ.

If with Laplace we assume that Y' is a function of μ' only, then by substituting for μ' by (7) we make Y' a function of p and of q' such that q' enters through $\cos q'$. In this case by the first principles of the Integral Calculus the limits for q' may be any that just comprise the range 2π, and so we may take them to be 0 and 2π. Hence if Y' is a function of μ' only the formula (2) is established.

But when Y' is a function of ϖ' as well as of μ' it does not seem to me that this passage from (8) to (2) can be always effected. In this case we must express ϖ' in terms of the new variables by (5) and (6). Hence we get by division

$$\varpi'-\varpi=\tan^{-1}\left\{\frac{2\sin p\cos p\cos q}{\sin\theta-2\sin^2 p\cos q\sin(q+\theta)}\right\}.$$

Thus when we put $q=\frac{q'}{2}-\frac{\theta}{2}$, we shall introduce $\sin\frac{q'}{2}$ and $\cos\frac{q'}{2}$; and we cannot assert that the limits of the integration with respect to q' can be changed from $-\pi+\theta$ and $\pi+\theta$ to 0 and 2π.

Although the truth of (2) is asserted by Liouville and admitted by Poisson, even in the case in which Y' is a function of ϖ' as well as μ', yet for the reason just given the result seems to me inadmissible. See pages 212 and 312 of the volume cited in Art. 1255.

1262. Admitting then the truth of (2) on Laplace's supposition, let us see how he applies it.

Let n denote the centrifugal force at the distance unity from the axis. Then for relative equilibrium we must have

$$V+\frac{a^2}{2}n(1-\mu^2)=\text{constant.}$$

But from (1) and (2) we have

$$V = \frac{4\pi a^3}{3}(1 - \alpha Y) + a^3\alpha \int_0^\pi \int_0^{2\pi} Y' \sin p\, dp\, dq' \ldots\ldots\ldots\ldots (9).$$

Hence, dividing by a^2, we get

$$\frac{4\pi}{3}\alpha Y - \alpha \int_0^\pi \int_0^{2\pi} Y' \sin p\, dp\, dq' - \frac{\kappa}{2}(1 - \mu^2) = C \ldots\ldots (10),$$

where C is a constant.

Differentiate (10) three times with respect to μ, and observe that by (7) we have $\frac{d\mu'}{d\mu} = \cos^2 p$. Thus

$$\frac{4\pi}{3}\frac{d^3 Y}{d\mu^3} - \int_0^\pi \int_0^{2\pi} \frac{d^3 Y'}{d\mu'^3} \sin p \cos^6 p\, dp\, dq' = 0,$$

that is

$$\int_0^\pi \int_0^{2\pi} \left(\frac{7}{3}\frac{d^3 Y}{d\mu^3} - \frac{d^3 Y'}{d\mu'^3}\right) \sin p \cos^6 p\, dp\, dq' = 0.$$

Laplace then says:

Cette équation doit avoir lieu, quel que soit μ; or il est clair que parmi toutes les valeurs comprises depuis $\mu = -1$, jusqu'à $\mu = 1$, il en existe une que nous désignerons par h, et qui est telle, qu'abstraction faite du signe, aucune des valeurs de $\frac{d^3 Y}{d\mu^3}$ ne surpassera pas celle qui est relative à h; en désignant donc par H, cette dernière valeur, on aura

$$\int_0^\pi \int_0^{2\pi} \left(\frac{7}{3} H - \frac{d^3 Y'}{d\mu'^3}\right) \sin p \cos^6 p\, dp\, dq' = 0.$$

La quantité $\frac{7}{3} H - \frac{d^3 Y'}{d\mu'^3}$ est évidemment du même signe que H, et le facteur $\sin p \cos^6 p$ est constamment positif dans toute l'étendue de l'intégrale; les élémens de cette intégrale sont donc tous du même signe que H; d'où il suit que l'intégrale entière ne peut être nulle, à moins que H ne le soit lui-même, ce qui exige que l'on ait généralement, $0 = \frac{d^3 Y}{d\mu^3}$, d'où l'on tire en intégrant,

$$Y = l + m\mu + n\mu^2;$$

l, m, n, étant des constantes arbitraires.

1203. This reasoning, says Liouville, is specious and might at the first glance deceive us; but on reflecting we see that it ceases to be applicable if the maximum of the function $\frac{d^n Y}{d\mu^n}$ could be infinite; and this would be the case if for example $Y = (1 - \mu^2)^{\frac{1}{2}}$ or $(1 - \mu^2)^{\frac{1}{3}}$.

In order to manifest the unsoundness of the principle on which this reasoning rests, Liouville takes a very simple example. Suppose it required to find a function $\phi(x)$ which satisfies the equation

$$\int_0^1 \phi(ax)\, da = \frac{3}{10}\phi(x) \text{ .... (11)},$$

x being an independent variable.

Differentiate three times with respect to x; thus

$$\int_0^1 a^3\phi'''(ax)\, da = \frac{3}{10}\phi'''(x),$$

or, which is the same thing,

$$\int_0^1 a^3 \left\{\frac{12}{10}\phi'''(x) - \phi'''(ax)\right\} da = 0.$$

Now if we apply to this equation and the function $\phi(x)$ the reasoning of Laplace, without changing a word, we shall arrive as before at the result $\phi'''(x) = 0$. But this is absurd, for the value of $\phi(x)$ which satisfies (11) is obviously of the form $\phi(x) = Ax^{\frac{7}{3}}$, where A is an arbitrary constant.

Similarly by putting the equation

$$\int_0^1 \phi(ax)\, da = 2\phi(x) \text{ .... (12)}$$

in the form

$$\int_0^1 [2\phi(x) - \phi(ax)]\, da = 0,$$

we might conclude by Laplace's reasoning that $\phi(x) = 0$. But it is obvious that (12) is satisfied by the more general value

$$\phi(x) = \frac{A}{\sqrt{x}}.$$

Thus in order that Laplace's demonstration should be sufficient it would be necessary to shew that $\frac{d^2 Y}{d\mu^2}$ is always finite. But this would amount to imposing additional restriction on the value of Y, when the only restriction that ought to be used is that Y is to be always finite. Therefore some method must be employed very different from Laplace's.

1264. Accordingly Liouville gives another demonstration; it is rather long, but interesting and satisfactory.

He restricts himself to the case in which the figure is assumed to be one of revolution; for the remainder of the problem he refers to the supplementary investigation which, as I have stated in Art. 1260, Laplace himself gave.

As I have stated in Art. 1261, Liouville asserts the universal truth of (2); yet he confines himself, as Laplace did, to the case in which Y' is a function of μ' only; and thus his process is not affected by my objection to the universal truth of (2).

The process used by Liouville does not admit of any convenient abbreviation; and I must therefore leave the student to consult the original paper.

1265. I now pass to a paper by Poisson, which is entitled *Note relative à un passage de la Mécanique céleste;* it is given in Liouville's *Journal de Mathématiques*, Vol. II. 1837, pages 312...316. I shall translate this paper, for the investigation which it contains is so brief and simple, that it ought to form part of any standard treatise on the subject.

In the translation, I shall continue the numbering of the equations which has been already used in the Chapter.

1266. In the twenty-sixth section of the Third Book [of the *Mécanique Céleste*] the author proposes to demonstrate, without recourse to the reduction into a series, that a homogeneous fluid turning uniformly round a fixed axis has only a single figure of equilibrium which differs very little from a sphere. The objection which M. Liouville has urged against the generality of this demonstration is real; see the number of this *Journal* for the month of

June last: but the demonstration which he has substituted for that of the *Mécanique Céleste* is very complicated, and we may arrive more simply at the result by the following considerations which differ less from those which Laplace used.

I retain without stating it here, all the notation of the memoir of M. Liouville, and the equation (10) cited at the beginning of the second article, that is

$$C = \frac{4\pi\alpha}{3\rho} Y - \alpha \int_0^{\pi}\int_0^{2\pi} Y' \sin p \, dp \, dq' - \frac{1}{2}\kappa (1 - \mu^2) \ldots (10).$$

The radius vector r of any point of the surface is represented by

$$r = a(1 + \alpha Y).$$

The unknown quantity Y may be any function of the two variables denoted by μ and ϖ, provided it is always finite. We do not assume that the surface is one of revolution, or that Y is independent of the angle ϖ; nor do we assume that the fluid has its centre of gravity on the axis of rotation: we assume only that the figure differs very little from a sphere which would have its centre on this axis. The constant a may differ from the radius of the sphere equivalent in volume to the fluid, provided that the difference is of the order of smallness of α, the same as that of κ, and the square of which we neglect.

The rigorous condition of equilibrium consists in this, that the sum of the elements of the fluid divided by their respective distances from any point of the surface, together with the quantity $\frac{1}{2}a^2\kappa(1 - \mu^2)$, which arises from the centrifugal force at the point, should be constant. The part of this constant relative to the sphere of radius a and independent of the centrifugal force, is equal to $\frac{4\pi a^2}{3}$; the part relative to this force and to the non-sphericity of the fluid is $-a^2C$, where C is the constant of the preceding equation. If we denote by γ its complete value we have

$$\gamma = \frac{4\pi a^2}{3} - a^2 C.$$

Now for each possible figure of equilibrium this constant γ is evidently a determinate quantity, which cannot depend on the radius that we take for a, that is to say, on the difference between this radius and that of the sphere which is equivalent to the given volume of the fluid. The constant C then is indeterminate like this difference; so that for any value we may take for a, the preceding equation will determine the corresponding value of C; and conversely, if we take for C a value which is of the order of smallness of a, this equation will determine the radius a.

Suppose then
$$Y = l\mu + m\mu^2 + X,$$
l and m being undetermined constants, and X a new unknown function of μ and ϖ, the values of which are always finite. Let c denote the greatest of these values, and put
$$c - X = Z;$$
then the unknown quantity Z can never be negative, and the expression for Y will become
$$Y = c + l\mu + m\mu^2 - Z.$$

Substitute this in the equation (10). Let μ' denote what μ becomes in Y', then
$$\mu' = \mu \cos^2 p - \sin^2 p \cos q'.$$

Hence we shall have
$$\int_0^{\pi}\int_0^{2\pi} \mu' \sin p \, dp \, dq' = \frac{4\pi\mu}{3},$$
$$\int_0^{\pi}\int_0^{2\pi} \mu'^2 \sin p \, dp \, dq' = \frac{4\pi}{5}\left(\mu^2 + \frac{4}{3}\right);$$
and also we have
$$\int_0^{\pi}\int_0^{2\pi} \sin p \, dp \, dq' = 4\pi.$$

The result of the substitution then will be
$$C = \left(\frac{8\pi a}{15} m + \frac{1}{2}\kappa\right)\mu^2 - \frac{16\pi a}{15} m - \frac{8\pi a}{3} c - \frac{1}{2}\kappa$$
$$- \frac{4\pi a}{3} Z + a\int_0^{\pi}\int_0^{2\pi} Z \sin p \, dp \, dq'.$$

where Z' denotes what Z becomes in Y'.

Now since the constants m and C can be taken arbitrarily, we may suppose that

$$\frac{8\pi a}{15} m + \frac{1}{2} \epsilon = 0,$$

$$C = -\frac{16\pi a}{15} m - \frac{8\pi a}{3} c - \frac{1}{2} \epsilon \text{ .... } (13),$$

which reduces the preceding equation to

$$\int_0^\pi \int_0^{2\pi} Z' \sin p \, dp \, dq' - \frac{4\pi}{3} Z = 0,$$

which may be written in this form,

$$\int_0^\pi \int_0^{2\pi} \left(Z' - \frac{1}{3} Z\right) \sin p \, dp \, dq' = 0.$$

Now let h and k denote the values of μ and ϖ, which correspond to the least of all the possible values of Z; and denote by L the least value; for $\mu = h$, and $\varpi = k$, the last equation will become

$$\int_0^\pi \int_0^{2\pi} \left(Z' - \frac{1}{3} L\right) \sin p \, dp \, dq' = 0 \text{ .... } (14).$$

But it is evident that Z' or Z being by hypothesis a positive quantity or zero, the difference $Z' - \frac{1}{3} L$ is also positive or zero. Then as all the elements of the double integral have the same sign the double integral cannot be zero unless the factor $Z' - \frac{1}{3} L$ is zero; and this condition cannot be satisfied unless Z' or Z is constantly zero.

From the preceding equations we obtain

$$m = -\frac{15\epsilon}{16\pi a},$$

$$c = \frac{3\epsilon}{16\pi a} - \frac{3}{8\pi a} C.$$

Substitute the values of m and c in the expression for Y, suppressing the term Z; and put this expression in the value of r: thus

$$r = a\left\{1 + \frac{3(\kappa - 2C)}{16\pi} + al\mu - \frac{15\kappa}{16\pi}\mu^2\right\}.$$

This result involves the indeterminate constant al, which depends on the origin of the coordinates on the axis of rotation. We may make it disappear by a convenient displacement of this origin on this straight line; or if we please we may suppose it zero, and write

$$r = a\left\{1 + \frac{3(\kappa - 2C)}{16\pi} - \frac{15\kappa}{16\pi}\mu^2\right\}.$$

We can also without difficulty make the constants a and C disappear from the value of r. In fact let

$$a\left\{1 + \frac{3(\kappa - 2C)}{16\pi}\right\} = b\left(1 + \frac{5\kappa}{16\pi}\right) \text{..........} (15);$$

then neglecting the squares and the product of κ and C, and putting for brevity,

$$\frac{15\kappa}{16\pi} = n,$$

we shall have finally

$$r = b\left\{1 + n\left(\frac{1}{3} - \mu^2\right)\right\}.$$

It is easy to see from this expression for r that b is the radius of the sphere equivalent in volume to the fluid, and so is given. Thus there is nothing unknown or indeterminate in this expression, and we conclude that the fluid has only one possible figure of relative equilibrium which differs but little from a sphere: which was to be proved.

This demonstration is more simple than that which is based on the reduction of r to a series of a certain form, and which supposes the properties of the terms of this development to be known, as well as the generality of this form of series which had been contested, but which I have placed beyond question in my memoir on the *Attraction of Spheroids* in the *Additions à la Connaissance des Tems*, 1829.

If we put

$$a = b, \qquad \alpha Y = n\left(\frac{1}{3} - \mu^2\right),$$

in the expression $a(1 + \alpha Y)$ for r, which makes it coincide with the final expression for the radius vector; and denote by D the value of the constant C, which corresponds to these values of a and of αY; and have regard to what n represents, we shall find without difficulty that the equation (10) reduces to

$$D = -\frac{1}{3}\kappa.$$

We shall have at the same time

$$\gamma = \frac{4\pi b^3}{3} + \frac{1}{3}\kappa b^3.$$

As we have said above, this quantity γ ought to be the same whatever radius, differing little from b, we take for a; thus we must have

$$\frac{4\pi b^3}{3} + \frac{1}{3}\kappa b^3 = \frac{4\pi a^3}{3} - a^3 C.$$

This result coincides in fact with equation (15), neglecting always the squares and the product of κ and C.

1267. Such is Poisson's treatment of the problem: I shall make two remarks on it.

In the first place it will be seen that Poisson assumes the truth of equation (2), which I do not allow. But on examination it will be found that equation (8), which has been strictly demonstrated, will be sufficient for his purpose; so that no objection can be taken on this ground to the demonstration.

Secondly, the quantity which Poisson designates by L is simply zero; for he assumes c to be the greatest value of X, and so when X is equal to c, the value of Z is least, namely zero.

1268. I venture to propose the following demonstration, which though less decisive than Poisson's may be found worthy of study.

Let us restrict ourselves to the case of figures of revolution. Then take the equation which has been already established,

$$\frac{4\pi\rho}{3}Y - \rho\int_0^{\pi}\int_0^{2\pi} Y' \sin p\, dp\, dq' - \frac{\kappa}{2}(1 - \mu^2) = \text{constant}.$$

Assume $$Y = -\frac{15\kappa}{16a\pi}\mu^2 + Z;$$

thus we obtain

$$\frac{4\pi a}{3} Z - a\int_0^{\pi}\int_0^{2\pi} Z' \sin p\, dp\, dq' = \text{constant} \ldots\ldots (16).$$

Differentiate with respect to μ; thus

$$\frac{4\pi}{3}\frac{dZ}{d\mu} - \int_0^{\pi}\int_0^{2\pi}\frac{dZ'}{d\mu'}\cos^2 p \sin p\, dp\, dq' = 0,$$

which may be written

$$\int_0^{\pi}\int_0^{2\pi}\left(\frac{dZ}{d\mu} - \frac{dZ'}{d\mu'}\right)\cos^2 p \sin p\, dp\, dq' = 0.$$

But this is impossible unless $\frac{dZ}{d\mu}$ is a constant. For if $\frac{dZ}{d\mu}$ be not constant take θ so that $\frac{dZ}{d\mu}$ has its greatest value, then $\frac{dZ}{d\mu} - \frac{dZ'}{d\mu'}$ is never negative; and the definite double integral must have some positive value, and not be zero. Hence $\frac{dZ}{d\mu}$ must be a constant; so that $Z = l\mu + h$, where l and h are constant.

This shews that for surfaces of revolution the figure must be that of an oblatum, which becomes a sphere if $\kappa = 0$.

Then by Laplace's supplementary investigation the solution of the problem may be extended to the case in which the figure is not assumed to be one of revolution.

It may be objected to the above process that it is not quite satisfactory, for $\frac{dZ}{d\mu}$ might be infinite. It would I think be a sufficient answer to say that if $\frac{dZ}{d\mu}$ can be infinite then $\frac{dr}{d\mu}$ will be infinite; and there would be a singular line on the surface of the nature of a ridge or chasm, or a mountain or valley with vertical sides; but it might be shewn by general reasoning that in such cases there would not be relative equilibrium. Moreover the approximate value of V which we have used throughout cannot be held to be safely established unless we admit that $\frac{dr}{d\mu}$ is never infinite.

1269. But it may I think be shewn that a value of $\phi(\mu)$ which itself always remains finite, but allows $\frac{d\phi(\mu)}{d\mu}$ to be infinite, cannot satisfy the equation

$$\frac{4\pi}{3}\phi(\mu) = \int_0^{\pi}\int_0^{\pi}\phi(\mu\cos^2 p - \sin^2 p\cos q)\sin p\, dq\, dp + C \ldots (17),$$

where C is a finite constant. This equation is equivalent to (16).

Let μ_1 and μ_2 be particular values of μ; put u for

$$\frac{\phi(\mu_2) - \phi(\mu_1)}{\mu_2 - \mu_1},$$

and v for

$$\frac{\phi(\mu_2\cos^2 p - \sin^2 p\cos q) - \phi(\mu_1\cos^2 p - \sin^2 p\cos q)}{\mu_2 - \mu_1},$$

then we deduce from (17)

$$\frac{4\pi}{3}u = \int_0^{2\pi}\int_0^{\pi} v\sin p\, dq\, dp \ldots\ldots\ldots\ldots\ldots\ldots (18),$$

and this is true however small $\mu_2 - \mu_1$ may be.

Now I say that if $\frac{d\phi(\mu)}{d\mu}$ could be infinite, we could make the left-hand member of (18) incomparably greater than the right-hand member; which is absurd.

For suppose μ_1 to denote the value of μ for which $\frac{d\phi(\mu)}{d\mu}$ becomes infinite; then we can make u as large as we please, by taking $\mu_2 - \mu_1$ small enough. But v will be very large indeed only over a very small part of the range of integration; in fact an infinitesimal part. However let σ denote a small but finite part of the whole spherical surface, over which the integration with respect to v may be supposed to extend, namely, that part within which the very large values of v occur. Then v will be numerically less than u, except p should happen to be zero. Thus the corresponding part of the right-hand member of (18) may be denoted by $\lambda u\sigma$, where λ denotes some proper fraction.

The rest of (18) may be denoted by $(4\pi - \sigma)\, w$, where w denotes some value of v intermediate between the greatest and least of a set of values which are all finite.

Thus instead of (18) we have

$$\frac{4\pi}{3} u = \lambda u\sigma + (4\pi - \sigma)\, w \text{ .... (19).}$$

We can of course take σ far less than $\frac{4\pi}{3}$; and then as u increases indefinitely, the left-hand member of (19) is obviously far greater than the right-hand.

The same reasoning holds if we suppose that $\frac{d\phi(\mu)}{d\mu}$ can become infinite, twice, or thrice, or any finite number of times.

The argument may be much strengthened by observing that v when very large is not always of the same sign; it may be said roughly to be as often positive as negative. This *attenuates* extremely the values of λ in (19).

1270. We may now conveniently introduce Laplace's supplementary investigation to which we have alluded in Art. 1260. We have already arrived at the following result: if a fluid mass in the form of a figure of revolution, nearly spherical, rotating with uniform angular velocity is in relative equilibrium, the form must be that of an oblatum; and if there is no rotation the oblatum reduces to a sphere. The result is then to be extended to the case where the figure is not assumed to be one of revolution.

1271. Return to equation (10) corrected as in (8). Suppose that besides the value $a(1 + \alpha Y')$ which belongs to an oblatum, the equation may be satisfied by another value of the radius vector which we will denote by $a(1 + \alpha Y' + \alpha v')$, where v' is some function of θ' and ϖ'. Thus

$$\frac{4\pi\alpha Y}{3} - \alpha\int_0^\pi\int_{-\pi+\varpi}^{\pi+\varpi} Y' \sin p\, dp\, dq' - N = \text{constant},$$

where N is the expression arising, as in Art. 1262, from the centrifugal force, or other small given external forces.

Also

$$\frac{4\pi 2}{3}(Y+v) - a\int_0^{\pi}\int_{-\pi+\theta}^{\pi+\theta}(Y'+v')\sin p\,dp\,dq' - N = \text{constant}.$$

Hence, by subtraction,

$$\frac{4\pi 2v}{3} - \int_0^{\pi}\int_{-\pi+\theta}^{\pi+\theta} v'\sin p\,dp\,dq' = \text{constant}.........\ (20).$$

Laplace himself does not state what the limits of the integration for q' are; I give them in accordance with my remarks in Art. 1261.

The equation (20) is obviously that of a homogeneous spheroid in equilibrium, of which the radius vector is $a(1+\alpha v)$, and in which there is no force acting besides the attractions of the molecules. Now as the equation is satisfied whatever be the value of ϖ by the radius vector $a(1+\alpha v)$, we may in this radius vector change ϖ successively into $\varpi + d\varpi$, $\varpi + 2d\varpi$, ... Denote by v_1, v_2, ... what v becomes by reason of these successive changes. Then the equation will also be satisfied by the radius vector

$$a\{1 + \alpha v_1 d\varpi + \alpha v_2 d\varpi + \alpha v_3 d\varpi + ...\}.$$

Hence we may take for the radius vector $a\left(1 + \alpha\int_0^{2\pi} v\,d\varpi\right)$, which will be the radius vector of a surface of revolution. Now as we have shewn in Art. 1268, this must be the surface of a sphere. Let us see what v must consequently be.

Let a be the shortest distance of the centre of gravity of the spheroid of which the radius vector is $a(1+\alpha v)$ from the surface; and let us place the pole, that is the origin of the angle θ, at the extremity of the shortest distance. Then v will be zero at the pole, and positive at every other place; and the same will be true for $\int_0^{2\pi} v\,d\varpi$. Now as the centre of gravity of the spheroid which has the radius vector $a(1+\alpha v)$ is at the centre of the sphere of radius a, so also is the centre of gravity of the spheroid of which the radius vector is $a\left(1 + \alpha\int_0^{2\pi} v\,d\varpi\right)$. The radii vectores drawn from the centre to the surface of the last spheroid are therefore unequal if v is not zero; and so the surface cannot be that of a sphere, unless v is zero. Hence we are certain that a homogeneous fluid which is acted on by very small external forces, and is nearly spherical, can be in equilibrium in only one way.

I may observe that there is an awkward note on Bowditch's page 275, numbered 1183. To make his process sound, instead of $a\left(1+\alpha\int_0^{2\pi} v d\varpi\right)$ for the radius he should use $a\left(1+\frac{\alpha}{2\pi}\int_0^{2\pi} v d\varpi\right)$.

1272. Another paper on the subject is given in the fourth volume of the *Journal de Mathématiques* 1839. This is entitled *Extrait d'une Lettre de M. Wantzel à M. Liouville*; it occupies pages 185...188 of the volume.

The paper commences thus:

Je me suis occupé de nouveau de la question d'Analyse qui a pour but de déterminer la figure d'équilibre d'une masse fluide soumise aux attractions de ses particules et animée d'une vitesse constante de rotation, lorsqu'on suppose cette figure peu différente de la sphère. Je crois avoir levé l'objection que vous avez faite à la seconde méthode de Laplace dans le tome II. du *Journal de Mathématiques* (juin 1837), ou plutôt avoir rendu cette méthode rigoureuse par une légère modification.

Wantzel's paper is not very clear, and it does not seem to me satisfactory. He introduces into some of the expressions a variable factor which is ultimately equal to unity; but the step seems to me not justified. The paper is not of sufficient interest to warrant me in devoting more space to it.

1273. It may be useful to give the process by which equation (8) can be obtained from the ordinary formulæ for the transformation of double integrals.

Take the two equations

$$\mu' - \mu\cos^2 p + \sin^2 p\cos(2q+\theta) = 0,$$

$$\sqrt{(1-\mu^2)}\sin(\varpi'-\varpi) - 2\sin p\cos p\cos q = 0;$$

denote the former by $F_1 = 0$, and the latter by $F_2 = 0$.

Then we know that the transformations consist in replacing $d\mu' d\varpi'$ by

$$\frac{\dfrac{dF_1}{dp}\dfrac{dF_2}{dq} - \dfrac{dF_1}{dq}\dfrac{dF_2}{dp}}{\dfrac{dF_1}{d\mu'}\dfrac{dF_2}{d\varpi'} - \dfrac{dF_1}{d\varpi'}\dfrac{dF_2}{d\mu'}}\, dp\, dq.$$

Now
$$\frac{dF_1}{dp} = \sin 2p\,\{\cos(2q+\theta) + \mu\},$$

$$\frac{dF_2}{dq} = \sin 2p \sin q,$$

$$\frac{dF_1}{dq} = -2\sin^2 p \sin(2q+\theta),$$

$$\frac{dF_2}{dp} = -2\cos 2p \cos q\,;$$

therefore
$$\frac{dF_1}{dp}\frac{dF_2}{dq} - \frac{dF_1}{dq}\frac{dF_2}{dp}$$

$$= \sin^2 2p \sin q\,\{\cos(2q+\theta) + \mu\} - 4\sin^2 p \cos 2p \cos q \sin(2q+\theta).$$

In this, the term involving $\cos\theta$

$$= 4\sin^2 p\,\{(1+\cos 2q)\sin q \cos^2 p - \cos 2p \cos q \sin 2q\}\cos\theta$$

$$= 8\sin^2 p \cos^2 q \sin q\,(\cos^2 p - \cos 2p)\cos\theta$$

$$= 8\sin^4 p \cos^2 q \sin q \cos\theta$$

$$= 2\rho^2 \sin^2 p \sin q \cos\theta.$$

The term involving $\sin\theta$

$$= -4\sin^2 p\,\{\cos^2 p \sin q \sin 2q + \cos 2p \cos q \cos 2q\}\sin\theta$$

$$= -4\sin^2 p\,\{\cos^2 p \cos q - \sin^2 p \cos q \cos 2q\}\sin\theta$$

$$= -4\sin^2 p \cos q \sin\theta + 8\sin^4 p \cos^3 q \sin\theta$$

$$= -2\rho \sin p \sin\theta + 2\rho^2 \sin^2 p \cos q \sin\theta.$$

Thus

$$\frac{dF_1}{dp}\frac{dF_2}{dq} - \frac{dF_1}{dq}\frac{dF_2}{dp} = -2\rho \sin p\,\{\sin\theta - \rho \sin p \sin(q+\theta)\}$$

$$= -2\rho \sin p \sin\theta' \cos(\varpi' - \varpi) \text{ by equation (6).}$$

And $$\frac{dF_1}{d\mu'}\frac{dF_2}{d\varpi'}-\frac{dF_1}{d\varpi'}\frac{dF_2}{d\mu'}=\sqrt{(1-\mu'^2)}\cos(\varpi'-\varpi).$$

Hence $d\mu'\,d\varpi'$ transforms to

$$\frac{-2\rho\sin p\sin\theta'\cos(\varpi'-\varpi)}{\sqrt{(1-\mu'^2)}\cos(\varpi'-\varpi)}\,dp\,dq,$$

that is to $$-2\rho\sin p\,dp\,dq.$$

And as $\sqrt{(2-2\lambda)}=\rho$ it follows that

$$\frac{d\mu'\,d\varpi'}{\sqrt{(2-2\lambda)}}\text{ transforms to }\frac{-2\rho\sin p\,dp\,dq}{\rho},$$

that is to $$-2\sin p\,dp\,dq.$$

CHAPTER XXXIII.

LAPLACE'S MEMOIRS.

1274. LAPLACE published various memoirs during the first quarter of the present century. We shall here give a brief notice of them, reserving a fuller account for the next Chapter, which will be devoted to the fifth volume of the *Mécanique Céleste*, where most of the investigations are reproduced.

1275. A paper entitled *Sur l'anneau de Saturne* occurs on pages 450...453 of the *Connaissance des Tems* for 1811, which was published in 1809. The paper occurs also on pages 426...428 of the *Nouveau Bulletin...par la Société Philomatique*, Vol. I. 1807.

Laplace adverts to the results which he had obtained in the third Book of the *Mécanique Céleste* as to Saturn's ring. He says that the fact of the rotation of the ring in about ten hours and a half had been suggested by himself from theory and confirmed by Herschel's observations. But some observations by Schroeter seemed to throw doubt on the fact of the rotation. Laplace makes some remarks with a view of explaining the discrepancy; that is, he accounts for Schroeter's phenomena without giving up the rotation.

Laplace gives some account of the labours of astronomers and mathematicians on Saturn's ring in the *Mécanique Céleste*, Vol. V. pages 288...291. He says nothing about Schroeter in these pages.

1276. A paper entitled *Sur la Rotation de la Terre* occurs on pages 53...60 of the *Annales de Chimie*... Vol. VIII., published in 1818. It is said to have been read to the Academy of Sciences on the 16th of May, 1818.

This paper constituted the preamble to a memoir with the same title printed in the *Connaissance des Tems* for 1821; it is entirely reproduced there.

1277. A paper entitled *Sur la Figure de la Terre, et la Loi de la pesanteur à sa surface*, occurs on pages 313...318 of the same volume of the *Annales de Chimie...* It is said to have been read to the Academy of Sciences on the 3rd of August, 1818.

The paper is also printed on pages 122...125 of the *Bulletin... par la Société Philomatique* for 1818. This seems to have been the preamble to the memoir which we shall notice in Art. 1280, and which was reproduced in the second Chapter of the Eleventh Book of the *Mécanique Céleste*. The preamble is not however reproduced with the memoir, though much of it is in the pages 11...16 of the first Chapter of the Eleventh Book of the *Mécanique Céleste*. Still there are some remarks in this paper which do not reappear in the *Mécanique Céleste*. For instance, take the following sentence:

Dans le nombre infini des figures que comprend l'expression analytique des surfaces de la mer et du sphéroïde terrestre, on peut en choisir une qui représente l'élévation et les contours des continens et des îles: ainsi, je trouve qu'un petit terme du troisième ordre, ajouté à la partie elliptique du rayon terrestre, suffit pour rendre conformément à ce que l'observation semble indiquer, la mer plus profonde et plus étendue vers le pôle austral que vers le pôle boréal, et même pour laisser ce dernier pôle à découvert.

There is nothing about this matter in the memoir.

1278. A memoir entitled *Sur la rotation de la Terre* occurs on pages 242...259 of the *Connaissance des Tems* for 1821, which was published in 1819. About half this memoir concerns us; the rest relates to the movement of the plane of the Earth's equator and to the movement of the plane of the Moon's orbit.

1279. The following interesting paragraph forms part of the preamble:

Le système du sphéroïde terrestre et des fluides qui le recouvrent, est troublé par les actions du Soleil et de la Lune, qui changent continuellement la position de son équateur. L'explication de ce changement observé sous les noms de *précession* et de *nutation* est, à mon sens, le résultat le plus frappant et le moins attendu de la découverte de la pesanteur universelle. Les anciens avaient bien connu que la cause du flux et du reflux de la mer réside dans ces deux astres. Képler avait conclu de ce phénomène et des lois des mouvemens célestes, l'attraction mutuelle de toutes les parties de la matière. Mais personne, avant Newton, n'avait soupçonné la cause de la précession des équinoxes, cause d'autant plus cachée, qu'elle dépend de l'aplatissement de la Terre, inconnu jusqu'alors. La manière dont ce grand géomètre a déduit la précession, de l'ellipticité du sphéroïde terrestre et de la théorie du mouvement rétrograde des nœuds de l'orbe lunaire, deux choses qu'il avait tirées de sa découverte; cette manière, dis-je, quoiqu'inexacte à plusieurs égards, est un des plus beaux traits de son génie.

1280. The part of the memoir with which we are concerned contains results which are reproduced in the fifth Volume of the *Mécanique Céleste*: see the pages 16, 17 and 57...67.

The mathematical investigation however is much simpler in the memoir than in the *Mécanique Céleste*; but I think not less satisfactory.

1281. Laplace establishes the following theorem: Suppose the density of every stratum of the Earth to be diminished by the density of the sea; take one of the principal axes of this imaginary spheroid passing through its centre of gravity; let the Earth be supposed to rotate uniformly round this axis; then supposing the sea to be in relative equilibrium, this axis will be a principal axis of the whole mass of the Earth and sea, and the centre of gravity of the imaginary spheroid will also be the centre of gravity of the whole mass. The sea is assumed to cover the whole surface of the Earth.

We will now give the mathematical investigation.

1282. Take the principal axis of the imaginary spheroid for the axis of rotation. Let V denote the potential of the Earth, and V' that of the sea at a point on the surface of the sea, which has for polar coordinates the usual r and μ. Let ω be the angular velocity of rotation.

Then for relative equilibrium we must have

$$V + V' + \frac{\omega^2 r^2}{2}(1 - \mu^2) = \text{constant} \;\ldots\ldots\ldots\ldots\; (1).$$

Let the radius vector of any stratum of the Earth be $a(1 + \alpha y)$, and suppose that the density ρ is a function of a. Let y be expanded in a series of Laplace's functions denoted by

$$Y_1 + Y_2 + Y_3 + \ldots\ldots$$

Let M denote the mass of the Earth. Then

$$V = \frac{M}{r} + \frac{4\pi\alpha}{r}\int_0^a \rho \frac{d}{da}\left\{\frac{a^4 Y_1}{3r} + \frac{a^5 Y_2}{5r^2} + \frac{a^6 Y_3}{7r^3} + \ldots\right\} da,$$

where the upper limit of the integral denotes the value of a at the surface of the solid part.

Suppose σ the density of the sea, and let the radius vector of the surface of the sea be denoted by $b(1 + \alpha z)$; and suppose z expanded in a series of Laplace's coefficients $Z_1 + Z_2 + Z_3 + \ldots$ Let $\bar{y}$ denote the value of y at the surface of the solid part. Then V' may be found by considering the sea to be the difference between two homogeneous spheroids of density σ, one having for radius vector $b(1 + \alpha z)$, and the other having for radius vector $a(1 + \alpha\bar{y})$.

Thus if M' denote the mass of the sea

$$V' = \frac{M'}{r} + \frac{4\pi\alpha\sigma}{r}\left\{\frac{b^4 Z_1 - a^4 \overline{Y_1}}{3r} + \frac{b^5 Z_2 - a^5 \overline{Y_2}}{5r^2} + \ldots\right\}.$$

Hence (1) may be expressed thus:

$$\frac{M + M'}{r} + \frac{4\pi\alpha}{r}\int_0^a (\rho - \sigma)\frac{d}{da}\left\{\frac{a^4 Y_1}{3r} + \frac{a^5 Y_2}{5r^2} + \frac{a^6 Y_3}{7r^3} + \ldots\right\} da$$

$$+ \frac{4\pi\alpha\sigma}{r}\left\{\frac{b^4 Z_1}{3r} + \frac{b^5 Z_2}{5r^2} + \frac{b^6 Z_3}{7r^3} + \ldots\right\}$$

$$- \frac{\omega^2 r^2}{2}(\mu^2 - 1) = \text{constant}.$$

Then approximate, rejecting α^2; thus

$$-\frac{M+M'}{b}\alpha\,(Z_1+Z_2+Z_3+\dots)$$

$$+\frac{4\pi}{b}\int_0^a(\rho-\sigma)\frac{d}{da}\left\{\frac{a^4Y_1}{3b}+\frac{a^5Y_2}{5b^2}+\frac{a^6Y_3}{7b^3}+\dots\right\}da$$

$$+4\alpha\pi b^2\sigma\left\{\frac{Z_1}{3}+\frac{Z_2}{5}+\frac{Z_3}{7}+\dots\right\}$$

$$-\frac{\omega^2b^2}{2}\left(\mu^2-\frac{1}{3}\right)=\text{constant}\ \dots\dots\dots\dots\dots\dots\ (2).$$

Then in the usual way we equate to zero the aggregate of Laplace's functions of each order. Now by supposition the origin is at the centre of gravity of the imaginary spheroid; hence by Livre III. § 31, we have

$$\int_0^a(\rho-\sigma)\frac{d}{da}(a^4Y_1)\,da=0,$$

and therefore from (2) we have $Z_1=0$.

Thus the origin is also the centre of gravity of the entire mass; for the condition that it should be so is

$$\int_0^a(\rho-\sigma)\frac{d}{da}(a^4Y_1)\,da+b^4\sigma Z_1=0,$$

and this condition is satisfied.

Next we have from (2)

$$\alpha\left\{\frac{4\pi b^2\sigma}{5}-\frac{M+M'}{b}\right\}Z_2+\frac{4\pi}{5b^3}\int_0^a(\rho-\sigma)\frac{d}{da}(a^5Y_2)\,da$$

$$-\frac{\omega^2b^2}{2}\left(\mu^2-\frac{1}{3}\right)=\text{constant}\ \dots\dots\dots\dots\dots\ (3).$$

And we have other equations by which in general Z_n is made to depend on Y_n; and so the figure of the sea necessary for relative equilibrium follows from the figure of the solid part when this is given.

Now we have supposed that the axis of rotation is a principal axis at the origin for the imaginary spheroid; hence it follows by Livre III. § 32, that $\int_0^a (\rho-\sigma)\frac{d}{da}(a^5 Y_2)\,da$ must be of the form

$$H\left(\mu^2-\frac{1}{3}\right)+H'(1-\mu^2)\cos 2\phi,$$

where H and H' are constants, and ϕ is the third polar coordinate of the point to which r and μ belong.

Hence by (3) it follows that

$$\int_0^a (\rho-\sigma)\frac{d}{da}(a^5 Y_2)\,da + b^5\sigma Z_2$$

is of the same form; and this ensures that the axis of rotation is a principal axis at the origin for the whole mass.

1283. A memoir entitled *Sur la loi de la pesanteur, en supposant le sphéroïde terrestre homogène et de même densité que la mer*, occurs on pages 284...290 of the *Connaissance des Tems* for 1821, which was published in 1819. This memoir is entirely embodied in that which I notice in Article 1280, and which was reproduced in the *Mécanique Céleste*, Livre XI. Chapitre II.

The memoir may be considered to be summed up in the formula which Laplace gives on page 40 of his fifth volume and calls the *expression remarquable*.

In the memoir itself Laplace demonstrated the formula on the supposition that the Earth has the same density as the sea. But in an Addition to the memoir on page 353 of the volume, Laplace states that it is true whatever may be the ratio of the density of the sea to that of the Earth, supposed homogeneous. Also he here states the six results which he draws from theory and observation, as he does at the commencement of the memoir noticed in Art. 1280. I quote these six results in my account of the *Mécanique Céleste*, Vol. V. Chapter II.: see Art. 1301.

1284. A paper entitled *Sur la Figure de la Terre* occurs on pages 97...100 of the *Bulletin...la Société Philomatique* for 1819.

This is the preamble to the memoir which we notice in the next Article, and is reproduced in that memoir.

1285. A memoir entitled *Sur la Figure de la Terre* occurs in pages 284...293 of the *Connaissance des Tems* for 1822, which was published in 1820. The memoir is said to have been read before the *Bureau des Longitudes* on the 26th of May, 1819.

The memoir is the same as that which appeared in the *Mémoires de l'Académie*...for 1818, under the title of an *Addition* to a memoir in the preceding volume of the Academy.

1286. A memoir by Laplace entitled *Mémoire sur la Figure de la Terre* is contained in the *Mémoires de l'Académie*... for 1817, published in 1819. The memoir occupies pages 137...184 of the volume. The memoir was read on August 4th, 1818.

An *Addition* to the memoir occupies pages 489...502 of the *Mémoires de l'Académie*...for 1818, published in 1820.

The memoir and the addition are substantially reproduced as the second Chapter of the Eleventh Book of the *Mécanique Céleste;* and will be discussed by us hereafter.

1287. A paper entitled *Sur la Diminution de la durée du jour par le refroidissement de la Terre* occurs on pages 410...417 of the *Annales de Chimie*, Vol. XIII., which was published in 1820.

This consists of the preamble of a memoir under the same title printed in the *Connaissance des Tems* for 1823, together with a sketch of the analysis employed. The preamble is reproduced almost identically in the memoir.

The paper concludes thus:

Je développerai dans la connaissance des temps de 1823, cette analyse, son extension aux sphéroïdes peu différens d'une sphère, et son application à la diminution de la durée du jour par le refroidissement de la terre.

But this intention was not completely carried out, for there is nothing in the memoir as to the extension of the analysis to spheroids differing but little from spheres.

1288. There is an addition to the preceding paper entitled *Addition au Mémoire sur la Diminution de la durée du jour par le refroidissement de la Terre, inséré dans le Cahier des Annales du mois d'avril* 1820. This addition occurs in pages 315 and 316 of the *Annales de Chimie*... Vol. XIV., which was published in 1820.

This addition gives a formula which it says will be demonstrated in the "*Connaissance des Temps* de 1823, qui paraîtra incessamment." The demonstration is contained in the pages 324...327 of the volume.

1289. A memoir entitled *Sur la Diminution de la durée du jour, par le refroidissement de la Terre,* occurs on pages 245...257 of the *Connaissance des Tems* for 1823, which was published in 1820. There is an *Addition* to the memoir on pages 324...327 of the volume.

The mathematical part of the memoir is reproduced with some additions in the fourth Chapter of the Eleventh Book of the *Mécanique Céleste;* we shall speak of it hereafter. The preamble of the memoir is reproduced substantially in pages 18...21 of the first Chapter of the Eleventh Book.

1290. In the preamble, as given in the *Connaissance des Tems,* Laplace after stating that the heat increases as we penetrate into the Earth, adds the following sentence:

...C'est ce que M. Daubuisson a fait voir, dans son excellent traité de Géognosie. MM. les rédacteurs des Annales de Chimie et de Physique, ont confirmé ce résultat, en ajoutant beaucoup d'observations, à celles que M. Daubuisson avait rapportées.

This sentence does not occur in the paper which I have noticed in Art. 1287, nor in the *Mécanique Céleste.*

In the paper which is noticed in Art. 1287, Laplace says that he had determined a certain constant by means of the annual variations of temperatures at different depths; and he refers then to observations made at the Observatory of Paris at the depth of 28 metres. In the memoir Laplace instead of these observations refers to experiments made by Saussure; and so also in the *Mécanique Céleste*, Vol. v. page 20.

On the page just cited there is a sentence which does not occur in the present memoir nor in the paper noticed in Art. 1287. Laplace is referring to the law of the diminution of heat from the centre to the surface of the Earth, and he says:

La loi dont il s'agit, que j'ai publiée en 1810 dans le recueil de la Connaissance des Tems, et que M. Poisson a confirmée depuis par une savante analyse,...

1291. A paper entitled *Sur la Densité moyenne de la Terre* occurs on pages 410...416 of the *Annales de Chimie...* Vol. XIV., which was published in 1820; it occurs also on pages 328...331 of the *Connaissance des Tems* for 1823, which was published in 1820.

There are no mathematical investigations. Laplace refers to the two operations undertaken for determining the mean density of the Earth; namely, that connected with the mountain Schehallien in Scotland, in which Maskelyne, Hutton, and Playfair were concerned; and the experiment suggested by Michell and executed by Cavendish. He considers that we may on the whole regard the density of the Earth to be about 5·48 times that of water.

1292. The paper is interesting; but I do not see why it appeared so long after the observations and experiments to which it refers. The following extracts from it may be read with pleasure.

The paper begins thus:

Un des points les plus curieux de la Géologie, est le rapport de la moyenne densité du sphéroïde terrestre à celle d'une substance connue. Newton, dans ses Principes mathématiques de la Philosophie naturelle, a donné le premier aperçu que l'on ait publié sur cela. Cet admirable Ouvrage contient les germes de toutes les grandes découvertes qui ont été faites depuis sur le système du monde: l'histoire de leur développement par les successeurs de ce grand géomètre, serait à la fois le plus utile commentaire de son Ouvrage, et le meilleur guide pour arriver à de nouvelles découvertes. Voici le passage de cet Ouvrage, sur l'objet dont il s'agit, tel qu'il se trouve dans la première édition et dans les suivantes :

Laplace then gives a translation of part of the tenth Proposition of Newton's Third Book: see Art. 17.

In referring to the operations at Schehallien Laplace describes Hutton as

...géomètre illustre, auquel les Sciences mathématiques sont redevables d'ailleurs d'un grand nombre de recherches importantes.

With respect to Cavendish's experiment Laplace says:

...En examinant avec une scrupuleuse attention, l'appareil de M. Cavendish et toutes ses expériences faites avec la précision et la sagacité qui caractérisent cet excellent physicien, je ne vois aucune objection à faire à son résultat qui donne 5·48 pour la densité moyenne de la Terre...

The paper concludes thus:

Ces expériences et ces observations mettent en évidence l'attraction réciproque des plus petites molécules de la matière, en raison des masses divisées par le carré des distances. Newton l'avait conclue du principe de l'égalité de l'action à la réaction, et de ses expériences sur la pesanteur des corps, qu'il trouva, par les oscillations du pendule, proportionnelle à leur masse. Malgré cette preuve, Huyghens, fait plus qu'aucun autre contemporain de Newton pour bien l'apprécier, rejeta cette attraction de la matière, de molécule à molécule, et l'admit seulement entre les corps célestes; mais sous ce dernier rapport, il rendit aux découvertes de Newton la justice qui leur était due. Au reste, la gravitation universelle n'avait pas pour les contemporains de Newton, et pour Newton lui-même, toute la certitude que les progrès des Sciences mathématiques, qui lui sont dus principalement, et les observations subséquentes lui ont donnée; et l'on peut justement appliquer à cette découverte, la plus grande qu'ait faite l'esprit humain, ces paroles de Cicéron : *opinionum commenta delet dies, naturae judicia confirmat.*

CHAPTER XXXIV.

FIFTH VOLUME OF THE *MÉCANIQUE CÉLESTE.*

1293. The fifth volume of the *Mécanique Céleste* was published in 1825. The volume consists of historical sketches of the progress of physical astronomy, and of various investigations which Laplace had made since the date of his former volumes, and had published in the Paris *Mémoires* and in the *Connaissance des Tems.*

1294. We are concerned with the Eleventh Book, which is entitled *De la Figure et de la Rotation de la Terre;* this extends from the beginning of the volume to page 85: it is divided into four Chapters.

1295. The first Chapter of the Eleventh Book is entitled *Notice historique des travaux des géomètres sur cet objet:* this occupies pages 2...21 of the volume. About half of the Notice is devoted to the period extending to the date of the second volume of the *Mécanique Céleste;* the other half gives an analysis of the results which Laplace himself had since obtained in various investigations which are reproduced in the following three Chapters of the Eleventh Book.

1296. In his pages 11...16 Laplace states the results obtained by him in the investigations which are reproduced in the second Chapter of the Eleventh Book. We may say in general terms that Laplace considers the hypotheses involved in his mathematical theory of the Figure of the Earth to be well confirmed by experiment and observation.

Some remarks on pages 14 and 15 may be noticed as specially interesting. Laplace will not admit that there has ever been any considerable displacement of the poles of the Earth. He refers to the elephant which had been found with his flesh well preserved in a mass of ice; and says

La découverte de cet animal a donc confirmé ce que la théorie mathématique de la Terre nous apprend...

This was a greatly honoured beast, to whom it was given to corroborate some of the profoundest investigations of the first of modern physical astronomers.

1297. I do not know whether our Geologists and Natural Historians will allow themselves to be annexed to the Mathematical Sciences as Laplace suggests; he says on his page 11:

...En se rapprochant ainsi de la nature, on entrevoit les causes de plusieurs phénomènes importans que l'Histoire naturelle et la Géologie nous offrent; ce qui peut répandre un grand jour sur ces deux sciences, en les rattachant à la théorie du Système du monde.

1298. On his pages 16...18 Laplace states the results obtained by him in the investigations which are reproduced in the third Chapter of the Eleventh Book. These investigations relate to the axis of rotation of the Earth. The importance of the subject is well indicated by the words with which Laplace begins the account:

Toute l'Astronomie repose sur l'invariabilité de l'axe de rotation de la Terre à la surface du sphéroïde terrestre, et sur l'uniformité de cette rotation.

1299. On his pages 18...21 Laplace states the results obtained by him in the investigations which are reproduced in the fourth Chapter of the Eleventh Book. These investigations relate to the heat of the Earth.

1300. We now proceed to the next three Chapters which involve the mathematical investigations. As no commentary has been published on this volume of the *Mécanique Céleste*, like that of Bowditch on the first four volumes, we shall find it expedient to give occasionally more detail than would otherwise have been necessary.

1301. The second Chapter of the Eleventh Book of the *Mécanique Céleste* is entitled *De la figure de la Terre:* it occupies pages 22...56. As we have already stated the Chapter is the reproduction of a memoir: see Art. 1286. The title is rather vague. We shall find that the most important subjects discussed are the form of the ocean, and the constitution of the interior of the Earth. An analysis of the contents of the Chapter is given by Laplace on pages 11...16 of the volume. The following summary occurs at the beginning of the memoir of 1817, but is not reproduced:

Les géomètres ont, jusqu'à présent, considéré la terre comme un sphéroïde formé de couches de densités quelconques, et recouvert en entier d'un fluide en équilibre. Ils ont donné les expressions de la figure de ce fluide, et de la pesanteur à sa surface; mais ces expressions, quoique fort étendues, ne représentent pas exactement la nature. L'Océan laisse à découvert une partie du sphéroïde terrestre; ce qui doit altérer les résultats obtenus dans l'hypothèse d'une inondation générale, et donner naissance à de nouveaux résultats. A la vérité, la recherche de la figure de la terre présente alors plus de difficultés; mais le progrès de l'analyse, sur-tout dans cette partie, fournit le moyen de les vaincre, et de considérer les continens et les mers, tels que l'observation nous les présente. C'est l'objet de l'analyse suivante, qui, comparée aux expériences du pendule, aux mesures des degrés et aux observations lunaires, conduit à ces résultats:

1° La densité des couches du sphéroïde terrestre croît de la surface au centre;

2° Ces couches sont à très-peu-près régulièrement disposées autour de son centre de gravité;

3° La surface de ce sphéroïde, dont la mer recouvre une partie, a une figure peu différente de celle qu'elle prendrait en vertu des loix de l'équilibre, si la mer cessant de la recouvrir, elle devenait fluide;

4° La profondeur de la mer est une petite fraction de la différence des deux axes de la terre;

5° Les irrégularités de la terre et les causes qui troublent sa surface, ont peu de profondeur;

6° Enfin, la terre entière a été primitivement fluide.

Ces résultats de l'analyse, des observations et des expériences, me semblent devoir être placés dans le petit nombre des vérités que nous offre la géologie.

1302. Let there be a point on the surface of the sea. Let r be its radius vector, and μ the cosine of the angle which the radius vector makes with the axis. Let V be the potential of the solid part, W the potential of the sea itself, ω the angular velocity. Then the condition of relative equilibrium is

$$V + W + \frac{\omega^2}{2} r^2 (1 - \mu^2) = \text{constant} \text{ ....(1).}$$

Now this equation is to be transformed into the notation generally employed by Laplace in these researches.

1303. The Earth is supposed to consist of nearly spherical strata. Let $a(1 + \alpha y)$ denote the radius vector of a stratum, where a is a parameter which particularises the stratum, α is a very small constant, and y a function of a and of the usual polar co-ordinates. Let ρ be the density of the stratum, ρ being a function of a.

Supposing y expanded in a series of Laplace's functions, so that

$$y = Y_0 + Y_1 + Y_2 + \ldots$$

Then by Arts. 900 and 1074

$$V = \frac{4\pi}{r}\int_0^a \rho a^2 da + 4\pi\alpha \int_0^a \rho \frac{d}{da}\left\{\frac{a^3 Y_0}{r} + \frac{a^4 Y_1}{3r^2} + \frac{a^5 Y_2}{5r^3} + \ldots\right\} da.$$

Here the upper limit of the integrals denotes the value of a at the surface. Laplace uses unity for it.

Let $\bar{y}$ denote the value of y at the surface of the solid part, and let the radius vector of the surface of the sea be $a(1 + \alpha\bar{y} + \alpha z)$, Laplace uses y' for z. Suppose that z can be expanded in a series of Laplace's functions, so that

$$z = Z_0 + Z_1 + Z_2 + \ldots$$

Then $a\alpha z$ expresses to our order of approximation the depth of the sea. Where the land rises above the sea z becomes negative, so that $-a\alpha z$ then is what we may call the height above the level of the sea: see Laplace's page 39.

Denote the density of the sea by σ; Laplace denotes it by unity.

Let W_1 denote the potential for a homogeneous spheroid of density σ and radius vector $a(1 + \alpha\bar{y} + \alpha z)$. Let W_2 denote the potential for a homogeneous spheroid of density σ and radius vector $a(1+\alpha\bar{y})$. Then

$$W = W_1' - W_2' + X,$$

where X denotes the potential for that part of the dry land which is above the level of the sea, and supposed to have the density σ; for all this occurs negatively in $W_1 - W_2$, and so we must allow for it in expressing W.

Now $W_1 = \dfrac{4\pi a^3\sigma}{3r}$

$$+ 4\alpha\pi\sigma\left\{\frac{a^3(Y_0+Z_0)}{r} + \frac{a^4(\bar{Y}_1+Z_1)}{3r^2} + \frac{a^5(\bar{Y}_2+Z_2)}{5r^3} + \dots\right\},$$

$$W_2 = \frac{4\pi a^3\sigma}{3r} + 4\alpha\pi\sigma\left\{\frac{a^3\bar{Y}_0}{r} + \frac{a^4\bar{Y}_1}{3r^2} + \frac{a^5\bar{Y}_2}{5r^3} + \dots\right\}.$$

Therefore $\quad W_1 - W_2 = 4\alpha\pi\sigma\left\{\dfrac{a^3Z_0}{r} + \dfrac{a^4Z_1}{3r^2} + \dfrac{a^5Z_2}{5r^3} + \dots\right\}.$

Hence finally (1) becomes

$$\text{constant} = \frac{4\pi}{r}\int_0^a \rho a^2 da$$

$$+ 4\alpha\pi\int_0^a \rho\frac{d}{da}L(a, r, Y)\,da$$

$$+ 4\alpha\pi\sigma L(a, r, Z)$$

$$+ X - \frac{\omega^2 r^2}{2}\left(\mu^2 - \frac{1}{3}\right) \text{.......................} (2),$$

where L is a functional symbol, such that

$$L(a, r, Y) = \frac{a^3Y_0}{r} + \frac{a^4Y_1}{3r^2} + \frac{a^5Y_2}{5r^3} + \dots$$

It will be seen that $\omega^2 r^3$ differs from a constant by a term of the order we reject; and thus we have modified the last term on the left-hand side of (1) for convenience.

1304. Let j denote the ratio of the centrifugal force at the equator to the attraction there; then

$$\frac{a\omega^2}{\dfrac{4\pi}{a^2}\displaystyle\int_0^a \rho\, a^2 da} = j \text{ very approximately,}$$

so that $\qquad \omega^2 = \dfrac{4\pi j}{a^3}\displaystyle\int_0^a \rho\, a^2 da.$

We may use this expression for ω^2 in (2) if we please; Laplace always uses it: but for the sake of simplicity we shall frequently retain ω^2.

1305. If we differentiate the right-hand side of (2) with respect to r, and change the sign, we obtain an expression for gravity at any point, Laplace's *pesanteur*. Denote it by p: thus

$$p = \frac{4\pi}{r^2}\int_0^a \rho\, a^2 da - 4\pi\int_0^a \rho \frac{d^2}{dr\, da} L(a, r, Y)\, da$$

$$- 4a\pi\sigma \frac{d}{dr} L(a, r, Z) - \frac{dX}{dr} + \omega^2 r\left(\mu^2 - \frac{1}{3}\right) \ldots\ldots (3).$$

This is the value of gravity at the surface of the sea.

1306. The preceding equations (2) and (3) are those which Laplace denotes by the same numbers, with slight differences of notation. At this point he interposes a discussion of his well-known equation; but we have given sufficient attention to this matter in Chapter XXX.

We will therefore assume with Laplace that

$$\frac{1}{2} X + a \frac{dX}{dr} = 0 \ldots\ldots\ldots\ldots\ldots\ldots\ldots\ldots (4).$$

Multiply (2) by $-\frac{1}{2a}$ and add to (3). Thus

$$p = \text{constant} + \left(\frac{4\pi}{r^2} - \frac{2\pi}{ar}\right)\int_0^a \rho\, a^2 da$$

$$- \frac{2a\pi}{a}\int_0^a \rho \frac{d}{da} L(a, r, Y)\, da - 4a\pi\int_0^a \rho \frac{d^2}{dr\, da} L(a, r, Y)\, da$$

$$- \frac{2a\pi\sigma}{a} L(a, r, Z) - 4a\pi\sigma \frac{d}{dr} L(a, r, Z)$$

$$+ \frac{r^2\omega^2}{4a}\left(\mu^2 - \frac{1}{3}\right) + r\omega^2\left(\mu^2 - \frac{1}{3}\right) \ldots\ldots\ldots\ldots (5).$$

Now put $a(1 + \alpha\bar{y} + \alpha s)$ for r; then to our order of approximation we obtain

$$p = \text{constant} - \frac{6\alpha\pi(\bar{y} + s)}{a^3}\int_0^a \rho\, a^2 da$$
$$+ 2\alpha\pi\int_0^a \rho\frac{d}{da}\left(\frac{a^3Y_0}{a} + \frac{a^4Y_1}{a^2} + \frac{a^5Y_2}{a^3} + \dots\right)da$$
$$+ 2\alpha\pi\sigma a s + \frac{5}{4}a\omega^2\left(\mu^2 - \frac{1}{3}\right) \dots\dots\dots\dots (6).$$

If the Earth is supposed homogeneous we have from (6)

$$p = \text{constant} - 2\alpha\pi\rho a(\bar{y} + s) + 2\alpha\pi\rho a\bar{y} + 2\alpha\pi\sigma a s + \frac{5}{4}a\omega^2\left(\mu^2 - \frac{1}{3}\right)$$
$$= \text{constant} - 2\alpha\pi(\rho - \sigma)as + \frac{5}{4}a\omega^2\left(\mu^2 - \frac{1}{3}\right).$$

And if the density of the sea is then supposed to be the same as that of the land, we have

$$p = \text{constant} + \frac{5}{4}a\omega^2\mu^2 = P\left(1 + \frac{5a\omega^2}{4P}\mu^2\right),$$

where P is the value of p at the equator.

Thus, by the definition of j in Art. 1304, we get

$$p = P\left(1 + \frac{5}{4}j\mu^2\right).$$

1307. It is very important to observe that equation (6) of the preceding Article, and those which follow from it, hold even when we suppose the surface of the dry land to be made irregular by elevated plains and mountains. For by reason of these bodies a term would be added to (1) expressing their potential, say X'; then for X' the equation corresponding to (4) would hold, so that X' would not appear in (6). This is a remarkable result of Laplace's process.

Here we arrive at the end of the second section of the Eleventh Book.

1308. Laplace says that to determine the figure of the sea when that of the Earth is given, the simplest method consists in arranging the approximations according to powers of the ratio of the density of the sea to the mean density of the Earth; this ratio is about $\frac{2}{11}$.

Take equation (2) and divide by $4\pi\int_0^a \rho a^2 da$, which we shall denote by $4\pi\phi(a)$. Thus

$$\text{constant} = \frac{1}{r} + \frac{\alpha}{\phi(a)}\int_0^a \rho \frac{d}{da} L(a, r, Y)\, da + \frac{\alpha\sigma}{\phi(a)} L(a, r, Z) + \frac{X}{4\pi\phi(a)} - \frac{\omega^2 r^2}{8\pi\phi(a)}\left(\mu^2 - \frac{1}{3}\right).$$

We have then to put $a(1 + \alpha\bar{y} + \alpha s)$ for r, and reject the square of α. Thus

$$\frac{\alpha}{a}(\bar{y} + s) = \frac{\alpha}{\phi(a)}\int_0^a \rho \frac{d}{da} L(a, a, Y)\, da + \frac{\alpha\sigma}{\phi(a)} L(a, a, Z) + \frac{X}{4\pi\phi(a)} - \frac{j}{2a}\left(\mu^2 - \frac{1}{3}\right) + \text{constant} \ldots\ldots\ldots (7).$$

Laplace says that he will consider the figure of the sea, neglecting the ratio just mentioned, that is, supposing the sea to be an infinitely rare fluid. This, as he allows, would amount to neglecting the terms in (7) which involve σ. But instead of neglecting these terms in (7), he says in his next sentence that he will neglect only the term X; this term of course involves σ by its definition: see Art. 1303. There is something not quite satisfactory in this process, for thus Laplace retains some terms, and neglects others, which may be comparable with these. We may say that he retains the sea, and neglects the dry land which is above the level of the sea, supposed homogeneous and of the same density as the sea.

1309. In equation (7) we neglect X and arrange both sides in a series of Laplace's functions; then we equate to zero the aggregate of the functions of the same order, supposing all the terms brought to one side. Thus we obtain in general

$$Z_i\left\{1 - \frac{a^2\sigma}{(2i+1)\,\phi(a)}\right\} = -Y_i + \frac{\int_0^a \rho \frac{d}{da}(a^{i+3} Y_i)\, da}{(2i+1)\, a^i \phi(a)}.$$

This holds for positive integral values of i except $i=2$. When $i=2$ we must add the term $-\frac{j}{2a}\left(\mu^2-\frac{1}{3}\right)$ to the right-hand side. It does not hold when $i=0$; for then a constant should be added to one side.

Now observation shews that $\bar{Y}_1$, $\bar{Y}_3$, $\bar{Y}_4$, ... are all small when compared with $\bar{Y}_2$; and that $\bar{Y}_2$ is very approximately $-\bar{h}\left(\mu^2-\frac{1}{3}\right)$, where $\bar{h}$ is a constant.

Put $-h\left(\mu^2-\frac{1}{3}\right)$ for Y_2. Then the equation for determining Z_2 is

$$Z_2\left\{1-\frac{a^3\sigma}{5\phi(a)}\right\}=\left\{\bar{h}-\frac{\int_0^a \rho\frac{d}{da}(a^5h)\,da}{5a^2\phi(a)}-\frac{j}{2a}\right\}\left(\mu^2-\frac{1}{3}\right).$$

Let h' stand for

$$\left\{\frac{j}{2a}-\bar{h}+\frac{\int_0^a \rho\frac{d}{da}(a^5h)\,da}{5a^2\phi(a)}\right\}\div\left\{1-\frac{a^3\sigma}{5\phi(a)}\right\};$$

then $$Z_2=-h'\left(\mu^2-\frac{1}{3}\right).$$

The equation which defines h' may be put in the form

$$\left\{a(h'+\bar{h})-\frac{j}{2}\right\}5a^2\phi(a)=ah'a^3\sigma+a\int_0^a \rho\frac{d}{da}(a^5h)\,da;$$

and thus it is seen to agree with what Clairaut had obtained. If we suppose ρ constant we have the result given in II. of Art. 324, neglecting there the difference between r_1 and r'; and if we do not suppose ρ constant, the result may be shewn to coincide with (2) of Art. 323. The ϵ_1 of those Articles is equivalent to the $a(h'+\bar{h})$ of the present Article.

1310. Laplace now says that

$$z=l-h'\mu^2,$$

where l is some constant. The constant l may be supposed to arise partly from the term $\frac{h'}{3}$ in Z_2, and partly from Z_0.

Thus, Laplace in fact takes Z_1 to be zero, and for this he gives the following reason: the origin of the radii vectores is supposed to be at the centre of gravity of the terrestrial spheroid, which makes Y_1 and Z_1 zero. I do not understand this; there may be some connexion, though I cannot trace it exactly, with the result established in the *Connaissance des Tems* for 1821, which is also investigated in the third Chapter of the Eleventh Book: see Art. 1281.

Plana in the *Astronomische Nachrichten*, Vol. XXXVIII. page 236, makes a remark with respect to this point which I will reproduce here in the notation of my present Chapter.

If we take the origin at the centre of gravity of the solid part we have

$$\int_0^a \rho \frac{d}{da}(a^4 Y_1)\, da = 0;$$

thus

$$Z_1 = -\frac{Y_1}{1 - \frac{a^3\sigma}{3\phi(a)}}.$$

But the phenomena of the tides demonstrate that the existence of the term aZ_1 in the depth of the sea is inadmissible; so that we must have $Y_1 = 0$, in order that we may have $Z_1 = 0$.

I do not think that this appeal to the phenomena of the tides is satisfactory when we are discussing the relative equilibrium of the fluid on the Earth's surface; so that I do not feel satisfied as to Plana's development of Laplace's statement.

1311. Thus Laplace takes for the depth of the sea the expression

$$a\alpha\,(l - h'\mu^2).$$

Now he says it is easy to see that h' will be zero if the sea being annihilated the surface of the spheroid should be in equilibrium on becoming fluid. I should prefer to put it thus: if $\bar{\lambda}$ has the value which would belong to the earth considered as a fluid then we may suppose $h' = 0$. Then Laplace says that if the surface is less flattened than in this case h' will be positive; and if the surface is more flattened h' becomes negative. I do not quite understand these statements. Consider the equations

$$\frac{j}{2a} - \bar{h} + \frac{\int_0^a \rho \frac{d}{da}(a^5 h)\,da}{5a^3\phi(a)} = 0,$$

and

$$\frac{j}{2a} - \bar{h} + \frac{\int_0^a \rho \frac{d}{da}(a^5 h)\,da}{5a^3\phi(a)} = C,$$

where C is some positive quantity.

Laplace's remark then would imply that the second equation *necessarily* requires $\bar{h}$ to be less than the first; but this seems to me not the case: for the value of h in terms of a may be so adjusted possibly as to allow $\bar{h}$ in the second equation to be less than in the first. In other words the sign of h' does not appear to depend solely on the ellipticity of the bounding surface of the solid part, but also on the law of ellipticity of the interior strata. It is obvious that Laplace does not assume the form of the earth to be that which corresponds to original fluidity, for if he did, then h' would be zero.

1312. It may happen that the volume of the sea is not sufficient to cover the entire surface of the earth: in this case, if h' be positive the equatorial part is covered, and if h' be negative the polar part is covered.

1313. We shall now obtain an expression for gravity.

Take (6) and omit Y_3, Y_4, ...; thus

$$p = \text{constant} - \frac{6a\pi\phi(a)}{a^3}(\bar{y} + s) + \frac{2a\pi}{a^5}\int_0^a \rho \frac{d}{da}(a^5 Y_2)\,da$$
$$+ 2a\pi\sigma a s + \frac{5}{4} a\omega^2\left(\mu^2 - \frac{1}{3}\right)$$
$$= \text{constant} - \frac{6a\pi\phi(a)}{a^3}(\bar{y} + s)$$
$$+ \frac{2a\pi}{a^5}\left\{\left(\frac{j}{2a} - \bar{h}\right)\left(\mu^2 - \frac{1}{3}\right) + s\left[1 - \frac{\sigma a^3}{5\phi(a)}\right]\right\} 5a^5\phi(a)$$
$$+ 2a\pi\sigma a s + \frac{5}{4} a\omega^2\left(\mu^2 - \frac{1}{3}\right)$$

$$= \text{constant} + \frac{6\alpha\pi\phi(\mathrm{a})}{\mathrm{a}^2}\left(\mu^2 - \frac{1}{3}\right)\left\{\bar{k} + k' + \frac{5}{3}\left(\frac{j}{2\alpha} - \bar{k}\right)\right\}$$

$$+ \frac{10\alpha\pi\phi(\mathrm{a})}{\mathrm{a}^2}s + \frac{5}{4}\mathrm{a}\omega^2\left(\mu^2 - \frac{1}{3}\right)$$

$$= \text{constant} + \frac{4\alpha\pi\phi(\mathrm{a})}{\mathrm{a}^2}\left(\mu^2 - \frac{1}{3}\right)\left\{\frac{5}{2}\frac{j}{\alpha} - \bar{k} - k'\right\}.$$

Thus $$p = P\left\{1 + \frac{4\alpha\pi\phi(\mathrm{a})}{\mathrm{a}^2 P}\left(\frac{5}{2}\frac{j}{\alpha} - \bar{k} - k'\right)\mu^2\right\},$$

where P denotes the gravity at the equator. And in the small term we may take $\frac{4\pi\phi(\mathrm{a})}{\mathrm{a}^2 P}$ as unity, so that

$$p = P\left\{1 + \left(\frac{5}{2}j - \alpha\bar{k} - \alpha k'\right)\mu^2\right\}.$$

In like manner we might put the expression for p thus,

$$p = P\left\{1 - \left(\frac{5}{2}j - \alpha\bar{k} - \alpha k'\right)\left(1 - \mu^2\right)\right\},$$

where P now denotes the gravity at the poles.

Laplace gives these two forms, taking the former in the case in which k' is positive, and the latter in the case in which k' is negative. It is of little importance, but it might seem more natural to use the first formula when the sea covers the poles, and the second when it covers the equator, that is to reverse Laplace's allotment.

Laplace uses the coefficient $\frac{5}{2}$ in the first formula, but the coefficient $\frac{5}{4}$ in the second. In the national edition $\frac{5}{4}$ is taken in both cases. It should be $\frac{5}{2}$ in both cases, as I give it; and in fact it is so in Laplace's original memoir. This example is one of many which reflect little credit on the editors of the national edition of Laplace's works.

1314. Laplace now digresses to some very remarkable investigations respecting Legendre's functions.

He gives an expression for the nth function by means of a definite integral, namely

$$P_n = \frac{1}{\pi}\int_0^\pi [x - \cos\phi \sqrt{(x^2-1)}]^n \, d\phi;$$

see Laplace's page 33. The investigation involves the use of imaginary symbols. For another investigation see Heine's *Handbuch der Kugelfunctionen*, pages 11...14.

If $x = 1$, the above formula shews that $P_n = 1$. Laplace says that if x is less than 1, the formula makes P_n less than 1; "comme il est facile de le prouver." I presume he would adopt some such process as this:

If x is less than 1, assume

$$x = k \cos\psi,$$

and
$$\cos\phi \sqrt{(1-x^2)} = k \sin\psi;$$

so that
$$k^2 = x^2 + (1-x^2)\cos^2\phi$$
$$= 1 - (1-x^2)\sin^2\phi.$$

Then
$$[x - \cos\phi \sqrt{(x^2-1)}]^n = k^{\frac{n}{2}} [\cos\psi - \sqrt{(-1)}\sin\psi]^n$$
$$= k^{\frac{n}{2}} [\cos n\psi - \sqrt{(-1)}\sin n\psi].$$

This expression is always less than unity; so that P_n is less than unity, for $\int_0^\pi Q d\phi$ is less than π if Q is always less than unity.

Laplace deduces from the definite integral for P_n the following approximate value of P_n when n is very large

$$\frac{\sqrt{2}}{\sqrt{(n\pi \sin\theta)}} \cos\left\{\left(n+\frac{1}{2}\right)\theta - \frac{1}{4}\pi\right\},$$

where $\cos\theta = x$.

The investigation is given rather more fully in the original memoir than in the *Mécanique Céleste*; in the latter place, instead of the details of the process, there is a vague reference to the

Mémoires de l'Académie...for 1782, and to the *Théorie des Probabilités*.

The investigation cannot be considered very satisfactory, for it does not supply us with any estimate of the amount of the error made in using this expression instead of the exact value of P_n. Moreover it is obvious that the result does not hold universally; for instance it is not true when $\theta = 0$, and we can have no confidence that it is true when θ is very small.

Laplace gives another investigation in the supplement to the fifth volume of the *Mécanique Céleste;* this investigation makes no use of imaginary quantities, but can be considered only as a very rude process of approximation.

Heine does not advert to this approximate value of P_n, and I do not know whether it has been discussed by any other writer than Laplace himself.

A misprint at the bottom of Laplace's page 33 is reproduced in the national edition.

Also in the investigation in the supplement to the fifth volume it will be found that in the two fundamental equations on the middle of page 3, we must put on the right-hand side $\frac{\cos\theta}{\sin\theta}$ instead of $\sin\theta$. The misprint is important; for if it were not corrected we should be at a loss to see why the result cannot be relied on when $\theta = 0$. The misprint is reproduced in the national edition.

1315. Laplace has a few words about a second approximation to the value of s; see his page 36: but he does not really work out his suggestion. Nor does he make any use of the results he obtains with respect to Legendre's coefficients; an apparent exception to this statement occurs in a paragraph on pages 36 and 37 relating to the case in which the earth is supposed to be a figure of revolution. This paragraph was not in the original memoir; it finishes with a formula which can command very little confidence.

Here we arrive at the end of the third section of the Eleventh Book.

1316. Laplace now proposes to consider the variations of the lengths of degrees and of the value of gravity at the surface of continents and islands; these are the only variations which we can observe. In order to obtain their analytical expression, imagine an atmosphere infinitely rare, of constant density, very little elevated, but sufficiently so as to cover all the mountains. Let $a\alpha\zeta$ represent the height above the surface of the terrestrial spheroid; Laplace uses y'' for ζ.

The equation (2) neglecting α^2 will apply to that part of the surface of the atmosphere which is above the sea; we must put $a(1+\alpha\bar{y}+\alpha\zeta)$ for r. But this equation also applies to the surface of the sea when we put $a(1+\alpha y+\alpha s)$ for r. Then if we subtract one of the results thus obtained from the other, we have

$$a(\alpha\zeta-\alpha s)=\text{constant}.$$

Therefore all the points of the surface of this atmosphere which correspond to the surface of the sea are equally elevated above the latter surface.

Then as to that part of the atmosphere which is above the solid part of the earth. Here again Laplace makes out that (2) holds; for the potential of the sea will be of the same form whether the point to which it relates be close to the surface of the sea, or close to the surface of the dry land.

Thus again we obtain

$$a(\alpha\zeta-\alpha s)=\text{constant}.$$

Then the constant must be the same in these two equations, as we see by considering the case of points just on the sea shore. This constant Laplace denotes by $a\alpha l$. Therefore

$$a(\alpha\zeta-\alpha s)=a\alpha l.$$

This equation then holds universally at the surface of the imaginary atmosphere.

1317. In the preceding Article I have given, I think, the meaning of Laplace; but I do not find him altogether clear. The surface determined by the radius vector $a(1+\alpha\bar{y}+\alpha s)$ is what he calls the level of the sea; where there is sea this equation represents the surface of the sea. Where the dry land appears, if we suppose broad canals cut across the continents, the water would I apprehend from his equation still rise to the level determined by the above radius vector. But he does not make this remark, though it seems to me necessary in order to have a clear conception of his process. But he does partially state this in his page 53, or something like it. See also the *Annales de Chimie*, Vol. VIII. 1818, page 310.

1318. We have now to find an expression for gravity at the surface of the supposed atmosphere. The equation (2) may be put in the form

$$\text{constant} = \frac{4\pi}{r}\int_0^a \rho a^2 da + 4a\pi\int_0^a \rho \frac{d}{da} L(a, r, Y)\, da$$
$$+ V_1 - \frac{\omega^2 r^2}{2}\left(\mu^2 - \frac{1}{3}\right) \ldots\ldots\ldots\ldots (8),$$

where V_1 denotes the potential of the sea. We may even suppose, for greater generality, that V_1 includes the potential arising from the mountains and cavities of the surface of the Earth, observing that the part of V_1 relative to the cavities is negative.

Hence denoting the gravity by p' we have

$$p' = \frac{4\pi}{r^2}\int_0^a \rho a^2 da - 4a\pi\int_0^a \rho \frac{d^2}{da\, dr} L(a, r, Y)\, da$$
$$- \frac{dV_1}{dr} + \omega^2 r\left(\mu^2 - \frac{1}{3}\right) \ldots\ldots\ldots\ldots\ldots\ldots\ldots (9).$$

Multiply (8) by $-\frac{1}{2a}$ and add to (9); thus to our order we obtain, as in Art. 1300,

$$p' = \text{constant} - \frac{6a\pi(\bar{y} + \zeta)}{a^3}\int_0^a \rho a^2 da$$
$$+ 2a\pi\int_0^a \rho \frac{d}{da}\left(\frac{a^3 Y_0}{a^3} + \frac{a^4 Y_1}{a^4} + \frac{a^5 Y_2}{a^5} + \ldots\right) da$$
$$+ \frac{5}{4} a\omega^2\left(\mu^2 - \frac{1}{3}\right) \ldots\ldots\ldots\ldots\ldots\ldots\ldots (10).$$

Now if p'' be the gravity at the surface of the spheroid corresponding to p' at the surface of the atmosphere, we have

$$p' = p'' - \frac{2a\alpha P\zeta}{a},$$

where P may denote the gravity at the equator at the level of the sea. Hence

$$p'' - 2\alpha P\zeta = \text{constant} - \frac{6\pi\pi\zeta}{a^2}\int_0^a \rho\, a^2 da$$

$$-\frac{6\pi\pi}{a^2}\bar{y}\int_0^a \rho\, a^2 da + 2\alpha\pi\int_0^a \rho \frac{d}{da}\left(\frac{a^3 Y_2}{a^2} + \frac{a^4 Y_3}{a^3} + \frac{a^5 Y_4}{a^4} + \dots\right) da$$

$$+\frac{5}{4}a\omega^2\left(\mu^2 - \frac{1}{3}\right).$$

By integrating by parts the terms in the second line we obtain

$$p'' = \text{constant} + \frac{1}{2}\alpha P\zeta$$

$$+\frac{2\pi\pi}{a^2}\bar{y}\int_0^a a^3 \frac{d\rho}{da}\, da - 2\alpha\pi\int_0^a \frac{d\rho}{da}\left(\frac{a^3 Y_2}{a^2} + \frac{a^4 Y_3}{a^3} + \dots\right) da$$

$$+\frac{5}{4}a\omega^2\left(\mu^2 - \frac{1}{3}\right) \dots\dots\dots\dots\dots\dots (11).$$

Laplace observes that this expression for p'' includes the attraction of the mountains, and generally all the effects of attraction due to the irregularities of the surface, provided that the attracted point is far removed from them; *for this condition is necessary to the existence of the equation*

$$0 = \frac{dV_1}{dr} + \frac{1}{2a}V_r$$

In the words which I have marked with Italics it seems to me that Laplace really treats his own much used equation with that caution which Ivory would have desired. See Chapter XXX.

1319. Suppose the earth homogeneous, so that $\frac{d\rho}{da}$ is zero; then from (11) we have

$$p'' = P\left[1 - \frac{1}{2}\alpha(l - \zeta) + \frac{5}{4}j\mu^2\right].$$

where P now denotes the gravity at the equator at the level of the sea; Laplace calls this an *expression remarquable*: see Art. 1283. He draws attention to it as one of his most interesting results on his page 11.

Laplace says that this formula may be used to test the hypothesis of homogeneity. For the atmosphere which we have hitherto imagined may be taken to be the real atmosphere reduced to its mean density. Then if to the value of p'' determined by the pendulum, we add the value of $\frac{1}{2} P\alpha (l - \zeta)$ determined by the barometer, the value of gravity thus corrected should become $P(1 + \frac{5}{4} j\mu^2)$.

Now $\frac{5}{4} j = \cdot 004325$.

Thus the increment of gravity would be $\cdot 004325 P\mu^2$.

But numerous experiments in both hemispheres agree in making this increment about $\cdot 0054 P\mu^2$. Hence Laplace concludes that the hypothesis of homogeneity is excluded by these experiments.

On his page 12 Laplace referred to pendulum experiments, and stated the result which is here obtained; and then he added on that page

On voit de plus, en les comparant à l'analyse, que les densités des couches terrestres vont en croissant de la surface au centre.

But I do not see where Laplace really establishes the statement thus made.

Moreover he says that the heterogeneity of the strata must extend from the surface beyond quantities of the order α, in order that the quantity in equation (11)

$$\frac{2\alpha\pi}{\mathrm{a}^5} \bar{y} \int_0^a a^5 \frac{d\rho}{da} da - 2\alpha\pi \int_0^a \frac{d\rho}{da} \left(\frac{a^5 Y_2}{\mathrm{a}^3} + \frac{a^7 Y_4}{\mathrm{a}^5} + \dots \right) da,$$

may be of the order α and become equal to

$$P (\cdot 0054 - \cdot 004325) \left(\mu^2 - \frac{1}{3} \right).$$

It seems to me that Laplace should not say that it must become *equal* to this, for it might differ from this by a constant.

Here we arrive at the end of the fourth section of the Eleventh Book.

1320. Laplace begins his next section thus: "Comparons maintenant l'analyse aux observations." As he has just made a very important comparison of this kind, he ought to have said: "let us proceed with our comparison of the analysis with observations."

In this section such comparison is made with respect to four different things, namely, pendulum experiments, certain terms in the lunar theory, measures of degrees of the meridian, and precession and nutation.

1321. Multiply (6) by $\frac{3}{2a}$, reduce it to our order of approximation and subtract it from (10); thus

$$p' = \text{constant} + 4\alpha\pi\int_0^a \rho\frac{d}{da}\left\{-\frac{a^3Y_0}{a^2}+\frac{a^5Y_2}{5a^4}+\frac{2a^6Y_3}{7a^5}+\dots\right\}da$$
$$-\frac{3}{2a}V_1+2a\alpha^2\left(\mu^2-\frac{1}{3}\right)\dots\dots(12).$$

Developing V_1 in powers of $\frac{1}{r}$, and ultimately putting a for r, we obtain an expression of this form

$$U_0+U_1+U_2+\dots$$

where U_n denotes a Laplace's function of the nth order. Then as the terms arising from Y_0 and U_0 may be included in the constant, we obtain from (12)

$$p' = \text{constant} + 4\alpha\pi\int_0^a \rho\frac{d}{da}\left\{\frac{a^5Y_2}{5a^4}+\frac{2a^6Y_3}{7a^5}+\dots\right\}da$$
$$-\frac{3}{2a}\left[U_1+U_2+U_3+\dots\right]+2a\alpha^2\left(\mu^2-\frac{1}{3}\right)\dots\dots(13).$$

It follows from numerous experiments with the pendulum that

$$p' = \text{constant} + aqP\left(\mu^2-\frac{1}{3}\right),$$

where aq is very nearly equal to ·0054, and P denotes the gravity at the equator, so that $a\omega^2 = jP$ approximately.

Hence it follows that

$$4a\pi \int_0^a \rho \frac{d}{da}\left\{\frac{2a^4 Y_2}{7a^2} + \frac{3a^7 Y_3}{9a^3} + \dots\right\} da - \frac{3}{2a}\left[U_1 + U_3 + U_4 + \dots\right]$$

is very small relatively to the term $aqP\left(\mu^2 - \frac{1}{3}\right)$; and that the function

$$\frac{4a\pi}{5a^2}\int_0^a \rho \frac{d}{da}(a^5 Y_2)\, da - \frac{3}{2} U_2$$

is very nearly equal to $(aq - 2j)\, P\left(\mu^2 - \frac{1}{3}\right)$.

The general expression of the above Laplace's function of the second order is

$$A\left(\mu^2 - \frac{1}{3}\right) + A_1 \mu \sqrt{(1-\mu^2)} \sin\phi + A_2 \mu \sqrt{(1-\mu^2)} \cos\phi$$
$$+ A_3 (1-\mu^2) \sin 2\phi + A_4 (1-\mu^2) \cos 2\phi.$$

Hence A_1, A_2, A_3, A_4 must be very small compared with A; and we have very approximately

$$A = (aq - 2j)\, P.$$

Now the pendulum experiments make $aq = $ ·0054 very nearly; and $j = \frac{1}{289}$: thus we obtain

$$A = -\cdot 00152P \quad \dots\dots\dots\dots(14).$$

1322. Next Laplace takes a certain term in the Lunar Theory, and compares it with observations.

Let $Q\left(\mu^2 - \frac{1}{3}\right)$ denote the part of U_2 which is independent of the angle ϕ. Then he finds that

$$A + \frac{5}{2} Q = -\cdot 001558P \dots\dots\dots\dots(15).$$

Laplace obtains $Q = -\cdot 00015P$ from equations (14) and (15); but it should be $-\cdot 0000152P$. There is a misprint, but not the same, in the original memoir at this point.

A second term in the Lunar Theory gives also the same result.

The errors of observations and of experiments would render this value very uncertain; but still we may safely infer that Q is very small. Hence we conclude that the sea is neither very deep nor very dense.

1323. The measures of degrees of the meridian reduced to the level of the supposed atmosphere are next considered.

Equation (8) is put by Laplace in the following form:

$$a\alpha(\bar{y}+\zeta)P = 4\alpha\pi \int_0^a \rho \frac{d}{da}\left(\frac{a^5 Y_2}{5a^2} + \frac{a^7 Y_4}{7a^4} + \dots\right) da$$

$$+ U_2 + U_4 + \dots - \frac{jPa}{2}\left(\mu^2 - \frac{1}{3}\right) + \text{constant} \dots\dots\dots\dots (16);$$

he says that the origin of coordinates is at the common centre of gravity of the sea and of the terrestrial spheroid, which makes the quantities Y_1 and U_1 and the other functions of the same nature disappear: see Art. 1310.

The comparison of degrees measured in distant parts of the world led Delambre to the result

$$a(\bar{y}+\zeta) = \text{constant} - \cdot 00324\left(\mu^2 - \frac{1}{3}\right) \dots\dots\dots\dots (17).$$

When ζ is supposed expanded in a series of Laplace's functions let $-h''\left(\mu^2 - \frac{1}{3}\right)$ be that part of the function of the second order which is independent of the angle ϕ. Let $-\bar{h}\left(\mu^2 - \frac{1}{3}\right)$ be the similar term in $\bar{y}$. Thus

$$a(\bar{h} + h'') = \cdot 00324 \dots\dots\dots\dots\dots\dots\dots\dots (18).$$

Laplace has a troublesome misprint as to this notation at the top of his page 45, which is reproduced in the national edition: the original memoir is correct.

By comparing (17) with (16) we see that Y_3, Y_4, ... U_3, U_4, U_5, ... are very small; as appears also from the pendulum experiments. Then from (16)

$$-Pa(\bar{h} + h'') = A + \frac{5}{2}Q - \frac{j}{2}P,$$

and from (18) we obtain

$$A + \frac{5}{2} Q = -\cdot 00151 P \ldots\ldots\ldots\ldots\ldots\ldots\ldots\ldots (19).$$

It is obvious that (19) agrees very closely with (15).

Laplace says he has supposed the degrees measured on the surface of the spheroid and reduced to the level of the atmosphere, to be the same as the degrees measured at the surface of the atmosphere. In order to justify this it must be shewn that rejecting α^2 the direction of gravity is the same at the surface of the spheroid as at the surface of the atmosphere. He proceeds to shew this briefly. What he seems to make out is, that the direction of gravity at the level of the sea is the same as at the level of the atmosphere; and this, I presume, is what really ought to be shewn, as the degrees measured on the Earth's surface are in general referred to the level of the sea. That the correspondence required between degrees referred to the surface of the sea and to the surface of the atmosphere really exists, follows from the fact, that the former surface *is at a small constant elevation above the latter:* see Art. 1316. Laplace's remarks are rather obscure; the real point seems to me to lie in the statement which I have just given in italics.

1324. Lastly, Laplace refers to Precession and Nutation. This however does not yield any very decisive result, as we are obliged to make some hypothesis respecting the density of the Earth. The treatment of this point in the original memoir and that in the *Mécanique Céleste* are rather different. But the matter belongs properly not to our subject, but to the theory of Precession and Nutation. It will be sufficient to say that in the *Mécanique Céleste* Laplace takes as an hypothesis that the density increases from the surface to the centre in arithmetical progression; and assuming that the mean density at the surface is three times that of the sea, he finds that the mean density of the Earth is 4·761 times that of the sea.

Plana has discussed this passage of the *Mécanique Céleste* in the *Astronomische Nachrichten*, Vol. XXXV. pages 177...192. Plana considers that the mean density of the surface cannot be so great as Laplace supposed. Plana also discusses this law of density in the *Astronomische Nachrichten*, Vol. XXXVI. pages 313...334. And a memoir by Plana in Vol. XXXVIII. of the *Astronomische Nachrichten* may be said to go over nearly the same extent of ground as the §§ 2...5 of Laplace's Eleventh Book.

Here we arrive at the end of the fifth section of the Eleventh Book.

1325. Laplace's next section is devoted to the discussion of a certain hypothetical law connecting the pressure and the density inside the Earth; namely, the law expressed by the equation $\frac{d\Pi}{d\rho} = 2k\rho$, where Π is the pressure, ρ the density, and k a constant. This section comes from the *Addition* to the original memoir: see Art. 1286. This section has now passed substantially into the elementary books, and has thus become familiar to us. See Airy's *Mathematical Tracts*, Pratt's *Figure of the Earth*, O'Brien's *Mathematical Tracts*, and Resal's *Traité Elémentaire de Mécanique Céleste*. I shall offer only a few remarks.

1326. Laplace arrives from his hypothesis at a law of density which Legendre had formerly given as an example: see Art. 942. Thus Laplace says on his page 51:

Je dois observer ici que M. Legendre a déterminé l'aplatissement de la Terre, dans le cas où la densité des couches est exprimée par $\frac{A}{a} \cdot \sin an$.

But this does not ascribe to Legendre the idea of the hypothetical law expressed by $\frac{d\Pi}{d\rho} = 2k\rho$; and I conclude that Resal is wrong in saying, as he does on his page 227, that this hypothesis was imagined and discussed by Legendre.

1327. Laplace arrives in his discussion at the following differential equation

$$\frac{d^2v}{da^2} + \left(n^2 - \frac{6}{a^2}\right) v = 0 \text{.....................(20)},$$

where v stands for $h\int_0^a \rho' a da$, and ρ' is such that

$$\frac{d^2\rho'}{da^2} + n^2\rho' = 0.$$

Laplace observes that it is easy to see that the equation for v is satisfied by

$$v = H\rho'\left(1 - \frac{3}{n^2a^2}\right) + \frac{3H}{n^2a}\frac{d\rho'}{da},$$

where H is an arbitrary constant.

. As ρ' may be supposed equal to $A \sin (na + B)$, where A and B are arbitrary constants, this value of v will involve, as it should do, two arbitrary constants. Thus in fact we may say that

$$v = C\left\{\left(1 - \frac{3}{n^2a^2}\right)\sin (na + B) + \frac{3}{na}\cos (na + B)\right\}.$$

For the integration of a general equation which includes (20) as a particular case see Art. 942, and Boole's *Differential Equations*, third edition, page 424.

1328. A troublesome misprint occurs on Laplace's page 52 in the fourth line. Instead of $D = \frac{3q}{n^2}$ we must read $\frac{D}{(\rho)} = \frac{3q}{n^2}$. This misprint occurs in the original *Addition*, as well as in the *Connaissance des Tems* for 1822, and in the national edition of Laplace's works.

A student who wishes to verify Laplace's numerical calculations must remember that the radius of the earth is assumed to be unity. The results obtained on page 52 should be compared with those given in Schmidt's *Lehrbuch der Mathematischen und Physischen Geographie*, Vol. I. page 387. Schmidt makes the mean density of the earth 1·814 times the density of the superficial stratum, and the mean density of the earth 4·785 times that of water. In Humboldt's *Cosmos*, Vol. I. Note 130, this number 4·785 is mentioned in connexion with 4·761 given by Laplace on his page 47; but there Laplace has a very different law of density, and there is no just ground for the connexion.

1329. The expression obtained by theory for the Precession of the Equinoxes involves various Astronomical elements, such as the ratio of the Moon's mass to the Earth's, and the ratio of the mean motion of the Moon round the Earth to that of the mean motion of the Earth round the Sun. But in connexion with our subject the most important element involved is the fraction $\frac{\int \rho a^4 da}{\int \rho a^2 da}$, where ρ denotes the density of the stratum having the parameter a; the integrals are taken between the limits zero and the extreme value of a. Suppose we calculate the value of this fraction on the assumption that ρ has the form which Laplace is here discussing; then we can make an interesting comparison of the theoretical expression with the results of observation.

Similar remarks apply to the expression for Lunar nutation.

Such a comparison is made in the three elementary works cited in Art. 1325; so that I need not enter upon it here.

I shall therefore only remark that Laplace himself treats on this comparison briefly in his original memoirs; see Arts. 1285 and 1286: but he does not reproduce his remarks in the *Mécanique Céleste.*

1330. Some general observations which form the first seven pages of the *Addition* are reproduced in substance in the first Chapter of the Eleventh Book of the *Mécanique Céleste.* One slight change may be noted. After remarking that geometers had not yet introduced in their researches on the figure of the Earth the compressibility of the strata Laplace says in the *Addition*:

M. Young vient d'appeller leur attention sur cet objet, par la remarque ingénieuse, que l'on peut expliquer de cette manière l'accroissement de densité des couches du sphéroïde terrestre.

This is omitted in the *Mécanique Céleste,* Vol. v. page 15, and we have instead:

...quoique Daniel Bernoulli, dans sa pièce sur le flux et le reflux de la mer, eût déjà indiqué cette cause de l'accroissement de densité des couches du sphéroïde terrestre.

The remark of Young is to be found in the *Philosophical Transactions* for 1819: see his *Works,* Vol. II. pages 19, 78, 82.

Some observations bearing on the subject of this section of Laplace by Plana will be found in the fifth volume of De Zach's *Correspondance Astronomique*.

Here we arrive at the end of the sixth section of the Eleventh Book.

1331. Take equations (13) and (16); subtract: neglect U_1, U_2, U_3,...on the ground that the action of the sea is small owing to its small density or small depth; and assume the strata to be elliptical so that Y_2, Y_3,...vanish. Thus

$$p' - P\alpha(\bar{y} + \zeta) = \text{constant} + \frac{5}{2} j P \left(\mu^2 - \frac{1}{3}\right).$$

Laplace omits the constant. The coefficients of $\mu^2 - \frac{1}{3}$ in $-\bar{y}$ and $-\zeta$ are supposed to be $\bar{h}$ and h'' respectively. Let αqP be the coefficient of this term in p'. Then from the above equation

$$\alpha q + \alpha (\bar{h} + h'') = \frac{5}{2} j.$$

This is in fact Clairaut's theorem applied to the supposed atmosphere surrounding the earth. It will be observed that $\alpha(\bar{h} + h'')$ is the ellipticity of the surface of the atmosphere, and therefore of the sea, since one of these surfaces is at a small constant distance from the other.

Let p denote the gravity at the surface of the earth corresponding to p'; thus, according to Laplace,

$$p = p' + 2\alpha(l - \zeta) P.$$

If then $\alpha\bar{q}P$ be the coefficient of the term $\mu^2 - \frac{1}{3}$ in p we must have

$$\bar{q} = q + 2h''.$$

But the correct formulæ should be

$$p = p' + 2\alpha\zeta P, \quad \bar{q} = q - 2h''.$$

The result obtained by Laplace would give, he remarks, the difference $\alpha h''$ of the ellipticities of the atmosphere and of the terrestrial spheroid, if we knew by the pendulum experiments the values of $\bar{q}$ and q. But he says it follows from the experiments which have been made for the most part at the level of the sea or a little above it that $\alpha h''$ is very small and almost insensible.

I presume he means that $\bar{q}$ is determined directly by observation, and q is deduced by allowing for the difference in elevation, which is ascertained by the aid of the barometer. But I find it difficult to catch precisely Laplace's train of thought. The words which immediately follow, on his page 54, "La surface de l'atmosphère supposée"... seem to me the commencement of a new paragraph, and they should, I think, have been so distinguished in printing. Laplace is actually about to investigate the effect of an elevated plateau, like that on which Quito is situated, on the value of gravity.

1332. Accordingly on his pages 55 and 56 Laplace considers the effect of the attraction of a mountain. He obtains the common result which is now in elementary books; namely, $2\pi\rho_1 k$ for the attraction, where ρ_1 is the density and k the height of the mountain. Laplace applies this to an experiment recorded by Bouguer, and infers that the density of the mountains near Quito is about one-fifth of the mean density of the Earth, that is about the density of water. Plana has touched on the subject in the first of his three memoirs cited in Art. 1324.

1333. The original memoir by Laplace contains in its pages 178...182 matter which is not reproduced in the *Mécanique Céleste*. This relates principally to the influence of the attraction of a mountain on the measure of the degrees of the meridian. It is interesting and not difficult.

1334. The memoir terminates with some remarks on the stability of the figure of the Earth; see the pages 182...184 of the memoir. These remarks relate to the subject discussed in the third Chapter of the Eleventh Book, namely, the axis of rotation of the Earth. The remarks bear upon the case in which the sea is not supposed to cover the whole earth; they are not reproduced in the *Mécanique Céleste*, though Laplace alludes to the same matter on his page 71.

1335. In leaving Chapter II. of the Eleventh Book of the *Mécanique Céleste*, I may state that there are numerous misprints; and most of them are reproduced in the national edition.

It will be seen from our analysis that the Chapter contains important matter, and that it is original.

1336. The third Chapter of the Eleventh Book of the *Mécanique Céleste* is entitled *De l'axe de rotation de la Terre:* it occupies pages 57...71.

1337. The pages 57...67 are devoted to the investigation of the theorem which we have given in Art. 1282. As I have already remarked, Laplace's original investigation is much simpler than that which he gives here.

But it is true he establishes something more here. We know that at any point of a given mass a system of principal axes can be found; Laplace himself gives a demonstration of this in his Livre I. § 27. It might seem at first sight that he is giving again a demonstration of this theorem; thus he says on his page 63, "L'existence d'un pareil axe est donc toujours possible...." However, what he really shews is something different, namely, that if we reject the square of the usual small quantity a, the conditions necessary for the existence of principal axes can be satisfied. But it does not seem to me that for his main purpose this result is of any importance.

1338. Let us pass on to his next point, which was not in the original memoir. Suppose the Earth covered by the sea to be in relative equilibrium, rotating round one of the principal axes through the centre of gravity; if it be made to rotate round one of the other principal axes instead of the actual axis the figure of the sea would change. The three figures which can thus be obtained by taking in succession the three principal axes, have between themselves some simple relations which are interesting to know.

Suppose the whole mass rotating round an axis, which we will call the first principal axis. With the notation of Art. 1303, we have for the radius vector of the surface of the sea

$$a\{1 + al + a(\overline{Y}_1 + Z_1) + a(Y_2 + Z_2) + a(Y_3 + Z_3) + \dots\},$$

where l is a constant, such that aal expresses the mean depth of the sea.

If αm denote the whole volume of the sea, we have

$$\alpha m = a^2 \int_0^{2\pi} \int_{-1}^{1} \alpha l \, d\phi \, d\mu ;$$

therefore $$\alpha m = 4\pi a^2 \alpha l \text{ .... } (21).$$

Now we have by Art. 1309,

$$Z_i = \frac{-\overline{Y}_i (2i+1) a^i \phi (a) + \int_0^a \rho \frac{d}{da} (a^{i+3} Y_i) \, da}{\{(2i+1) \phi (a) - a^3 \sigma\} a^i},$$

except when $i = 2$, and then we must add a certain term to the numerator on the right-hand side.

Hence $$Z_i + \overline{Y}_i = \frac{\int_0^a (\rho - \sigma) \frac{d}{da} (a^{i+3} Y_i) \, da}{\{(2i+1) \phi(a) - a^3 \sigma\} a^i} \text{ .... } (22),$$

except when $i = 2$; and then we have

$$Z_2 + \overline{Y}_2 = \frac{\int_0^a (\rho - \sigma) \frac{d}{da} (a^5 Y_2) \, da - \frac{5}{2} \frac{j}{a} \left(\mu_1^2 - \frac{1}{3}\right) a^2 \phi (a)}{\{5\phi (a) - a^3 \sigma\} a^2}.$$

Let u denote the sum of the values of the expression on the right-hand side of (22) from $i = 1$ to $i =$ infinity. Then the radius vector of the sea becomes

$$a + a\alpha l + a\alpha u - \frac{\frac{5}{2} aj \left(\mu_1^2 - \frac{1}{3}\right) \phi (a)}{5\phi (a) - a^3 \sigma}.$$

Now let us suppose that the whole mass turns round the second principal axis. What we denoted by μ_1 may now be denoted by μ_2, so that we have for the radius vector

$$a + a\alpha l + a\alpha u - \frac{\frac{5}{2} aj \left(\mu_2^2 - \frac{1}{3}\right) \phi (a)}{5\phi (a) - a^3 \sigma}.$$

Laplace says it is clear that αl and αu corresponding to the same point of the sea are the same as before. This is obvious with respect to l from equation (21). With respect to u, I suppose we see them to be the same by the following argument:

it is obvious that u does not involve the angular velocity, and if this angular velocity be zero, the two expressions denoted by u must be the same; hence they must always be the same.

In like manner for the third principal axis we find that the radius vector of the surface of the sea will be

$$a + a\alpha l + a\alpha u - \frac{\frac{5}{9}\alpha j\left(\mu_3^2 - \frac{1}{3}\right)\phi(a)}{5\phi(a) - a^3\sigma}.$$

Now we know that

$$\mu_1^2 + \mu_2^2 + \mu_3^2 = 1 \qquad (23),$$

so that if we take the mean of the three values of the radius vector we obtain

$$a + a\alpha l + a\alpha u,$$

which is independent of the angular velocity of rotation, and is the same as the radius vector of the sea, supposed in equilibrium on the earth without any rotation.

Laplace does not use equation (23), but proceeds in a less simple manner. By a misprint, followed in the national edition, he omits the term $\phi(a)$ in the numerator in the expression for the radius vector in the second and third cases.

I shall give some remarks which may perhaps be of service to a student of this Chapter of Laplace. The Chapter is original; but it does not seem to me very important.

1339. On page 57, at the beginning, Laplace says that the origin is supposed to be at the centre of gravity of the spheroid. By spheroid here he means the solid part of the Earth. But he really does not make any use of the supposition that the origin is at the centre of gravity of the solid part: the origin may be at any fixed point very near this centre of gravity.

1340. Let μ_1 and ϕ_1 denote the usual angular polar coordinates. Let there be a fixed radius for which μ_1 is $\cos\Theta$, and ϕ_1 is Φ. Let μ and ϕ be the polar coordinates referred to this fixed radius as a new axis, the angle ϕ being counted from the meridian which contains the original axis and the new axis.

Then Laplace gives, on his page 58, the following formulæ which connect the old and the new polar coordinates:

$$\mu = \cos\Theta\,\mu_1 + \sin\Theta\,\sqrt{(1-\mu_1^2)}\cos(\phi_1 - \Phi),$$

$$\sqrt{(1-\mu^2)}\sin\phi = \sqrt{(1-\mu_1^2)}\sin(\phi_1 - \Phi).$$

These are obvious from Spherical Trigonometry. He says they lead to

$$\sqrt{(1-\mu^2)}\cos\phi = \cos\Theta\,\sqrt{(1-\mu_1^2)}\cos(\phi_1 - \Phi) - \sin\Theta\,\mu_1.$$

The truth of this last result may be shewn thus. If we square and add our three equations we shall obtain an identity; so that as the other two are known to hold this must hold. But in this way we are left in doubt whether the sign on the left-hand side should not be negative. The best way of verifying the formula is to use Spherical Trigonometry. If we employ the ordinary notation the formula becomes

$$-\sin c\cos B = \cos a\sin b\cos C - \sin a\cos b.$$

If we substitute for $\cos B$ and $\cos C$ their known values, we shall find that this is always true.

1341. On his page 61 Laplace says: "Pour que le centre de gravité de la Terre soit libre, et dans l'axe principal de rotation,"... I see no meaning in the words *soit libre.*

1342. On his page 63 Laplace says: "L'existence d'un pareil axe est donc toujours possible...." These words seem to me premature; for it is not until page 65 that Laplace discusses the equations he has obtained, and shews that they always have a real solution.

1343. On his page 63 Laplace quotes the equation we have given in Art. 1338, connecting what we call $\overline{Y}_2$ and Z_2. Laplace adds "les quantités Z_2, $\overline{Y}_2$, Y_2, se rapportant ici à l'axe des μ. Mais rapportées à l'axe des μ_1, elles restent les mêmes:..." I cannot understand what is meant by the last four words I have quoted. It seems to me that when we change our axes, Z_2, $\overline{Y}_2$ and Y_2 *do not remain the same;* but that they are transformed by the aid of such formulæ as we have given in Art. 1340.

1344. Some troublesome misprints which occur in Laplace's edition and are preserved in the national edition may be noticed.

On page 63 in line 4 for *H* read Π.

On page 65 at the bottom there is a letter q which carries a bar, a dash, and another letter: the dash should be omitted.

On page 67 Laplace says "on aura $q^{(i)} = 0$." He ought to say "on aura $\int (\rho - 1) \cdot \frac{d}{da} (a^4 q^{(i)})\, da = 0$."

1345. The fourth Chapter of the Eleventh Book of the *Mécanique Céleste* is entitled *De la chaleur de la Terre, et de la diminution de la durée du jour par son refroidissement*; it occupies pages 72...85.

1346. This Chapter belongs rather to the researches on the theory of Heat, by Fourier, Poisson, and others, than to our proper subject. I do not profess to have verified the numerical calculations, but I have gone over the analysis, and shall make a few remarks which may be of service to the student. The Chapter is substantially reproduced from the *Connaissance des Tems* for 1823; see Art. 1289: but the last page is new.

1347. Laplace starts with two fundamental equations given by Fourier; he says "j'en donnerai la démonstration, dans un autre livre." I do not find that this intention was carried out.

1348. Laplace arrives at the differential equation

$$\frac{d^2 q'}{dx^2} + q' - \frac{i(i+1)}{x^2} q' = 0,$$

and gives a process of solution.

In the original memoir he omitted the process and referred to Legendre's memoir of 1789: see Art. 942.

1349. As I have stated in Art. 942, the solution of the equation is now known to take the following compact form,

$$\frac{C}{x^i} \frac{d^i}{da^i} \frac{\sin (x\sqrt{a} + H)}{\sqrt{a}},$$

where unity is to be substituted for a after the differentiation.

Laplace requires the expression to be finite when x is zero; this cannot be unless $B=0$.

We shall put y_n for $\frac{d^n}{da^n}\frac{\sin x\sqrt{a}}{\sqrt{a}}$, it being supposed that unity is substituted for a after the differentiations.

Now y_n vanishes when $x=0$. Suppose x to increase from zero, it is important for Laplace to know when y_n first vanishes after $x=0$.

It is obvious that y_0 first vanishes when $x=\pi$. Laplace says that y_1 first vanishes when x is between π and $\frac{3\pi}{2}$, that y_2 first vanishes when x is between $\frac{3\pi}{2}$ and 2π, that y_3 first vanishes when x is between 2π and $\frac{5\pi}{2}$, "et ainsi du reste": see his page 76.

1350. It will be found that

$$y_1=-\frac{1}{2}(\sin x-x\cos x),$$

$$y_2=\frac{3}{4}\left\{\sin x\left(1-\frac{x^2}{3}\right)-x\cos x\right\},$$

$$y_3=-\frac{15}{8}\left\{\sin x\left(1-\frac{2x^2}{5}\right)-x\cos x\left(1-\frac{x^2}{15}\right)\right\};$$

and thus Laplace's statements with respect to y_1, y_2 and y_3 may be verified. The following formula will be useful:

$$\begin{aligned}\frac{dy_n}{dx}&=\frac{d^n}{da^n}\frac{d}{dx}\frac{\sin x\sqrt{a}}{\sqrt{a}}\\&=\frac{d^n}{da^n}\cos x\sqrt{a}\\&=-\frac{x}{2}\frac{d^{n-1}}{da^{n-1}}\frac{\sin x\sqrt{a}}{\sqrt{a}}\\&=-\frac{x}{2}y_{n-1}.\end{aligned}$$

Thus y_n continually increases numerically as x changes from zero to the value for which y_{n-1} first vanishes.

1351. At the bottom of his page 82 Laplace gives a formula which reduced to his degree of approximation amounts to this:

$$\int_0^a \frac{r^s}{a^{s+1}} \sin \frac{\pi r}{a}\, dr = \frac{1}{\pi}\left\{1 - \frac{s(s-1)}{\pi^2} + \frac{s(s-1)(s-2)(s-3)}{\pi^4} - \dots\right\}.$$

But it ought to be stated that if s is *even*, the last term within the brackets must be doubled.

Thus, for example, $\int_0^a \frac{r^2}{a^3} \sin \frac{\pi r}{a}\, dr$ is not $\frac{1}{\pi}\left(1 - \frac{2}{\pi^2}\right)$, but $\frac{1}{\pi}\left(1 - \frac{4}{\pi^2}\right)$. Laplace seems to have gone wrong here. Thus a coefficient at the top of his page 83, which he makes to be $3 - e\left(1 - \frac{8}{\pi^2}\right)$, should be $3 - e\left(1 - \frac{16}{\pi^4}\right)$. This correction will affect some of his numerical results on his page 84.

1352. We may observe that Plana, by using a different value of the quantity denoted by e, strengthens Laplace's conclusions as to the permanence of the length of a day: see the *Astronomische Nachrichten*, Vol. XXXV. page 183.

1353. There are some misprints in this Chapter which do not occur in the original memoir nor in the national edition.

1354. The contributions made by Laplace to our subject, which are contained in his fifth volume, fall below those of his earlier years in interest and importance; but they are not unworthy of his eminent reputation. Those in the second Chapter seem the most remarkable, and may be said to consist of three parts. We have the process by which, instead of supposing fluid to cover the whole surface of the Earth, an investigation is given which may apply to the actual constitution of the Earth and sea; we have the important theorems respecting the approximate values of Laplace's coefficients; and lastly, there is the discussion of a certain hypothetical law connecting the pressure with the density. The later volume seems more obscure than the earlier volumes, and is certainly more disfigured by misprints; these defects may probably be attributed to the infirmity of advancing age, and may well be excused in the closing years of a life so full of great scientific achievements.

CHAPTER XXXV.

POISSON.

1335. I HAVE undertaken to carry the history of the theories of Attraction and of the Figure of the Earth down to the researches of Laplace, so that I shall not in general pass beyond the end of the first quarter of the present century. But I propose to make exceptions with respect to Poisson, Ivory, and Plana, and to give an account of all the contributions of these mathematicians to our subject. The labours of all three connect themselves closely and naturally with the matters we have already discussed; Poisson and Plana especially may be regarded as disciples and successors of Laplace, and may be conveniently and justly associated with him in mathematical history.

The present Chapter will be devoted to Poisson.

1336. The writings of Poisson, arranged chronologically, which may be considered as belonging to our subject, are the following, according to the list of his works and memoirs drawn up by himself and published at Paris in 1851:

I. Leçons de Mécanique. One volume in 4to. I have never seen this.

II. Traité de Mécanique. First edition in two volumes in 8vo. I have not seen this edition, which appeared I think in 1811.

III. Mémoire sur la Distribution de l'électricité à la surface des corps conducteurs.

Second Mémoire sur le même sujet.

IV. Extrait d'un Mémoire de M. Yvori sur l'attraction des Ellipsoïdes homogènes.

V. Addition à l'article précédent.

I have already noticed IV. and V.: see Art. 1160.

VI. Remarques sur une Equation qui se présente dans la théorie des attractions des sphéroïdes.

I have already noticed this: see Art. 1237.

VII. Mémoire sur la distribution de la Chaleur dans les corps solides.

I have already noticed this: see Art. 1223.

VIII. Observations relatives à un Mémoire de M. Ivory, sur l'Equilibre d'une masse fluide. *Annales de Chimie*...1824.

This criticises an assumption made by Ivory which will be noticed hereafter.

IX. Annonce de mon Mémoire sur l'Attraction des sphéroïdes. *Nouveau Bulletin*...*Philomatique* 1826.

This is a notice, extending to about a dozen pages of the memoir numbered X.

X. Mémoire sur l'Attraction des sphéroïdes. *Connaissance des Tems* 1829.

XI. Note sur une formule relative à l'Attraction des sphéroïdes *Philosophical Magazine* 1827.

XII. Additions au Mémoire sur l'Attraction des sphéroïdes. *Connaissance des Tems* 1831.

XIII. Traité de Mécanique. Second edition 1833.

XIV. Mémoire sur l'Attraction d'un ellipsoïde homogène. Paris *Mémoires* for 1835.

XV. Note sur l'Attraction d'un ellipsoïde hétérogène. *Connaissance des Tems* 1837.

XVI. Note sur un passage de la *Mécanique céleste*.

I have already noticed this: see Art. 1265.

XVII. Remarques à l'occasion d'un Rapport relatif à l'Attraction des ellipsoïdes. *Comptes Rendus*...Vol. VI.

Addition à ces Remarques. *Comptes Rendus*...Vol. VII.

XVIII. Note sur une Propriété générale des formules relatives aux attractions des sphéroïdes. *Comptes Rendus*...Vol. VII.

I proceed to give an account of such of these writings as have not been already noticed; the first of these is that numbered III.

1357. Two memoirs by Poisson entitled *Sur la distribution de l'électricité à la surface des corps conducteurs* are contained in the *Mémoires...de l'Institut* for 1811. The subject of the distribution of electricity is connected with that of attraction; but it is too extensive and important to be included in the present work: I must therefore content myself with expressing the hope that it may soon find its own historian.

Here I will only just notice the proof of Coulomb's theorem which was supplied by Laplace to Poisson, and inserted by the latter in his first memoir: see pages 5 and 29 of the memoir.

I have already explained the theorem in Art. 993. The present proof resembles that, in dividing the film into the two parts which I call S and S'; but differs in another respect. Here it is observed that if the film is spherical we have obviously $4\pi\rho$ for the joint action of S and S' at P'; *this is independent of the radius of the sphere.*

Now whatever be the form of the film we may consider the part S cut into elements by planes which all pass through the common normal, the angle between two consecutive planes being infinitesimal. Then we may admit that the action of any element will be the same at P' as the action of a spherical element of the same curvature. And in this way we obtain $2\pi\rho$ for the whole action of S. Then as in Coulomb's proof we obtain $4\pi\rho$ for the action of S and S', since the two parts will exert equal actions.

Some remarks for the purpose of rendering the demonstration more rigorous are given by Plana in his *Mémoire sur la distribution de l'électricité*...Turin, 1845: see page 23 of the memoir.

1358. In the *Connaissance des Tems* for 1829, published in 1826, there is a memoir by Poisson entitled *Mémoire sur l'attraction des Sphéroïdes.* The memoir occupies pages 329...379 of the volume. There is an *Addition* to the memoir in the *Connaissance des Tems* for 1831, published in 1828; this occupies pages 49...57 of the volume.

1359. The memoir is divided into three sections. The first section is entitled *Formules préliminaires;* it occupies pages 329...353. The second section is entitled *Formules relatives aux attractions des corps quelconques;* it occupies pages 354...364. The third section is entitled *Formules relatives aux sphéroïdes très-peu différens d'une sphère;* it occupies pages 364...379. The memoir may be said to form a new edition, with important improvements, of Laplace's researches on the subject.

1360. The first section constitutes a treatise on *Laplace's functions.* Poisson discussed these functions in a peculiar manner; he seems to have attached great importance to his process, and repeats it in various places. He refers to the *Journal de l'École Polytechnique,* 19ᵉ cahier, page 145. Poisson shews that a function of two variables, θ and ψ, can be expanded in a series of Laplace's coefficients; the expansion holds for values of θ between 0 and π, and of ψ between 0 and 2π. Exceptions may occur at the limiting values of the variables.

It is unnecessary to enter on Poisson's method here, because it is not our principal subject, but belongs rather to the history of the theory of Laplace's functions. Moreover it is readily accessible; for instance Poisson repeats it in the eighth Chapter of his *Théorie Mathématique de la Chaleur.* Some account of it will be found in Pratt's treatise on the *Figure of the Earth.*

1361. Supposing that a function has been expanded in a series of Laplace's coefficients, it will have to be investigated whether we shall continue to obtain equalities if we integrate or differentiate both members of the equation with respect to either of the variables θ and ψ. Poisson discusses this important point with great care. He shews that the formulæ obtained by integration are subject to no restriction; but those obtained by differentiation are liable to exceptions at the extreme values of the variables. He refers on page 337 to researches of his own on the subject "dans les derniers cahiers du *Journal de l'École Polytechnique,...*"

1362. Poisson's second section gives formulæ for the attraction of any body, expressed by means of the potential function as we call it.

Let r', θ', ψ' be the coordinates of any element of the attracting body; let ρ' be the density at that point. Let r, θ, ψ be the coordinates of the attracted particle. Let P_n' be Laplace's n^{th} coefficient, a function of θ' and ψ'. Let u denote the radius of the surface corresponding to θ' and ψ'.

Poisson uses $d\omega$ for $\sin\theta'\, d\theta'\, d\psi'$, so that $d\omega$ may be considered to be the element of the surface of a sphere of radius unity which is described with the origin as centre. The double integration with respect to θ' and ψ' may be replaced by a symbol of single integration; the integration will extend over the whole or some definite part of the spherical surface.

1363. It is almost unnecessary to write down the formula for V when r is greater than any value of u, as it has been given before. But for convenience we may repeat it.

$$V = \Sigma\left[\frac{1}{r^{n+1}}\int\left(\int_0^u \rho' r'^{n+2}\, dr'\right) P_n'\, d\omega\right] \text{ .... (1).}$$

Here, and throughout, Σ denotes summation with respect to n from $n = 0$ to $n = \infty$.

Next consider such an internal point that r is less than any value of u. Then

$$V = \Sigma\left[\frac{1}{r^{n+1}}\int\left(\int_0^r \rho' r'^{n+2}\, dr'\right) P_n'\, d\omega\right]$$
$$+ \Sigma\left[r^n\int\left(\int_r^u \frac{\rho'\, dr'}{r'^{n-1}}\right) P_n'\, d\omega\right] \text{ .... (2).}$$

In the last place suppose the external or internal point is such that r is greater than some values of u, but less than other values. Then

$$V = \Sigma\left[\frac{1}{r^{n+1}}\int'\left(\int_0^u \rho' r'^{n+2}\, dr'\right) P_n'\, d\omega\right]$$
$$+ \Sigma\left[\frac{1}{r^{n+1}}\int_{,}\left(\int_0^r \rho' r'^{n+2}\, dr'\right) P_n'\, d\omega\right]$$
$$+ \Sigma\left[r^n\int_{,}\left(\int_r^u \frac{\rho'\, dr'}{r'^{n-1}}\right) P_n'\, d\omega\right] \text{ .... (3).}$$

The integral indicated by $\int$ extends to those directions of r' in which u is less than r; and the integrals indicated by $\int_{,}$ to those directions of r' in which u is greater than r.

The limits of the integrals $\int$ and $\int_{,}$ relative to θ' and ψ' depend implicitly on the position of the attracted point, that is on its coordinates r, θ, and ψ; and it will be necessary to remember this when we differentiate V with respect to these variables. But we shall find that this will not affect the differential coefficients of V of the *first* order.

In fact we have identically

$$\int_0^u \rho' r'^{n+2}\, dr' = \int_0^r \rho' r'^{n+2}\, dr' + \int_r^u \rho' r'^{n+2}\, dr';$$

$$\int\left(\int_0^u \rho' r'^{n+2}\, dr'\right) P_n'\, d\omega + \int_{,}\left(\int_0^r \rho' r'^{n+2}\, dr'\right) P_n'\, d\omega$$

$$= \int\!\!\left(\int_0^r \rho' r'^{n+2}\, dr'\right) P_n'\, d\omega,$$

where the symbol $\int$ indicates the complete integral extended to all the directions of r'.

Therefore the last formula for V may be expressed thus:

$$V = \Sigma\left[\frac{1}{r^{n+1}}\int\!\!\left(\int_0^r \rho' r'^{n+2}\, dr'\right) P_n'\, d\omega\right]$$

$$+\,\Sigma\left[\frac{1}{r^{n+1}}\int\left(\int_r^u \rho' r'^{n+2}\, dr'\right) P_n'\, d\omega\right]$$

$$+\,\Sigma\left[r^n\int_{,}\left(\int_r^u \frac{\rho'\, dr'}{r'^{n-1}}\right) P_n'\, d\omega\right] \qquad (4).$$

Now the differential coefficient of V with respect to r, θ, or ψ will not involve any term arising from the variability of the limits in the integrals denoted by $\int$ and $\int_{,}$. For consider the former integral; a differential coefficient, so far as it depends on the variability of the limits, will have as a factor the value, *at the limit* considered, of $\int_r^u \rho' r'^{-n+2} dr'$; but at this limit by supposition $r = u$; and so this factor vanishes.

1364. Poisson says, and quite correctly, that commonly only two formulæ for V had been given, namely (1) and (2); and it had been assumed that (1) held for *all external* points, and (2) for *all internal* points. It is however obvious that there are external points, and there are internal points, for which the correct form is (3) or its equivalent (4). For example, if we consider a homogeneous ellipsoid, and place the origin at the centre, the formula (1) applies only to such external points as have the radius vector r greater than the greatest of the three semiaxes of the ellipsoid; and the formula (2) applies only to such internal points as have the radius vector r less than the least of the three semiaxes.

1365. The terms in (1) and (2) considered as functions of θ and ψ will be Laplace's functions; this depends on the fact that the limits of the integrations are independent of θ and ψ. But this will not be the case with the terms in (3) and (4).

With respect to an internal particle we may always take the origin so that r does not exceed any value of u; then the formula (2) will be applicable.

Accordingly Poisson makes an application of (2) to establish the correction which he had introduced in Laplace's fundamental equation for V with respect to an internal particle: see Art. 1237.

Since P_n satisfies the equation

$$\frac{1}{\sin\theta}\frac{d}{d\theta}\left(\sin\theta\frac{dP_n}{d\theta}\right) + \frac{1}{\sin^2\theta}\frac{d^2P_n}{d\psi^2} = -n(n+1)P_n,$$

we have from (2)

$$\frac{1}{\sin\theta}\frac{d}{d\theta}\left(\sin\theta\frac{dV}{d\theta}\right) + \frac{1}{\sin^2\theta}\frac{d^2V}{d\psi^2}$$

$$= -\Sigma\left[\frac{n(n+1)}{r^{n+1}}\int\left(\int_0^r \rho' r'^{n+2} dr'\right) P_n' \, d\omega\right]$$

$$-\Sigma\left[n(n+1)r^n\int\left(\int_r^u \frac{\rho' dr'}{r'^{n-1}}\right) P_n' \, d\omega\right] \ldots\ldots\ldots\ldots(5).$$

If we differentiate rV once with respect to r we shall find that no term arises from the variability of the limits in the integrations with respect to r'. This result depends on the fact that

$$\frac{d}{dr}\int_0^r \rho' r'^{n+2}\, dr' = \rho r^{n+2},$$

and

$$\frac{d}{dr}\int_r^a \frac{\rho' dr'}{r'^{n-1}} = -\frac{\rho}{r^{n-1}};$$

ρ being the density at the point corresponding to r, θ, ψ.

Hence we have simply

$$\frac{d(rV)}{dr} = -\Sigma\left[\frac{n}{r^{n+1}}\int\left(\int_0^r \rho' r'^{n+2}\, dr'\right) P_n'\, d\omega\right]$$
$$+\Sigma\left[(n+1)\, r^n \int\left(\int_r^a \frac{\rho' dr'}{r'^{n-1}}\right) P_n'\, d\omega\right].$$

But when we differentiate this again with respect to r, we shall obtain terms from the variability of the limits. Thus we shall have

$$r\frac{d^2(rV)}{dr^2} = \Sigma\left[\frac{n(n+1)}{r^{n+1}}\int\left(\int_0^r \rho' r'^{n+2}\, dr'\right) P_n'\, d\omega\right]$$
$$+\Sigma\left[n(n+1)\, r^n \int\left(\int_r^a \frac{\rho' dr'}{r'^{n-1}}\right) P_n'\, d\omega\right]$$
$$-r^2\Sigma\,(2n+1)\int \rho' P_n'\, d\omega \;\ldots\ldots\ldots\ldots\ldots\ldots\ldots (6).$$

Hence from (5) and (6) we have for an internal particle

$$r\frac{d^2(rV)}{dr^2} + \frac{1}{\sin\theta}\frac{d}{d\theta}\left(\sin\theta\frac{dV}{d\theta}\right) + \frac{1}{\sin^2\theta}\frac{d^2V}{d\psi^2}$$
$$= -r^2\Sigma\,(2n+1)\int \rho' P_n'\, d\omega \ldots\ldots\ldots\ldots\ldots\ldots (7).$$

Now suppose ρ' developed in a series of Laplace's functions; so that

$$\rho' = Q_0' + Q_1' + Q_2' + \ldots + Q_n' + \ldots$$

Then, by Art. 1009, we have

$$\int \rho' P_n' \, d\omega = \frac{4\pi}{2n+1} Q_n.$$

Therefore the right-hand side of (7) becomes

$$-4\pi r^2 \Sigma Q_n,$$

that is $$-4\pi r^2 \rho.$$

Hence too if V be expressed as a function of the rectangular coordinates x, y, z, we shall have for an internal particle

$$\frac{d^2 V}{dx^2} + \frac{d^2 V}{dy^2} + \frac{d^2 V}{dz^2} = -4\pi\rho.$$

1366. Poisson on his pages 362...364 determines the value of V relative to a sphere for any point external or internal; the method which he uses has now passed into the elementary books: see *Statics*, Art. 240.

1367. In his third section Poisson applies his formulæ for V to the case of spheroids which differ but little from spheres. He begins with supposing the body homogeneous.

The radius of the surface is denoted by $a(1 + \alpha y')$ where a is a constant, being the radius of a sphere which differs but little from the spheroid, and α is very small; and the peculiarity of Poisson's investigation is that he does not limit himself to the first power of α, but retains in general all the powers of α.

If r is greater than the greatest value of u we take the formula (1); we can separate the integrals relative to r' into two parts, one extending from $r' = 0$ to $r' = a$, and the other from $r' = a$ to $r' = u$. In the first part the integrals will be constants; and by reason of the properties of the function P_n' we shall have simply

$$\Sigma \left[\frac{1}{r^{n+1}} \int \left(\int_0^a \rho' r'^{n+2} dr'\right) P_n' \, d\omega\right] = \frac{4\pi\rho a^3}{3r}.$$

Hence the complete value of V will be in this case

$$V = \frac{4\pi\rho a^3}{3r} + \rho\Sigma \left[\frac{1}{r^{n+1}} \int \left(\int_a^u r'^{n+2} dr'\right) P_n' \, d\omega\right] \text{.........} (8).$$

If r is less than the least value of u we take the formula (2). Then we have

$$\Sigma\left[\frac{1}{r^{n+1}}\int\left(\int_0^r \rho' r'^{n+2}\,dr'\right) P_n'\,d\omega\right] = \frac{4\pi\rho r^2}{3}.$$

The integrals with respect to r' which are taken between $r' = r$ and $r' = u$ we separate into two parts, one extending from $r' = r$ to $r' = a$, and the other from $r' = a$ to $r' = u$. For the first part we shall have

$$\Sigma\left[r^n\int\left(\int_r^a \frac{\rho'\,dr'}{r'^{n-1}}\right) P_n'\,d\omega\right] = 2\pi\rho\,(a^2 - r^2).$$

Hence the complete value of V in this case is

$$V = 2\pi\rho a^2 - \frac{2\pi\rho r^2}{3} + \rho\Sigma\left[r^n\int\left(\int_a^u \frac{dr'}{r'^{n-1}}\right) P_n'\,d\omega\right] \text{.........(9).}$$

Let y denote the value of y' when for θ' and ψ' we put θ and ψ respectively; and suppose y and its powers developed in series of Laplace's functions.

Let $$y = Y_0 + Y_1 + Y_2 + \ldots + Y_n + \ldots$$
and generally
$$y^i = Y_0^{(i)} + Y_1^{(i)} + Y_2^{(i)} + \ldots + Y_n^{(i)} + \ldots$$

Since $\int_a^u r'^m\,dr' = a^{m+1}\left(\alpha y' + \frac{m}{2}\alpha^2 y'^2 + \frac{m(m-1)}{2.3}\alpha^3 y'^3 + \ldots\right)$,

we shall have

$$\int\left(\int_a^u r'^{n+2}\,dr'\right) P_n'\,d\omega$$
$$= \frac{4\pi a^{n+3}}{2n+1}\left\{\alpha Y_n + \frac{n+2}{2}\alpha^2 Y_n^{(2)} + \frac{(n+2)(n+1)}{2.3}\alpha^3 Y_n^{(3)} + \ldots\right\},$$

$$\int\left(\int_a^u \frac{dr'}{r'^{n-1}}\right) P_n'\,d\omega$$
$$= \frac{4\pi a^{-n+2}}{2n+1}\left\{\alpha Y_n - \frac{n-1}{2}\alpha^2 Y_n^{(2)} + \frac{(n-1)n}{2.3}\alpha^3 Y_n^{(3)} - \ldots\right\}.$$

Hence the preceding values of V will present themselves as series arranged in powers of α; we shall have from (8)

$$V = \frac{4\pi\rho a^3}{3r} + \frac{4\pi\rho a^3}{r}\left\{\alpha\Sigma\,\frac{1}{2n+1}\,\frac{a^n}{r^n}\,Y_n\right.$$

$$\left. + \frac{\alpha^2}{2}\Sigma\,\frac{n+2}{2n+1}\,\frac{a^n}{r^n}\,Y_n^{(1)} + \frac{\alpha^3}{2\,.\,3}\Sigma\,\frac{(n+2)(n+1)}{2n+1}\,\frac{a^n}{r^n}\,Y_n^{(2)} + \dots\right\} \dots (10),$$

and we shall have from (9)

$$V = 2\pi\rho a^2 - \frac{2\pi\rho r^2}{3} + 4\pi\rho a^2\left\{\alpha\Sigma\,\frac{1}{2n+1}\,\frac{r^n}{a^n}\,Y_n\right.$$

$$\left. - \frac{\alpha^2}{2}\Sigma\,\frac{n-1}{2n+1}\,\frac{r^n}{a^n}\,Y_n^{(1)} + \frac{\alpha^3}{2\,.\,3}\Sigma\,\frac{(n-1)\,n}{2n+1}\,\frac{r^n}{a^n}\,Y_n^{(2)} - \dots\right\} \dots. (11).$$

1368. Poisson says that the formula (10) holds for external points, and the formula (11) for internal points, *provided the point is not too near the surface.* It had however been usual to neglect this condition, and to apply (10) for any external point, and (11) for any internal point. The matter requires examination, and accordingly Poisson proceeds to consider it, starting from equation (4), which has been rigorously demonstrated. As we have stated in Art. 843, he does not consider it sufficient that the series finally obtained are convergent; he holds that the series employed throughout the investigation should be convergent; see his page 360.

1369. Since the spheroid is supposed homogeneous, the first series contained in (4) is reduced to its first term; and is equal to $\frac{4\pi\rho r^2}{3}$.

Also whatever Q' may denote we have identically

$$\int' Q'\,d\omega + \int_{,} Q'\,d\omega = \int Q'\,d\omega.$$

Hence if we eliminate successively each of the partial integrals denoted by $\int'$ and $\int_{,}$ and put for abbreviation

$$\frac{1}{r^{n+1}}\int_r^a r'^{n+2}\,dr' - r^n\int_r^a \frac{dr'}{r'^{n-1}} = U,$$

the equation (4) will take either of the two following equivalent forms:

$$V=\frac{4\pi\rho r^2}{3}+\rho\Sigma\left[\frac{1}{r^{n+1}}\int\left(\int_r^a r'^{n+2}\,dr'\right)P_n'\,d\omega\right]-\rho\Sigma\int_{,} UP_n'\,d\omega,$$

$$V=\frac{4\pi\rho r^2}{3}+\rho\Sigma\left[r^n\int\left(\int_r^a \frac{dr'}{r'^{n-1}}\right)P_n'\,d\omega\right]+\rho\Sigma\int' UP_n'\,d\omega \;\ldots\ldots(12).$$

It will be sufficient to consider the first of these formulæ; the reasoning will apply without difficulty to the second.

Let $u=r-z'$, so that z' represents a function of θ' and ψ', the value of which is very small, and of the same order of magnitude as α, for those values of r which we have to consider.

If we effect the integrations in U we shall find that

$$U=\frac{2n+1}{2}z'^2-\frac{2n+1}{3r}z'^3+\frac{(2n+1)(n+1)n}{2.3.4r^2}z'^4-\ldots$$

The integrals denoted by $\int_{,}$ extend only to negative values of z'; but if we denote by ζ' a discontinuous function of θ' and ψ', such that we have $\zeta'=z'$, or $\zeta'=0$, according as z' is negative or positive, we can change $\int_{,}$ into the complete integral $\int$; and then we shall have

$$\int_{,} UP_n'\,d\omega=\frac{2n+1}{2}\int\zeta'^2P_n'\,d\omega-\frac{2n+1}{3r}\int\zeta'^3P_n'\,d\omega+\ldots\ldots\ldots(13).$$

At this stage Poisson limits the approximation to the order α^3 inclusive, by rejecting powers of ζ' above the third.

Now suppose ζ'^2 and ζ'^3 are developed in series of Laplace's functions; and it must be observed that discontinuous functions may be so developed: let then

$$\zeta'^2=X_0+X_1+X_2+\ldots+X_n+\ldots$$

$$\zeta'^3=Z_0+Z_1+Z_2+\ldots+Z_n+\ldots$$

Then by the known properties of Laplace's functions we shall have from (13),

$$\int_{,} UP_n'\,d\omega=2\pi X_n-\frac{4\pi}{3r}Z_n;$$

and therefore

$$\Sigma \int_{\prime} UP_n' \, d\omega = 2\pi\zeta^2 - \frac{4\pi}{3r}\zeta^3.$$

Now by the nature of ζ this is zero whenever the attracted point is outside the spheroid. Thus for all such points the first equation (12) reduces to

$$V = \frac{4\pi\rho r^2}{3} + \rho\Sigma\left[\frac{1}{r^{n+1}}\int\left(\int_r^a r'^{n+2}\,dr'\right)P_n'\,d\omega\right].$$

This will agree with (8) if we observe that

$$\int_r^a r'^{n+2}\,dr' = \int_0^a r'^{n+2}\,dr' - \int_0^r r'^{n+2}\,dr',$$

and that the part of the sum denoted by Σ which corresponds to the second integral becomes $\frac{4\pi r^2}{3} - \frac{4\pi a^3}{3r}$.

In the same manner it may be shewn that the sum $\Sigma\int' UP_n'\,d\omega$ which occurs in the second of equations (12) is zero whenever the attracted point is within the spheroid; so that for all internal points we shall have

$$V = \frac{4\pi\rho r^3}{3} + \rho\Sigma\left[r^n\int\left(\int_r^a \frac{dr'}{r'^{n-1}}\right)P_n'\,d\omega\right].$$

This will agree with (9) if we observe that

$$\int_r^a \frac{dr'}{r'^{n-1}} = \int_0^a \frac{dr'}{r'^{n-1}} + \int_r^0 \frac{dr'}{r'^{n-1}},$$

and that the part of the sum denoted by Σ which corresponds to the second integral becomes $2\pi a^2 - 2\pi r^2$.

1370. Poisson then says that it has been shewn that the two formulæ (8) and (9), or the two formulæ (10) and (11), which are the developments of them in convergent series, will apply to all positions of the attracted particle, namely (8) and (10) to all external positions, and (9) and (11) to all internal positions. If we differentiate the expression for V with respect to r, θ, and ψ we obtain in the usual way expressions for the resolved attraction.

For example, suppose we require the attraction resolved along the radius. We thus get from (10)

$$\frac{dV}{dr} = -\frac{4\pi\rho a^3}{3r^2} - \frac{4\pi\rho a^3}{r^2}\left[\alpha\Sigma\frac{n+1}{2n+1}\frac{a^n}{r^n}Y_n\right.$$

$$+\frac{\alpha^2}{2}\Sigma\frac{(n+2)(n+1)}{2n+1}\frac{a^n}{r^n}Y_n^{(1)}$$

$$\left.+\frac{\alpha^3}{2.3}\Sigma\frac{(n+1)^2(n+2)}{2n+1}\frac{a^n}{r^n}Y_n^{(2)}+\ldots\right]\ldots\ldots\ldots(14);$$

this holds for external particles.

And we get from (11)

$$\frac{dV}{dr} = -\frac{4\pi\rho r}{3} + \frac{4\pi\rho a^3}{r}\left[\alpha\Sigma\frac{n}{2n+1}\frac{r^n}{a^n}Y_n\right.$$

$$-\frac{\alpha^2}{2}\Sigma\frac{n(n-1)}{2n+1}\frac{r^n}{a^n}Y_n^{(1)}$$

$$\left.+\frac{\alpha^3}{2.3}\Sigma\frac{n^2(n-1)}{2n+1}\frac{r^n}{a^n}Y_n^{(2)}+\ldots\right]\ldots\ldots\ldots\ldots(15);$$

this holds for internal particles.

1371. I do not quite understand the view which Poisson takes of his results. Both here and in the latter part of his memoir he seems to imply that they are true for all powers of α, whereas he has only demonstrated them so far as α^2 inclusive. In a paper published in the *Proceedings of the Royal Society*, Vol. xx. 1872, I have extended Poisson's investigation to all powers of α.

1372. When the point considered is on the surface of the spheroid the values of V ought to coincide, as well as those of $\frac{dV}{dr}$. We will verify this coincidence as far as the order α^2 inclusive.

We put then $r = a(1+\alpha y)$. Thus (10) becomes

$$V = \frac{4\pi\rho a^2}{3} + 4\pi\rho a^2\alpha\left\{\Sigma\frac{1}{2n+1}Y_n - \frac{y}{3}\right\}$$

$$+4\pi\rho a^2\alpha^2\left\{\frac{1}{2}\Sigma\frac{n+2}{2n+1}Y_n^{(1)} - y\Sigma\frac{n+1}{2n+1}Y_n + \frac{y^2}{3}\right\}\ldots\ldots(16).$$

In the same way (11) becomes

$$V = \frac{4\pi\rho a^3}{3} + 4\pi\rho a^3\alpha\left\{\Sigma\frac{1}{2n+1}Y_n - \frac{y}{3}\right\}$$

$$+ 4\pi\rho a^3\alpha^2\left\{-\frac{1}{2}\Sigma\frac{n-1}{2n+1}Y_n^{(1)} + y\Sigma\frac{n}{2n+1}Y_n - \frac{y^2}{6}\right\}\text{......}(17).$$

Now it is obvious that in these two values of V the term without α is the same in both; so also the term involving α is the same in both. The terms involving α^2 will agree provided

$$\frac{1}{2}\Sigma\frac{n+2}{2n+1}Y_n^{(1)} - y\Sigma\frac{n+1}{2n+1}Y_n + \frac{y^2}{3}$$

$$= -\frac{1}{2}\Sigma\frac{n-1}{2n+1}Y_n^{(1)} + y\Sigma\frac{n}{2n+1}Y_n - \frac{y^2}{6};$$

this leads to $\qquad \frac{1}{2}\Sigma Y_n^{(1)} - y\Sigma Y_n + \frac{y^2}{2} = 0,$

that is $\qquad \frac{1}{2}\Sigma Y_n^{(1)} - y^2 + \frac{y^2}{2} = 0;$

and this is obviously true.

Now let us compare the values of $\frac{dV}{dr}$ for a point on the surface. We put then $r = a(1+\alpha y)$. Thus (14) becomes

$$\frac{dV}{dr} = -\frac{4\pi\rho a}{3} + 4\pi\rho a\alpha\left\{-\Sigma\frac{n+1}{2n+1}Y_n + \frac{2}{3}y\right\}$$

$$+ 4\pi\rho a\alpha^2\left\{-\frac{1}{2}\Sigma\frac{(n+2)(n+1)}{2n+1}Y_n^{(1)} + y\Sigma\frac{(n+2)(n+1)}{2n+1}Y_n - y^2\right\}\text{...}(18).$$

In the same way (15) becomes

$$\frac{dV}{dr} = -\frac{4\pi\rho a}{3} + 4\pi\rho a\alpha\left\{\Sigma\frac{n}{2n+1}Y_n - \frac{y}{3}\right\}$$

$$+ 4\pi\rho a\alpha^2\left\{-\frac{1}{2}\Sigma\frac{n(n-1)}{2n+1}Y_n^{(1)} + y\Sigma\frac{n(n-1)}{2n+1}Y_n\right\}.$$

Now it is obvious that in these two values of $\frac{dV}{dr}$ the term without α is the same in both. The terms involving α agree, for by equating them we arrive at the identity $-\Sigma Y_n + y = 0$. The terms involving a^2 agree, for by equating them we arrive at the identity $-\Sigma Y_n^{(1)} + 2y\Sigma Y_n - y^2 = 0$.

1373. Some of the coefficients which occur in the preceding Article admit of transformations which may be occasionally useful. Thus the coefficient of a^2 in (16) is transformed by Poisson in the following manner:

$$\frac{1}{2}\Sigma\frac{n+2}{2n+1}Y_n^{(1)} - y\Sigma\frac{n+1}{2n+1}Y_n + \frac{y^2}{3}$$

$$= \frac{1}{2}\Sigma\frac{n+2}{2n+1}Y_n^{(1)} - y\Sigma\frac{n+\frac{1}{2}+\frac{1}{2}}{2n+1}Y_n + \frac{y^2}{3}$$

$$= \frac{1}{2}\Sigma\frac{n+2}{2n+1}Y_n^{(1)} - \frac{y}{2}\Sigma Y_n - \frac{y}{2}\Sigma\frac{1}{2n+1}Y_n + \frac{y^2}{3}$$

$$= \frac{1}{2}\Sigma\frac{n+2}{2n+1}Y_n^{(1)} - \frac{y^2}{6} - \frac{y}{2}\Sigma\frac{1}{2n+1}Y_n$$

$$= \frac{1}{2}\Sigma\left(\frac{n+2}{2n+1} - \frac{1}{3}\right)Y_n^{(1)} - \frac{y}{2}\Sigma\frac{1}{2n+1}Y_n$$

$$= \frac{1}{6}\Sigma\frac{n+5}{2n+1}Y_n^{(1)} - \frac{y}{2}\Sigma\frac{1}{2n+1}Y_n.$$

1374. Let us proceed with Poisson to an application of the formulæ in Art. 1372.

Take the value of V from (16) and the value of $\frac{dV}{dr}$ from (18); thus we shall find that

$$\frac{dV}{dr} + \frac{1}{2a}V + \frac{2\pi\rho a}{3} = 2\pi\rho a\alpha\left\{\Sigma\frac{1}{2n+1}Y_n - \frac{y}{3} - \Sigma\frac{2n+2}{2n+1}Y_n + \frac{4}{3}y\right\}$$

$$+ 4\pi\rho a\alpha^2\left\{\frac{1}{4}\Sigma\frac{n+2}{2n+1}Y_n^{(1)} - \frac{y}{2}\Sigma\frac{n+1}{2n+1}Y_n + \frac{y^2}{6}\right.$$

$$\left. - \frac{1}{2}\Sigma\frac{(n+2)(n+1)}{2n+1}Y_n^{(1)} + y\Sigma\frac{(n+2)(n+1)}{2n+1}Y_n - y^2\right\}.$$

The term involving α disappears; that involving α^2 is

$$\pi\rho a\alpha^2\left\{-\Sigma(n+2)Y_n^{(1)} + 2y\Sigma(n+1)Y_n + 4y\Sigma\frac{n+1}{2n+1}Y_n - \frac{10y^2}{3}\right\}.$$

which may be reduced to

$$\pi\rho a\alpha^2\left\{-\Sigma n Y_n^{(1)} + 2y\,\Sigma n Y_n + 4y\,\Sigma \frac{n+1}{2n+1} Y_n - \frac{10y^2}{3}\right\}.$$

Poisson does not work out the term α^2 as it is given here; and I do not know that any previous writer has put it explicitly in this form. If we neglect the term in α^2 we arrive at Laplace's equation: see Art. 1196.

1375. Poisson for an application of his formulæ discusses the relative equilibrium of a mass of homogeneous fluid in rotation. His method presents some novelty.

In order that the figure may be nearly spherical, the centrifugal force must be small compared with the attraction. Accordingly he supposes that α expresses the ratio of the centrifugal force at the distance a from the axis to the mean attraction at the same distance; so taking a as the radius of a sphere equal in volume to that of the fluid, the mean attraction is $\frac{4\pi\rho a}{3}$, and the centrifugal force is $\frac{4\pi\rho a\alpha}{3}$. Hence at a distance x from the axis of rotation, the centrifugal force will be $\frac{4\pi\rho x\alpha}{3}$. Therefore, by the principles of Hydrostatics, the surface of the fluid will be determined by the equation

$$\frac{2\pi\alpha\rho x^2}{3} + V = \text{constant} \qquad\ldots\ldots\ldots\ldots(19),$$

where V and x relate to the same point of the surface.

Now Poisson does not assume as usual that the axis from which his θ is reckoned coincides with the axis of rotation, but only that the two straight lines are parallel. Let ϵ denote the distance between them. The plane containing these two straight lines is the plane from which the angle ψ is reckoned.

Since the distance from the axis of θ of a point on the surface is $a(1+\alpha y)\sin\theta$, we see that

$$x^2 = a^2(1+\alpha y)^2\sin^2\theta - 2a\epsilon(1+\alpha y)\sin\theta\cos\psi + \epsilon^2 \ldots\ldots(20).$$

In (19), substitute the value of x^2 from (20), and the value of V from (16); then collecting the constant terms, we have

$$\text{constant} = \frac{a}{2}(1+\alpha y)^2\sin^2\theta - \frac{\alpha e}{a}(1+\alpha y)\sin\theta\cos\psi + 3\alpha\left(\Sigma\frac{1}{2n+1}Y_n - \frac{y}{3}\right)$$
$$+ 3\alpha^2\left\{\frac{1}{2}\Sigma\frac{n+2}{2n+1}Y_n^{(1)} - y\Sigma\frac{n+1}{2n+1}Y_n + \frac{y^2}{3}\right\} + \dots\dots\dots (21).$$

Since a is the radius of a sphere of equal volume, we have

$$\frac{a^3}{3}\iint(1+\alpha y)^3\sin\theta\, d\theta\, d\psi = \frac{4\pi a^3}{3} \dots\dots\dots\dots (22),$$

the integrals being taken for θ from 0 to π, and for ψ from 0 to 2π.

If we expand y, y^2, and y^3 in a series of Laplace's functions, we shall obtain by the aid of the fundamental properties of the functions

$$Y_0 + \alpha Y_0^{(1)} + \frac{1}{3}\alpha^2 Y_0^{(2)} = 0 \dots\dots\dots\dots\dots\dots (23),$$

where the notation is that of Art. 1367.

Then in order that nothing may be left undetermined it is convenient to fix the position of the origin of coordinates; let us take it at the centre of gravity of the mass. Then we have the three conditions

$$\iint(1+\alpha y)^4\cos\theta\sin\theta\, d\theta\, d\psi = 0,$$

$$\iint(1+\alpha y)^4\sin^2\theta\sin\psi\, d\theta\, d\psi = 0,$$

$$\iint(1+\alpha y)^4\sin^2\theta\cos\psi\, d\theta\, d\psi = 0.$$

Now if we substitute in these integrals the expansions of y, y^2, y^3, and y^4, all the terms disappear from the integrals except those which involve Laplace's coefficients of the first order. Moreover, each of these coefficients of the first order is of the form

$$h\cos\theta + h'\sin\theta\sin\psi + h''\sin\theta\cos\psi,$$

where h, h' and h'' are constants.

Hence the integrations can be completely effected; and by adding the results we obtain

$$Y_1 + \frac{3\alpha}{2}Y_1^{(1)} + \alpha^2 Y_1^{(2)} + \frac{\alpha^3}{4}Y_1^{(3)} = 0 \dots\dots\dots (24).$$

The equations (23) and (24) shew that Y_0 and Y_1 are both of the order α, on our hypothesis as to the value of a and the position of the origin; and as these terms are multiplied by α in the expression $a(1+\alpha y)$, they are to be neglected when we neglect quantities of the order α^2.

The quantity ϵ is unknown; but as only whole positive powers of α occur in equation (21) we see that ϵ may be represented thus

$$\epsilon = e + e'\alpha + e''\alpha^2 + \ldots,$$

where the coefficients e, e', e'',... are quantities independent of α, which have to be determined.

1376. Thus we see that the novelties in the process are these: Poisson has the *accurate* equations (23) and (24), of which Laplace used the approximate forms. Also Poisson does not *assume* that the axis of rotation passes through the centre of gravity; but takes the distance of the centre of gravity from the axis of rotation as one of the quantities to be determined.

1377. Let us proceed with Poisson's solution. Take the equation (21) and retain only the first powers of α. Then

$$\text{constant} = \frac{\alpha}{2}\sin^2\theta - \frac{\alpha e}{a}\sin\theta\cos\psi + 3\alpha\Sigma\left(\frac{1}{2n+1}Y_n - \frac{y}{3}\right),$$

that is,

$$\text{constant} = \frac{\alpha}{2}\sin^2\theta - \frac{\alpha e}{a}\sin\theta\cos\psi - 2\alpha\Sigma\frac{n-1}{2n+1}Y_n.$$

Equate to zero in the right-hand member the sum of the terms which relate to each index n, except $n=0$; then since $\sin^2\theta - \frac{2}{3}$ is of the nature of Y_2, and $\sin\theta\cos\psi$ is of the nature of Y_1, we see that Y_n must be zero for every value of n greater than 2. And also

$$e = 0,$$

$$Y_2 = \frac{5}{4}\left(\sin^2\theta - \frac{2}{3}\right).$$

Hence the radius of the surface, which is $a(1+\alpha y)$, becomes

$$a\left\{1+\frac{5\alpha}{4}\left(\sin^2\theta-\frac{2}{3}\right)\right\}.$$

If we wish to proceed to a second approximation we may put

$$y=\frac{5}{4}\left(\sin^2\theta-\frac{2}{3}\right)+\alpha z,$$

and suppose that z is expanded in a series of Laplace's Functions; so that $z=\Sigma Z_n$.

Neglect in (21) the powers of α above the second. It will be easily seen that Z_n vanishes if n be greater than 4. Also it will be found that $e'=0$, and $Z_3=0$; and the values of Z_2 and Z_4 will be obtained. The values of Z_0 and Z_1 must be obtained from (23) and (24); which will give, to the order of approximation with which we are concerned,

$$Z_0=-Y_0^{(1)},$$

$$Z_1=-\frac{3}{2}Y_1^{(1)};$$

and $Y_0^{(1)}$ and $Y_1^{(1)}$ are known by our first approximation, which indeed gives $Y_1^{(1)}=0$, neglecting α as we may here.

1378. We will work out the approximation to the second order, which Poisson only sketches, as by comparison with what we gave from Legendre's fourth memoir, the two processes will afford mutual verification.

The equation (21) gives us to the second order

$$\text{constant}=\frac{a}{2}(1+2\alpha Y_2)\sin^2\theta-\frac{\alpha e}{a}(1+\alpha Y_2)\sin\theta\cos\psi$$

$$-\frac{2\alpha}{5}Y_2-2\alpha^2\Sigma\frac{n-1}{2n+1}Z_n$$

$$+3\alpha^2\left\{\frac{1}{6}\Sigma\frac{n+5}{2n+1}Y_n^{(1)}-\frac{Y_2}{2}\cdot\frac{Y_2}{5}\right\},$$

where Y_2 stands for $\frac{5}{4}\left(\sin^2\theta-\frac{2}{3}\right)$.

The transformation of Art. 1373 has been used here.

Thus the equation reduces to

$$\text{constant} = a^2 Y_2 \sin^2\theta - \frac{a^2 e'}{a} \sin\theta \cos\psi - 2a^2 \Sigma \frac{n-1}{2n+1} Z_n$$
$$+ 3a^2 \left\{\frac{1}{6} \Sigma \frac{n+5}{2n+1} Y_n^{(1)} - \frac{1}{10} (Y_2)^2\right\}.$$

Now $\sin^2\theta = \frac{4}{3} Y_2 + \frac{2}{3}$. Hence

$$\text{constant} = \frac{2}{3} a^2 Y_2 - \frac{a^2 e'}{a} \sin\theta \cos\psi - 2a^2 \Sigma \frac{n-1}{2n+1} Z_n$$
$$+ \frac{a^2}{2} \Sigma \left(\frac{n+5}{2n+1} + 1\right) Y_n^{(1)}.$$

We may divide by a^2 which is constant. Since there is no term to balance Z_n when n is greater than 4, we see that then Z_n must vanish. Also $e' = 0$, for there is no term of the first order in Laplace's functions, except that in which e' occurs. In like manner $Z_3 = 0$. The terms Z_0 and $Y_0^{(1)}$ may be included in the constant. Hence finally

$$\text{constant} = \frac{2}{3} Y_2 - 2\left\{\frac{1}{5} Z_2 + \frac{1}{3} Z_4\right\} + \frac{6}{5} Y_2^{(1)} + Y_4^{(1)}.$$

Therefore
$$\frac{2}{5} Z_2 = \frac{2}{3} Y_2 + \frac{6}{5} Y_2^{(1)},$$

and
$$\frac{2}{3} Z_4 = Y_4^{(1)}.$$

But $Y_2 = -\frac{5}{6} P_2$, so that $(Y_2)^2 = \frac{25}{36} (P_2)^2$; and it will be found that $35 (P_2)^2 = 18 P_4 + 10 P_2 + 7$. See Art. 913.

Thus
$$Z_4 = \frac{3}{2} \cdot \frac{5}{14} P_4,$$

$$Z_2 = \frac{5}{3} Y_2 + \frac{3 \cdot 25}{7 \cdot 18} P_2 = -\frac{4}{7} \cdot \frac{25}{18} P_2.$$

And from Art. 1377 we find that

$$Z_0 = -\frac{5}{36}, \quad Z_1 = 0.$$

Hence we obtain for the radius vector of the surface

$$a\left\{1+\alpha Y_2+\alpha^2\left[\frac{3}{2}\cdot\frac{5}{14}P_4-\frac{4}{7}\cdot\frac{25}{18}P_2-\frac{5}{36}\right]\right\}.$$

It will be found that the term involving α^2 reduces to

$$\alpha^2\left[\frac{75}{32}\cos^4\theta-\frac{43}{7}\cdot\frac{25}{48}\cos^2\theta+\frac{25.37}{7.9.32}\right];$$

so that the radius vector of the surface becomes

$$a\left\{1+\frac{5}{4}\alpha\left(\sin^2\theta-\frac{2}{3}\right)+\alpha^2\left[\frac{75}{32}\cos^4\theta-\frac{43}{7}\cdot\frac{25}{48}\cos^2\theta+\frac{25.37}{7.9.32}\right]\right\}.$$

Let b denote the polar radius; then

$$b=a\left\{1-\frac{5}{4}\cdot\frac{2}{3}\alpha+\alpha^2\left[\frac{75}{32}-\frac{43}{7}\cdot\frac{25}{48}+\frac{25.37}{7.9.32}\right]\right\}.$$

Substitute for a in terms of b in the expression for the radius vector; and to the second order we shall find that the radius vector

$$=b\left\{1+\frac{5}{4}\alpha\sin^2\theta+\alpha^2\sin^2\theta\left[-\frac{75}{32}(1+\cos^2\theta)+\frac{43}{7}\cdot\frac{25}{48}+\frac{25}{24}\right]\right\}$$

$$=b\left\{1+\frac{5}{4}\alpha\sin^2\theta+\alpha^2\sin^2\theta\left(\frac{75}{32}\sin^2\theta-\frac{25}{56}\right)\right\}.$$

We may now compare this with Legendre's result.

The expression in equation (16) of Art. 914 becomes, when the body is homogeneous,

$$b\left\{1+\frac{5}{4}\epsilon\sin^2\theta+\frac{3.25}{4.28}\epsilon^2\sin^2\theta\left(4+\frac{7}{2}\sin^2\theta\right)\right\}.$$

Now ω being the angular velocity we have Legendre's $\epsilon=\dfrac{b^3\omega^2}{M}$, and Poisson's $\alpha=\dfrac{a^3\omega^2}{M}$; so that

$$\frac{\epsilon}{\alpha}=\frac{b^3}{a^3};$$

and therefore $\qquad \epsilon=\alpha-\dfrac{5}{2}\alpha^2$ approximately.

Using this value of κ in terms of α we shall find that Legendre's expression for the radius vector coincides with Poisson's.

To complete the comparison of the two results we may determine the ellipticity furnished by Poisson's process; this of course will agree with Legendre's. Let the equatorial radius be denoted by $b(1+\epsilon)$; then

$$b(1+\epsilon) = a\left\{1 + \frac{5}{12}\alpha + \frac{25.37}{7.9.32}\alpha^2\right\},$$

and
$$b = a\left\{1 - \frac{5}{6}\alpha - \frac{25}{63}\alpha^2\right\};$$

hence by division we find that to the order α^2

$$\epsilon = \frac{5}{4}\alpha + \frac{25.17}{7.32}\alpha^2.$$

Now Legendre's value, given in equation (17) of Art. 914, becomes when the body is homogeneous

$$\epsilon = \frac{5}{4}\kappa + \frac{75.15}{8.28}\kappa^2.$$

Put $\alpha - \frac{5}{2}\alpha^2$ for κ, and it will be found that this coincides with the value obtained by Poisson's process.

1379. Poisson concludes his discussion of the problem with some remarks; one passage has been quoted in Art. 1084.

1380. We will now return to equations (10) and (11). Suppose in succession two spheroids of the same matter, very little different from the same sphere. Let $a(1+\alpha y)$ denote the radius vector of one, and let $a(1+\alpha y+\alpha z)$ denote the radius vector of the other. Here z is supposed to be a given function of θ and ψ, as y is; and we suppose also that z may be developed in a series of Laplace's functions, which may be denoted by ΣZ_n. Let ΔV represent that part of V which arises from the matter between these two surfaces. Then if we neglect powers of α superior to the first, we have from (10) and (11) respectively

$$\Delta V = \frac{4\pi\rho a^3\alpha}{r}\Sigma\frac{1}{2n+1}\frac{a^n}{r^n}Z_n,$$

$$\Delta V = 4\pi\rho a^2\alpha\,\Sigma\frac{1}{2n+1}\frac{r^n}{a^n}Z_n.$$

The first formula supposes the attracted particle to be outside the outer surface, and the second formula supposes the attracted particle to be inside the inner surface. Let R_1 denote the action towards the centre in the first case, and R_2 that in the second case. Then

$$R_1 = \frac{4\pi\rho a^2\alpha}{r^2}\Sigma\frac{n+1}{2n+1}\frac{a^n}{r^n}Z_n,$$

$$R_2 = -\frac{4\pi\rho a^2\alpha}{r}\Sigma\frac{n}{2n+1}\frac{r^n}{a^n}Z_n.$$

If we make $r = a$ in these formulæ we see that

$$R_1 - R_2 = 4\pi\rho a\alpha\Sigma Z_n = 4\pi\rho a\alpha u.$$

This shews that if two particles are situated on the same radius, one at the outer surface of the stratum and the other at the inner surface of the stratum, the difference of the actions at these points in the direction of the radius is proportional to the thickness of the stratum and is the same as if the stratum were spherical.

Poisson adds:

On trouve une démonstration synthétique et plus générale de cette même proposition, dans mon premier Mémoire sur l'Electricité; M. Cauchy l'a aussi démontrée d'une autre manière dans le Bulletin de la Société Philomatique.

The synthetical demonstration is, I presume, more general from not assuming the form to be nearly spherical: see Art. 1357.

1381. The transition is easy and obvious from the formulæ that have been given to those which will apply to a heterogeneous body, in which the density is a function of the parameter a by which each stratum is determined.

Let c be the value of a at the surface. Then from (10) we see that for an external particle

$$V = \frac{4\pi}{r}\int_0^c \rho a^2 da + \frac{4\pi}{r}\left[\alpha\Sigma\frac{1}{(2n+1)r^n}Q_n + \frac{\alpha^2}{2}\Sigma\frac{n+2}{(2n+1)r^n}Q_n^{(1)} + \dots\right],$$

where $Q_n^{(i)}$ stands for $\int_0^c \rho\frac{d\,.\,a^{n+3}\,Y_n^{(i)}}{da}da.$

For an internal particle situated on the stratum of which the parameter is a_0 we shall have for V a formula consisting of two parts; one part is derived from (10) applied to the body so far as it is comprised between $a=0$, and $a=a_0$; and the other part is derived from (11) so far as it is comprised between $a=a_0$ and $a=c$. Thus

$$V=\frac{4\pi}{r}\int_0^{a_0}\rho a^2 da+\frac{4\pi}{r}\left[a\Sigma\frac{1}{(2n+1)r^n}A_n+\frac{a^2}{2}\Sigma\frac{n+2}{(2n+1)r^n}A_n^{(1)}+\dots\right]$$

$$+4\pi\int_{a_0}^{c}\rho a da+4\pi\left[a\Sigma\frac{r^n}{2n+1}B_n-\frac{a^2}{2}\Sigma\frac{(n-1)r^n}{2n+1}B_n^{(1)}+\dots\right]\dots(25),$$

where $A_n^{(i)}$ stands for $\int_0^{a_0}\rho\frac{d\,.\,a^{n+3}Y_n^{(i)}}{da}da$,

and $B_n^{(i)}$ stands for $\int_{a_0}^{c}\rho\frac{d\,.\,a^{2-n}Y_n^{(i)}}{da}da$.

Moreover we must remember that $r=a_0(1+\alpha y)$, so that if we substitute this value V will become a function of θ and ψ.

The subscript may be omitted from a_0 without danger in the use we shall make of the formula for V.

1362. We shall have to be careful in determining the values of the differential coefficients of V for an internal particle. Poisson says that the differential coefficients of V with respect to θ and ψ are to be taken before the substitution of the value of r; and hence neglecting α^2 we shall have from the fundamental equation of Laplace's functions applied to V,

$$\frac{1}{\sin\theta}\frac{d}{d\theta}\left(\sin\theta\frac{dV}{d\theta}\right)+\frac{1}{\sin^2\theta}\frac{d^2V}{d\psi^2}=$$

$$-\frac{4\pi a}{a}\Sigma\frac{n(n+1)}{(2n+1)a^n}A_n-4\pi a\Sigma\frac{n(n+1)a^n}{2n+1}B_n\dots(26).$$

It seems to me that Poisson ought also to have stated that the differential coefficients of V with respect to θ and ψ are formed on the supposition that r is constant. Hence when we require $\frac{dV}{d\theta}$

we must vary a_1 in such a manner that $\frac{d}{d\theta} a_1(1+\alpha y)=0$: but this will not have any influence to the order he has retained.

To find $\frac{dV}{dr}$ Poisson substitutes for r its value: thus neglecting α^2 we have

$$V=\frac{4\pi}{a}(1-\alpha y)\int_0^a \rho a^2 da + 4\pi\int_a^1 \rho a\, da$$
$$+\frac{4\pi\alpha}{a}\Sigma\frac{A_n}{(2n+1)a^n}+4\pi\alpha\Sigma\frac{a^n B_n}{2n+1} \text{ ...... (27).}$$

If we differentiate with respect to a we shall find that the terms which arise from the variation in the limits of the integrals cancel; and so we get

$$\frac{dV}{da}=-\frac{4\pi}{a^2}\left\{1-\alpha y+\alpha a\frac{dy}{da}\right\}\int_0^a \rho a^2 da$$
$$-\frac{4\pi\alpha}{a^2}\Sigma\frac{n+1}{(2n+1)a^n}A_n+\frac{4\pi\alpha}{a}\Sigma\frac{na^n}{2n+1}B_n.$$

But we have $\frac{dV}{da}=\frac{dV}{dr}\frac{dr}{da}=\frac{dV}{dr}\left(1+\alpha y+\alpha a\frac{dy}{da}\right)$; and hence, putting for $\frac{dV}{da}$ its value, it follows that

$$\frac{dV}{dr}=-\frac{4\pi}{a^2}(1-2\alpha y)\int_0^a \rho a^2 da$$
$$-\frac{4\pi\alpha}{a^2}\Sigma\frac{n+1}{(2n+1)a^n}A_n+\frac{4\pi\alpha}{a}\Sigma\frac{na^n}{2n+1}B_n \text{ ...(28).}$$

Poisson says that this is the same result as we should have obtained if we had differentiated V before the value of r was substituted, and had not varied a; but we should go wrong if we formed the value of $\frac{d^2V}{dr^2}$ in this other way.

The equations (27) and (28) give

$$\frac{d}{dr}(rV)=4\pi\int_a^1 \rho a\, da-\frac{4\pi\alpha}{a}\Sigma\frac{n}{(2n+1)a^n}A_n+4\pi\alpha\Sigma\frac{(n+1)a^n}{2n+1}B_n.$$

Differentiate with respect to a, and multiply by $r\frac{da}{dr}$, that is by $a\left(1-\alpha a\frac{dy}{da}\right)$. Then we shall get

$$r\frac{d^2(rV)}{dr^2} = -4\pi\rho a^2(1+2\alpha y)$$
$$+\frac{4\pi\alpha}{a}\Sigma\frac{n(n+1)}{(2n+1)a^n}A_n + 4\pi\alpha\Sigma\frac{n(n+1)}{2n+1}\frac{a^n}{}B_n \ldots (29),$$

after suppressing the terms which vanish since

$$y = \Sigma Y_n \text{ and } \frac{dy}{da} = \Sigma\frac{dY_n}{da}.$$

The formula in (29) includes the term $-4\pi\rho a^2(1+2\alpha y)$ which would not have appeared if we had differentiated twice with respect to r without varying a.

From (26) and (29) we have by addition

$$r\frac{d^2(rV)}{dr^2} + \frac{1}{\sin\theta}\frac{d}{d\theta}\left(\sin\theta\frac{dV}{d\theta}\right) + \frac{1}{\sin^2\theta}\frac{d^2V}{d\psi^2} = -4\pi\rho a^2(1+2\alpha y).$$

The right-hand member is $-4\pi\rho r^2$, since we have neglected α^2. Thus the result agrees with one already found, namely Poisson's correction of Laplace's fundamental equation for the case of an internal particle: see Art. 1365.

1383. There is nothing inadmissible in the way in which Poisson finds the value of $\frac{dV}{dr}$ for an internal particle; but I prefer another way. It seems to be more natural to take equation (25) and to put

$$\frac{dV}{dr} = \left(\frac{dV}{dr}\right) + \left(\frac{dV}{da}\right)\frac{da}{dr},$$

where $\left(\frac{dV}{dr}\right)$ means that r alone varies, and $\left(\frac{dV}{da}\right)$ means that a alone varies. Then after differentiation substitute for r its value.

Thus

$$\left(\frac{dV}{dr}\right) = -\frac{4\pi}{r^2}\int_0^a \rho a^2 da - \frac{4\pi}{r^2}\alpha\Sigma\frac{n+1}{(2n+1)r^n}A_n + 4\pi\alpha\Sigma\frac{nr^{n-1}}{2n+1}B_n,$$

$$\left(\frac{dV}{da}\right) = \frac{4\pi\rho a^2}{r} - 4\pi\rho a + \frac{4\pi\alpha}{r}\Sigma\frac{1}{(2n+1)r^n}\frac{dA_n}{da} + 4\pi\alpha\Sigma\frac{r^n}{2n+1}\frac{dB_n}{da}.$$

When we develop the expression for $\left(\frac{dV}{da}\right)$ we find that

$$\left(\frac{dV}{da}\right) = 4\pi\rho a\left(\frac{a}{r}-1\right) + 4\pi\rho a\alpha\Sigma Y_n = -4\pi\rho a\alpha y + 4\pi\rho a\alpha y = 0.$$

Thus to our order of approximation

$$\frac{dV}{dr} = \left(\frac{dV}{dr}\right) = -\frac{4\pi}{a^2}(1-2\alpha y)\int_0^a \rho a^2 da$$

$$-\frac{4\pi\alpha}{a^2}\Sigma\frac{n+1}{(2n+1)a^n}A_n + \frac{4\pi\alpha}{a}\Sigma\frac{na^n}{2n+1}B_n.$$

In like manner

$$\frac{d^2V}{dr^2} = \left(\frac{d^2V}{dr^2}\right) + 2\left(\frac{d^2V}{da\,dr}\right)\frac{da}{dr} + \left(\frac{d^2V}{da^2}\right)\left(\frac{da}{dr}\right)^2 + \left(\frac{dV}{da}\right)\frac{d^2a}{dr^2}.$$

Now it will be found that to our order

$$\left(\frac{d^2V}{da\,dr}\right) = -4\pi\rho\left(1+\alpha y + a\alpha\frac{dy}{da}\right) = -4\pi\rho\frac{dr}{da},$$

so that

$$\left(\frac{d^2V}{da\,dr}\right)\frac{da}{dr} = -4\pi\rho.$$

And

$$\left(\frac{d^2V}{da^2}\right) = 4\pi\rho\left(1+2\alpha y+2a\alpha\frac{dy}{da}\right) = 4\pi\rho\left(\frac{dr}{da}\right)^2,$$

so that

$$\left(\frac{d^2V}{da^2}\right)\left(\frac{da}{dr}\right)^2 = 4\pi\rho.$$

Hence to our order

$$\frac{d^2V}{dr^2} = \left(\frac{d^2V}{dr^2}\right) - 8\pi\rho + 4\pi\rho = \left(\frac{d^2V}{dr^2}\right) - 4\pi\rho.$$

1384. That part of Poisson's memoir which relates to the expansion of a series in terms of Laplace's functions was criticised by Ivory in the *Philosophical Magazine* for May 1827: I do not see any fresh matter of importance.

Poisson replied in a paper inserted in the *Philosophical Magazine* for July 1827, entitled *Observations relatives à un Article de Mr. Ivory, inséré dans le No. 5. du Philosophical Magazine*...This is numbered XI. in the list of Art. 1356; the title there given was probably quoted by Poisson from memory, as it is not quite accurate; also the paper is assigned to June 1827, instead of to July 1827. Poisson's reply seems to me sufficient. He states here without demonstration the general theorem in the first section of the work numbered XII. in Art. 1356, to which we now proceed.

1385. We have to notice the *Addition* to Poisson's memoir which was published in the *Connaissance des Tems* for 1831: see Art. 1358.

This *Addition* consists of four Articles, and is mainly occupied with the theory of Laplace's functions.

1386. In his first Article, Poisson finds the value, when $1-\alpha$ is infinitesimal, of the double integral

$$\frac{c}{\pi 2^{1+c}} \int_0^{\pi} \int_0^{2\pi} \frac{(1-\alpha^2)^c f(\theta', \psi') \sin \theta' \, d\theta' \, d\psi'}{(1-2\alpha p+\alpha^2)^{1+\frac{c}{2}}},$$

where $$p = \cos \theta \cos \theta' + \sin \theta \sin \theta' \cos (\psi - \psi').$$

This is a more general process than that in the original memoir, where he had confined himself to the case of $c = 1$.

The value of the double integral is found to be $f(\theta, \psi)$.

1387. In his second Article, Poisson gives a particular case of the general investigation of his first Article; namely that in which $f(\theta', \psi') = p$.

1388. In his third Article, Poisson combats the notions of Ivory on Hydrostatics; these notions will come before us in the next Chapter. Poisson says on his page 53:

...M. Ivory a persisté dans son opinion, et en a pris occasion de la développer dans plusieurs articles du *Philosophical Magazine*. Je persiste également dans la mienne, et j'abandonne au jugement des géomètres les motifs que j'en ai donnés; je demande toutefois la permission d'ajouter à la note qui les renferme, une observation dont j'ai lieu d'espérer que mon honorable adversaire sera frappé.

The argument which is thus introduced is given again by Poisson in his *Traité de Mécanique*, Vol. II. page 549, but there he does not mention Ivory.

We may observe that the French writers or printers have been very unfortunate in their efforts to spell Ivory's name. Poisson has Yvory in the *Connaissance des Tems* for 1829, and Yvori in his *Traité de Mécanique*, Vol. I. page 194. Laplace has Ivori in the *Mécanique Céleste*, Vol. V. page 10.

1389. In his fourth Article Poisson treats of the convergence of the series obtained when a function is expanded in a series of Laplace's functions. This Article is reproduced by Poisson in his *Théorie de la Chaleur*, pages 222 and 223. A few lines at the end respecting approximate values of Laplace's coefficients, in which Poisson refers to the third page of the Supplement to the fifth volume of the *Mécanique Céleste*, are not reproduced.

1390. The second edition of Poisson's *Traité de Mécanique* was published in 1833 in two octavo volumes. There is nothing new in the work with regard to our subject. The first volume contains a Chapter on the attraction of bodies, which occupies pages 169...202; and also a calculation of the attraction of a mountain, on pages 492...496: this arrives at the result which, as we stated in Art. 363, was first given by Bouguer. The second volume contains, on pages 538...549, a brief account of the problem of the Figure of the Earth considered as a homogeneous fluid rotating with uniform angular velocity.

1391. We pass now to the memoir which is numbered XIV. in Art. 1356.

In the *Mémoires...de l'Institut de France*, Vol. XIII., published in 1835, there is a memoir entitled *Mémoire sur l'attraction d'un ellipsoïde homogène*. The memoir occupies pages 497...545 of the volume. The memoir was read to the Academy on October 7, 1833.

1392. The memoir may be described as consisting essentially of a new and easy demonstration of the final result obtained by Legendre in his remarkable but most difficult memoir of 1788:

Poisson's memoir is a fine specimen of his great mathematical powers; admirable alike for simplicity and profundity. He treats of the attraction both on an internal and an external particle; but it is only in the treatment of the latter that the novelty of the method consists.

1393. Poisson's introduction is very interesting, giving a brief sketch of the labours of preceding writers; I have quoted a passage from it in Art. 887.

1394. The main principle of the memoir is the mode of decomposition of the ellipsoid into elements. Poisson decomposes the ellipsoid into films bounded by similar, similarly situated, and concentric ellipsoids. He determines the attraction of such a film, and demonstrates the remarkable result that the attraction it exerts on an external particle is directed along the axis of the cone which has its vertex at the attracted particle and envelopes the film. An elementary demonstration of this result was given by Steiner in Crelle's *Journal für Mathematik*, Vol. XII. See *Statics*, Chapter XIII.

1395. We have in this memoir expressions for the components of the attraction of an ellipsoid under a form slightly different from that which had been previously given by all the writers on the subject, except Rodrigues.

Let a, b, c be the semiaxes of an ellipsoid; let f, g, h be the corresponding coordinates of an attracted particle. Then the resolved attraction parallel to the direction of a is

$$2\pi fabc\int_{\tau}^{\infty}\frac{dt}{(t+a^2)\sqrt{T}},$$

where $$T=(t+a^2)(t+b^2)(t+c^2),$$

and τ is found from the equation

$$\frac{f^2}{\tau+a^2}+\frac{g^2}{\tau+b^2}+\frac{h^2}{\tau+c^2}=1,$$

for the case of any external particle.

For a particle on the surface or within the body we put 0 for τ.

Poisson's own notation is not symmetrical like this; but his result is substantially the same.

The resolved attractions parallel to the directions of b and c can be immediately deduced by symmetry from the formula which has just been given.

1396. The expression given in the preceding Article may be easily obtained from the older form by transformation. In the formula of Art. 885 suppose

$$x^2 = \frac{\tau + a^2}{t + a^2} \text{ and } k^2 = \tau + a^2;$$

then we arrive at the new expression for the case of an external particle.

And conversely from the new expression given by Poisson we can pass as he does to the older form.

The expression given in the preceding Article may also be readily obtained from the value of the potential V which is investigated in Art. 1184; Rodrigues himself brings out results which are practically equivalent to Poisson's expression.

1397. It will be found on examination that Poisson's first three sections contain nothing that is really new, except the pages 508 and 509, which are used in his fourth section. The fourth section which occupies pages 533...545 is the important part.

1398. Poisson has followed Legendre's memoir of 1812 in expressing the attractions on an external particle by means of elliptic integrals.

Let X, Y, Z be the resolved attractions on the external point, f, g, h; then Poisson shows that

$$\frac{X}{f} + \frac{Y}{g} + \frac{Z}{h} = \frac{4\pi abc}{\sqrt{\{(\tau + a^2)(\tau + b^2)(\tau + c^2)\}}},$$

where τ is the same as in Art. 1395.

It could also be shewn that

$$\frac{X(\tau+a^2)}{f} + \frac{Y(\tau+b^2)}{g} + \frac{Z(\tau+c^2)}{h} = \frac{4\pi abc}{\sqrt{(c^2 - a^2)}} F(k, \phi).$$

where $F(k, \phi)$ is a certain elliptic integral of the first kind, and a is the least semiaxis and c the greatest.

These results are due to Legendre: see his memoir of 1812.

It follows that

$$\frac{Xa^2}{f} + \frac{Yb^2}{g} + \frac{Zc^2}{h} = \frac{4\pi abc}{\sqrt{(c^2 - a^2)}} F(k, \phi) - \frac{4\pi abc\tau}{\sqrt{[(\tau + a^2)(\tau + b^2)(\tau + c^2)]}}.$$

See Arts. 1157 and 1158 for the case of an internal particle.

1399. In the *Supplément au Livre* v. of his *Théorie Analytique du Système du Monde*, Pontécoulant reproduces the substance of Poisson's memoir of 1835. Pontécoulant confines himself to what is new in the memoir, and thus condenses it into pages 1...20 of his supplement.

Pontécoulant makes some changes in the notation which I think are not improvements; he has a few misprints, which are not serious except on his pages 20 and 21, where he gives two results which were obtained by Legendre in his memoir of 1812. The second of these results Pontécoulant states incorrectly both for the internal and external point: see Art. 1398.

1400. In the *Connaissance des Tems* for 1837, which was published in 1834, there is a note by Poisson entitled *Note relative à l'attraction d'un ellipsoïde hétérogène.* The note occupies pages 93...102 of the volume: it was read to the French Academy on Nov. 24th, 1834. This note may be considered as an Appendix to the memoir in the *Mémoires...de l'Institut* for 1835.

1401. Poisson begins by referring to a letter recently sent by Jacobi to the French Academy, in which two results were enunciated. One was what we call Jacobi's theorem, namely that an ellipsoid is a possible form of relative equilibrium for rotating fluid. The other related to the attraction of a heterogeneous ellipsoid; the components of this attraction might be expressed in certain cases in a finite form, by arcs of circles and logarithms, without the aid of elliptic functions. Poisson's note relates to the second result; Jacobi had not published his demonstration, and meanwhile Poisson proposed to shew that the integration could be readily deduced from the formulæ which he had given in his memoir.

1402. Suppose an ellipsoid to consist of infinitesimally thin shells, each shell being bounded by similar, similarly situated, and concentric ellipsoids. Let the principal semiaxes of a shell be denoted by k, $k\sqrt{m}$, and $k\sqrt{n}$, where m and n are constant for all the shells. Let the density of the shells be expressed by a function of k. Then Poisson gives formulæ for determining the components of the attraction of the ellipsoid at a given point, external or internal.

1403. Poisson works out fully the particular case in which the density varies inversely as k. In this case although the density is infinite at the centre, yet the components of the attraction are finite quantities. If the attracted point is within the ellipsoid, the components remain constant along a given direction from the centre to the surface.

This particular case is also discussed by Pontécoulant in pages 22...26 of the work named in Art. 1399. Pontécoulant follows Poisson closely, though with rather less detail.

Poisson said in his note that it would be difficult to discover from the ancient formulæ for the attraction of an ellipsoid, when the integration could be effected in finite terms; but Pontécoulant does not admit this. In fact the ancient formulæ and those which Poisson prefers are connected, as we have seen in Art. 1396, by a very simple transformation. Thus practically what could be derived from Poisson's formulæ could also be derived from the ancient formulæ.

1404. It will be convenient to notice here the controversy in 1837 between Poisson and Poinsot concerning the history of the problem of the attraction of an ellipsoid on an external particle. See the *Comptes Rendus*...Vol. VI. pages 808...812, 837...840, 869...872; and Vol. VII. pages 1...3, 23 and 24.

Poisson's share in the controversy forms the articles which are numbered XVII. in the list of Art. 1356.

Chasles presented to the Academy a memoir entitled *Solution synthétique du problème de l'attraction des ellipsoïdes, dans le cas général d'un ellipsoïde hétérogène, et d'un point extérieur.* The memoir was referred by the Academy to Libri and Poinsot; and the report on the memoir was made by Poinsot.

1405. In this report Poinsot gave no reference to Poisson's memoir. Poisson made some remarks on the report; in these remarks, after stating the nature of Legendre's memoir of 1788, he proceeds to his own researches. He lays great stress on the fact that he had decomposed the ellipsoid into shells indefinitely thin and bounded by homothetical surfaces, and had determined the attraction of such a shell on an external particle. He does not hesitate to say that this is the only mode of decomposition by which the double integrals occurring in the problem can be reduced to single integrals. He thinks that the title of his memoir might have been mentioned in the report respecting Chasles's memoir, in which the same method of decomposition was in fact adopted. It seems to me that Poisson is both just and reasonable in all he says.

The following passage from page 839 is of sufficient interest to be reproduced:

Si quelqu'un se fût avisé de différentier les expressions que Laplace a donné le premier, des composantes de l'attraction d'un ellipsoïde sur un point extérieur, en faisant varier les trois axes suivant un même rapport, il aurait vu que les intégrales disparaissent dans le résultat, et que les composantes de l'attraction d'une couche elliptique s'expriment sous forme finie. Cette remarque, que je n'ai faite qu'après coup, aurait mis sur la voie de la solution directe du problème, en montrant que pour réduire les intégrales doubles à des intégrales simples, il suffisait de déterminer *à priori*, en grandeur et en direction, par des considérations géométriques ou par l'analyse, l'attraction sur un point extérieur d'une couche infiniment mince, comprise entre deux surfaces elliptiques semblables.

1406. I may observe that Poisson in his remarks speaks of the *theorem of Laplace;* and I am glad to have his authority for this title, which I had adopted before I had read this passage, or that cited from Ivory in Art. 1142.

1407. Poisson draws attention to a slight want of accuracy in a phrase used by Poinsot, who spoke in fact of an infinitely thin ellipsoidal shell, without explicitly stating that the inner surface was homothetical with the outer. Poisson is right; but Poinsot probably assumed that his context made the matter clear.

1408. Poinsot replied to Poisson's remarks. In the reply Poinsot insists strongly that Legendre's solution is a *direct* solution, and the *first* direct solution. He also holds that the merit of decomposing the ellipsoid into films in the manner of Poisson's memoir belongs to Rodrigues. Poinsot allows on his page 870 that Maclaurin established a particular case of the theorem which I call Laplace's; thus he is more correct than many other French writers: see Art. 260.

1409. Thus far we have been consulting the sixth volume of the *Comptes Rendus*...; the last words on the subject are contained in the seventh volume, which we will reproduce, and then add a few remarks.

1410. The first paper is by Poisson; it occurs on the first three pages of the volume:

Addition aux Remarques insérées dans le Compte rendu de la séance du 18 *juin; par* M. Poisson.

Ces remarques ayant été l'objet d'une Note qui fait partie du *Compte rendu* de la séance suivante, je me trouve obligé d'y faire une très-courte addition.

Ainsi que je l'ai dit dans cet article, j'abandonne mon analyse au jugement des géomètres. Il ne me conviendrait pas d'en faire moi-même la comparaison avec celle de Legendre, ni de tout autre. Je ferai seulement remarquer la différence essentielle qui existe entre la méthode que j'ai suivie et celle qu'avait employée cet illustre géomètre; différence qui ne résulte pas des progrès de l'analyse; car je n'ai fait usage d'aucun procédé de calcul qu'il n'ait pu également employer, et même Lagrange, en 1773, à l'époque de son premier Mémoire. J'ai décomposé l'ellipsoïde en couches terminées par des surfaces elliptiques et semblables; ce qu'on n'avait pas fait auparavant, et ce qui m'a conduit à un théorème nouveau sur l'attraction d'une pareille couche, qui trouve une application immédiate dans la théorie de l'électricité.

Legendre a divisé ce corps en couches coniques dont le sommet est au point attiré. Mais à raison de la complication du calcul qui en est résulté, il a été contraint, à la page 480 de son Mémoire, de recourir à une considération particulière et d'abandonner le procédé direct d'intégration qu'il avait suivi jusque là, et qui n'aurait pu le conduire, comme il le dit lui-même, *presque à aucune conclusion après d'aussi longs calculs*.

Souvent il est arrivé qu'une idée très simple a fourni la solution d'une difficulté qui avait long-temps arrêté; mais relativement à la décomposition des couches elliptiques et semblables, je dois dire que cette idée, quel que soit le peu d'importance qu'on y veuille attacher, ne s'est présentée à moi qu'après plusieurs autres tentatives, et que j'y ai été conduit par la considération attentive des formules, ainsi qu'on peut le voir dans le n° 4 de mon Mémoire. Il y a plus; Legendre dit, à la fin du sien, que la décomposition du sphéroïde en couches coniques, lui paraît être la seule que l'on puisse employer; et il faut observer que ce Mémoire avait précisément pour objet général, le choix des variables le plus propres à la réduction des intégrales doubles, ou en d'autres termes, la manière la plus convenable de décomposer les corps auxquels elles se rapportent. Legendre ajoute que l'attraction d'une couche conique exigeant une intégration très-difficile, le problème est vraisemblablement au-dessus des moyens ordinaires de la synthèse, ce qui serait effectivement vrai en suivant la marche qu'il avait adoptée; mais, au contraire, l'intégration relative à une couche elliptique est assez simple, pour qu'on ait pu facilement l'effectuer par des considérations géométriques, dès que le résultat en a été connu.

Enfin, dans la Note à laquelle je réponds, il est dit que M. Rodrigues, en soutenant, il y a vingt ans, une thèse pour le doctorat, avait employé bien avant moi cette décomposition de l'ellipsoïde en couches infiniment minces, pour le calcul même de l'attraction sur les points extérieurs: cela n'est aucunement vrai; et il est même évident, pour tous ceux qui comprennent la question, que M. Rodrigues n'aurait point atteint le but qu'il se proposait, par la considération de couches pareilles à celles dont il s'agit. L'erreur où est tombé l'auteur de la Note, vient, sans doute, de ce qu'il n'a point eu égard à la condition de similitude des deux surfaces, externe et interne, de chaque couche elliptique, qui en est cependant le caractère essentiel. En aucun endroit de sa thèse, d'ailleurs fort remarquable, M. Rodrigues n'a considéré l'attraction d'une couche elliptique terminée par des surfaces semblables. Dans l'endroit où il démontre le théorème de Maclaurin ou de Laplace, il différentie, relativement aux trois axes de l'ellipsoïde et en supposant constantes les deux distances focales, le rapport de son attraction à son volume, afin de faire voir que cette différentielle se réduit alors à zéro. S'il eût différentié, sous ce point de vue, l'attraction même, il aurait obtenu celle d'une couche elliptique dont les deux surfaces ont les mêmes foyers, et, par conséquent, ne sont pas semblables. Les signes d'intégration n'auraient pas disparu dans son expression, et la considération de cette force n'eût pas été plus simple que celle de l'attraction de l'ellipsoïde entier; au lieu que l'attraction d'une couche elliptique, terminée par deux surfaces semblables, s'exprime sous forme finie; ce qui, quand on a déterminé sa valeur *à priori*, réduit ensuite à une intégrale simple, l'attraction de l'ellipsoïde entier, homogène ou hétérogène. Au reste, la démonstration que M. Rodrigues a rapportée dans sa thèse, est celle que M. Gauss a donnée en 1813, et qui est fondée sur la transformation des variables employées par M. Ivory, et sur une propriété générale des surfaces fermées.

1411. Next we have Poinsot's reply on pages 23 and 24 of the volume.

Note de M. Poinsot, *en réponse à l'auteur des Remarques insérées dans le* Compte rendu *de la séance du 2 juillet.*

Le dissentiment qui existe entre cet auteur et moi, au sujet de la partie historique du problème de l'attraction d'un ellipsoïde sur un point extérieur, roule sur les trois propositions suivantes:

J'ai avancé:

1°. Que M. Legendre avoit résolu la question *directement*, c'est-à-dire, sans passer par le théorème de Maclaurin. (*Compte rendu*, page 869.)

2°. Que M. *Rodrigues*, pour la démonstration du théorème de Maclaurin, auquel il ramène le cas des points extérieurs, a fait usage de la considération d'*une couche infiniment mince, comprise entre deux surfaces semblables entre elles, et semblables à la surface de l'ellipsoïde dans laquelle la couche est prise.*

3°. Enfin, que la phrase de notre Rapport, où l'auteur a cru voir une *inexactitude* (*Compte rendu*, page 840), est *géométriquement* et *grammaticalement* exacte, et qu'il n'y a rien à y changer.

Je maintiens ces trois propositions.

Je les soumets à l'attention des géomètres, et j'espère qu'après un nouvel examen, l'auteur des *Remarques* lui-même se rendra à l'évidence, sans que j'aie besoin de lui signaler les *erreurs* sur lesquelles il a fondé sa prétendue réfutation de l'opinion que j'avais émise au sujet du travail de M. *Rodrigues*.

There are two notes at the foot of the pages; one relates to the first of Poinsot's three propositions, and the other to the second. They stand thus:

L'auteur, au contraire, avait avancé que, "pour le cas général, M. Legendre s'était contenté de donner une démonstration du théorème de Laplace (*lisez* de Maclaurin), encore plus compliquée que celle de l'auteur (*lisez* de Laplace)." *Compte rendu*, page 836.

Voyez, tome III. de la *Correspondance sur l'Ecole Polytechnique*, le commencement de la page 367, où l'on trouve ces mots: *Considérons une couche elliptique*, etc., et voyez si cette couche n'est pas bien précisément celle qu'on vient de définir, et si la considération de cette même couche n'entre pas essentiellement dans la démonstration.

1412. Let us take the points in the order adopted by Poinsot.

I. As to the value of Legendre's solution. Perhaps Poisson rather underrates, and Poinsot rather overrates this. Legendre, as we see from Art. 1150, claims for it the merit of being *direct*, and Poinsot lays great stress also on this merit. But the term *direct* ought to be carefully defined if so much importance is attached to it; and it does not appear to me that it can be applied in any very strict sense to the whole of Legendre's process. In the note Poinsot elaborately corrects Poisson's phrase, *the theorem of Laplace*, into *the theorem of Maclaurin*; it is of no great importance by what name we call the theorem, provided we understand what theorem is meant, but I consider that Laplace's name and not Maclaurin's is the proper one.

II. As to what had been accomplished by Rodrigues. Here I hold Poisson to be right. It is true that in order to effect a certain integration Rodrigues decomposed the ellipsoid in the manner which Poinsot indicates; but Rodrigues did not determine the attraction of one of the infinitesimal shells: and this was the important novelty which Poisson claimed, and justly, for himself.

III. As to the charge of inexactness. The matter is of small account, but Poisson was certainly right: see Art. 1407.

1413. We now arrive at the last of Poisson's contributions. It is entitled *Note sur une propriété générale des formules relatives aux attractions des sphéroides.* This is given in the *Comptes Rendus...* Vol. VII. 1838, pages 3...5.

Let there be a sphere in which the density is any function of the distance from the centre. Let a, b, c be the coordinates of the centre. Let x, y, z be the coordinates of any other point; and let dm denote the element of mass at that point. Suppose a body entirely external to the sphere; and let $dm\,\phi_1(x, y, z)$ denote the attraction of this body on dm parallel to the axis of x; similarly let $dm\,\phi_2(x, y, z)$ and $dm\,\phi_3(x, y, z)$ denote the attractions parallel to the axes of y and z respectively. Then will

$$\int \phi_1(x, y, z)\, dm = \mu\phi_1(a, b, c),$$

$$\int \phi_2(x, y, z)\, dm = \mu\phi_2(a, b, c),$$

$$\int \phi_3(x, y, z)\, dm = \mu\phi_3(a, b, c),$$

where μ denotes the mass of the sphere, and the integrations extend throughout the sphere.

Poisson demonstrates the equations thus: let P denote any element of the external body. Then the attraction of the sphere on P is the same as if the sphere were collected at its centre. Hence the attraction of P on the sphere will be the same as if the sphere were collected at its centre. Hence the attraction of the whole external body on the sphere will be the same in magnitude and direction as that of the attraction of this body on a particle of mass μ at the centre of the sphere. This result is the translation of the three equations which were to be demonstrated.

Also if $dm\,f(x, y, z)$ denote the *potential* of the external body on dm we shall have

$$\int f(x, y, z)\, dm = \mu f(a, b, c).$$

Poisson says that this is a remarkable example of the rare cases in which simple reasoning, or what may be called the synthetical method, has a great advantage over analysis; for it would be very difficult to demonstrate, in all their generality, the preceding equations by mathematical analysis. But Liouville shewed that the equations could be easily obtained by analysis; see pages 84...86 of the same volume.

1414. Let us now give Liouville's process. He takes the last equation for example. Then expressing dm in the usual polar coordinates we have to shew that

$$\int_0^l\int_0^\pi\int_0^{2\pi} f(x, y, z)\, \rho r^2 \sin\theta\, dr\, d\theta\, d\psi = \mu f(a, b, c),$$

where ρ is the density, and l the radius of the sphere.

Denote the left-hand member by U.

By the definition of the function $f(x, y, z)$ we have

$$f(x, y, z) = \iiint \frac{\rho'\, dx'\, dy'\, dz'}{[(x-x')^2 + (y-y')^2 + (z-z')^2]^{\frac{1}{2}}},$$

where x', y', z' denote the coordinates of an element $\rho'\, dx'\, dy'\, dz'$ of the external body.

Let $$R = \int_0^l\int_0^\pi\int_0^{2\pi} \frac{\rho r^2 \sin\theta\, dr\, d\theta\, d\psi}{[(x-x')^2 + (y-y')^2 + (z-z')^2]^{\frac{1}{2}}}.$$

Then $$U = \iiint R\rho'\, dx'\, dy'\, dz'.$$

Also $x = a + r\cos\theta$, $y = b + r\sin\theta\sin\psi$, $z = c + r\sin\theta\cos\psi$;

and as we assume that $(x-x')^2 + (y-y')^2 + (z-z')^2$ cannot vanish, the common methods give

$$R = \frac{\mu}{[(a-x')^2 + (b-y')^2 + (c-z')^2]^{\frac{1}{2}}}.$$

Hence $$U = \mu \iiint \frac{\rho'\, dx'\, dy'\, dz'}{[(a-x')^2 + (b-y')^2 + (c-z')^2]^{\frac{1}{2}}},$$

that is, $$U = \mu f(a, b, c),$$

which was to be demonstrated.

This analytical demonstration is founded on principles like those which M. Poisson himself employs in the fifth Article of his Memoir *Sur la propagation du mouvement dans les milieux élastiques*. It corresponds exactly to the synthetical demonstration; we may say that they substantially coincide; at least they differ only in language.

1415. It will be seen that Poisson holds a distinguished place in the history of our subject. The correction which he supplied to Laplace's differential equation for the potential, has become a permanent part of the theory; so also has the extension of Ivory's theorem to any law of attraction.

The two great memoirs, which I have numbered X. and XIV. in my list, still deserve the careful study of those who wish to obtain a profound knowledge of the subject; the latter memoir may be justly considered to be the immediate preparation for the researches of Chasles.

Poisson himself appears to have attached great importance to his method of treating the theory of Laplace's functions; for he repeated it in various places. But this method does not seem to find favour with later writers; I doubt whether it is even alluded to in Heine's work, cited in Art. 784.

We may well concur with Legendre in thinking that the task of improving the *Mécanique Céleste* seemed to devolve naturally on Poisson: see Pontécoulant's *Système du Monde*, Vol. III. at the beginning.

CHAPTER XXXVI.

IVORY.

1416. The writings of Ivory, arranged chronologically, which may be considered as connected with our subject are the following:

I. On the Attractions of Homogeneous Ellipsoids. *Philosophical Transactions* for 1809. I have noticed this in Chapter XXIX.

II. On the Grounds of the Method which Laplace has given in the second Chapter of the third Book of his *Mécanique Céleste* for computing the Attractions of Spheroids of every Description. *Philosophical Transactions* for 1812. I have noticed this in Chapter XXX.

III. On the Attractions of an extensive Class of Spheroids. *Philosophical Transactions* for 1812.

IV. On the Expansion in a Series of the Attraction of a Spheroid. *Philosophical Transactions* for 1822.

V. On the Figure requisite to maintain the Equilibrium of a Homogeneous Fluid Mass that revolves upon an axis. *Philosophical Transactions* for 1824.

VI. The article *Attraction* for the *Supplement to the Encyclopædia Britannica.*

VII. Remarks on the Theory of the Figure of the Earth. *Philosophical Magazine,* May 1824.

VIII. Investigations connected with the Properties of the Geodetic Line on an Oblatum will be found in the *Philosophical Magazine* for July 1824, April 1825, April 1826, and May 1826. Towards the end Ivory compares some results which Bessel had obtained with his own, and expresses himself in a tone of dissatisfaction. But the matter belongs rather to Analytical Geometry than to our subject, and so I shall not notice it further.

IX. On the Theory of the Figure of the Earth. *Philosophical Magazine*, April 1825.

X. On the Variation of Density and Pressure in the interior Parts of the Earth. *Philosophical Magazine*, November 1825.

XI. On the Theory of the Figure of the Planets contained in the Third Book of the *Mécanique Céleste*. *Philosophical Magazine*, December 1825, January 1826, and February 1826.

XII. Notice relating to the Theory of the Equilibrium of Fluids. *Philosophical Magazine*, June 1826.

XIII. On the Equilibrium of a Fluid attracted to a fixt Centre. *Philosophical Magazine*, July 1826.

XIV. Six papers of various titles, but all relating to pendulum experiments, are published in the volume of the *Philosophical Magazine* which extends from July to December 1826.

XV. Notice respecting the Seconds Pendulum at Port Bowen. *Philosophical Magazine*, March 1827.

XVI. Some Remarks on a Memoir by M. Poisson, read to the Academy of Sciences at Paris, Nov. 20, 1826, and inserted in the *Conn. des Tems*, 1829. *Philosophical Magazine*, May 1827.

XVII. Six papers of various titles, but all relating to Laplace's Functions, or to the conditions of fluid equilibrium, are published in the volume of the *Philosophical Magazine* which extends from July to December 1827.

XVIII. Three papers on the Ellipticity of the Earth, as deduced from Experiments with the Pendulum, and two papers on the Figure of the Earth, as deduced from Measurements of the different Portions of the Meridian, are published in the volume of the *Philosophical Magazine* which extends from January to June 1828.

XIX. Some Remarks on an Article in the Bulletin des Sciences Mathématiques Physiques et Chimiques, for March 1828. *Philosophical Magazine*, October 1828.

XX. Four papers of various titles, but all relating to the measurement of an arc perpendicular to the meridian, are published in the volume of the *Philosophical Magazine* which extends from July to December 1828; and two papers connected with these are published in the volume which extends from January to June 1829.

XXI. Some Arguments tending to prove that the Earth is a Solid of Revolution. *Philosophical Magazine*, March 1829.

XXII. Some Remarks on an Article in the "Bulletin des Sciences Mathématiques" for June 1829, § 269. *Philosophical Magazine*, October 1829.

XXIII. Letter relating to the Figure of the Earth. *Philosophical Magazine*, April 1830.

XXIV. On the Figure of the Earth. *Philosophical Magazine*, June 1830.

XXV. Two papers relating to the Shortest Distance between Two Points on the Earth's Surface are published in the volume of the *Philosophical Magazine* which extends from July to December 1830.

XXVI. On the Equilibrium of Fluids and the Figure of a Homogeneous Planet in a Fluid State. *Philosophical Transactions* for 1831.

XXVII. On the Equilibrium of a Mass of Homogeneous Fluid at liberty. *Philosophical Transactions* for 1834.

XXVIII. Of such Ellipsoids consisting of Homogeneous Matter as are capable of having the Resultant of the Attraction of the Mass upon a Particle in the Surface, and a Centrifugal Force caused by revolving about one of the Axes, made perpendicular to the Surface. *Philosophical Transactions* for 1838, with a note in the Volume for 1839.

XXIX. Three papers of various titles, but all relating to the subject of fluid equilibrium, are published in the volume of the *Philosophical Magazine* which extends from July to December 1838.

XXX. On the Conditions of Equilibrium of an Incompressible Fluid, the Particles of which are acted on by Accelerating Forces. *Philosophical Transactions* for 1839.

I proceed to give an account of such of these writings as have not been already noticed; the first of these is that numbered III.

1417. A memoir entitled *On the Attractions of an extensive Class of Spheroids* is contained in the *Philosophical Transactions* for 1812, published in that year. The memoir occupies pages 46...82 of the volume; it was read on November 14, 1811.

1418. The class of spheroids to which this memoir relates consists of those which have their radii vectores rational integral functions of the angular coordinates.

By a rational integral function Ivory seems to mean, at least sometimes, any function which can be expanded in a series of rational integral terms: see his page 75, and also pages 43 and 44 of the memoir II. in the list of Art. 1416.

Ivory arrives at results equivalent to those given by Laplace in his treatment of the problem in the third Book of the *Mécanique Céleste.* Ivory does not use any property of Laplace's functions, but carries on his process so far as to shew how the requisite integrations can be theoretically effected.

On his page 48 he repeats an objection which he had given on page 33 of his memoir II.: see Art. 1215.

The memoir seems to me of small importance now; it might have been of some service perhaps as establishing various formulæ rigorously, so as to liberate an early student from any doubts left on his mind by Laplace's process.

1419. A memoir entitled *On the expansion in a series of the attraction of a Spheroid* is contained in the *Philosophical Transactions* for 1822, published in that year. The memoir occupies pages 99...112 of the volume; it was read January 17, 1822.

1420. Ivory has doubts as to the statement that *any* function can be expanded in a series of Laplace's functions, though he allows that any rational integral function of the three rectangular coordinates of a point can be so expanded. He holds that there is a real distinction to be made between the case in which the function proposed for expansion is an explicit function of the three rectangular coordinates of a point, and the case in which it is not. He says on his page 106:

A method of calculation which is clear, exact and elegant, when it is confined to the first case, becomes clouded with obscurity, if not merely symbolical, when it is extended to the other case. To say the least, there are certainly great difficulties which are not explained; and if there be any geometers who hesitate, and have doubts, they are not without their excuse, and ought not to be entirely condemned.

I have already adverted to one of the topics considered in this memoir; see Art. 1224. I do not attach any importance to the memoir. Perhaps Ivory is less confident in his condemnation of the proposition about the expansion of any function than he was ten years earlier.

1421. A memoir entitled *On the figure requisite to maintain the equilibrium of a homogeneous fluid mass that revolves upon an axis* is contained in the *Philosophical Transactions* for 1824, published in that year. The memoir occupies pages 85...150 of the volume; it was read December 18, 1824.

1422. This memoir *assumed* a new principle to be necessary for fluid equilibrium, namely the following: in order that a mass of fluid may be in equilibrium it is necessary that the arrangement of the strata be such that the matter comprised between any two level surfaces should exercise no attraction on a particle within the inner boundary. Ivory attempts to justify this assumption; but his efforts seem to me quite in vain.

We shall find that Ivory continued to advocate his peculiar notions in subsequent memoirs; he supposed that he modified them slightly in the memoirs XXVI. and XXVII., as we shall see hereafter.

Poisson criticised Ivory's assumption; see the *Annales de Chimie*... Vol. XXVII. 1824, pages 225...236, and the *Connaissance des Tems* for 1831, page 53. See also a paper by Robert Leslie Ellis in the *Cambridge Mathematical Journal*, Vol. II. pages 18...22. We shall notice some miscellaneous topics in the memoir.

1423. On his page 93 Ivory demonstrates the formula for Legendre's coefficients, which had been previously given by Rodrigues: see Art. 1187. We may infer that Ivory obtained the formula independently, as he adds no reference.

1424. A theorem is given on page 94, which may be reproduced. Let V denote the potential of an attracting mass at a point of which the radius vector is r, let dm' denote an element of the attracting mass of which the radius vector is r'; then

$$V = \int \frac{dm'}{\sqrt{(r^2 - 2rr'\gamma + r'^2)}},$$

where γ is the cosine of the angle between the directions of r and r'. Therefore

$$-r\frac{dV}{dr} = \int \frac{(r^2 - rr'\gamma)\,dm'}{(r^2 - 2rr'\gamma + r'^2)^{\frac{3}{2}}}.$$

Put s for $\sqrt{(r^2 - 2rr'\gamma + r'^2)}$; thus $r^2 - rr'\gamma = s^2 + rr'\gamma - r'^2$; and

$$-r\frac{dV}{dr} = \int \frac{dm'}{s} + \int \frac{rr'\gamma - r'^2}{s^3}\,dm'.$$

Therefore $$2V - r\frac{dV}{dr} = 3\int \frac{dm'}{s} + \int \frac{rr'\gamma - r'^2}{s^3}\,dm'.$$

Now substitute for dm' the usual expression $\rho r'^2\,d\mu'\,d\phi'\,dr'$, where ρ denotes the density; thus

$$2V - r\frac{dV}{dr} = \iint d\mu'\,d\phi' \int \rho \left\{\frac{3r'^2}{s} + \frac{r'^2(rr'\gamma - r'^2)}{s^3}\right\} dr'.$$

This may be expressed thus,

$$-r^3\frac{d}{dr}\left(\frac{V}{r^2}\right) = \iint d\mu'\,d\phi' \int \rho\,\frac{d}{dr'}\left(\frac{r'^3}{s}\right) dr'.$$

If the body is homogeneous so that ρ is constant, we obtain

$$-r^3\frac{d}{dr}\left(\frac{V}{r^2}\right) = \rho \iint \frac{r'^3\,d\mu'\,d\phi'}{s},$$

where r' now represents the radius vector of a point on the surface of the body corresponding to the other polar coordinates μ' and ϕ'.

1425. On his page 09 Ivory says that in a heterogeneous fluid body it is easy to perceive that the densities must decrease in approaching the outer surface. His reason for this furnishes a good specimen of the vagueness and inconclusiveness of his language; he says:

For, in two contiguous strata of different densities, if we take two molecules equal in volume, and placed at the same point of the separating surface; the common gravity acting upon both will produce a greater pressure in the denser molecule. Wherefore, if the denser matter were nearer the outer surface, it would penetrate into the rarer matter below it; which is contrary to the perfect separation of the strata of different densities.

1426. Ivory enunciates on his pages 111 and 112 his first Proposition in these words:

If a homogeneous fluid body revolving about an axis, be in equilibrio by the attraction of its particles in the inverse proportion of the square of the distance; any other mass of the same fluid having a similar figure, and revolving with the same rotatory velocity about an axis similarly placed, will likewise be in equilibrio, supposing that its particles attract one another by the same law.

This he establishes in four pages of general reasoning.

1427. On his page 115 Ivory enunciates his second Proposition in these words:

If a homogeneous fluid mass revolve about an axis, and be in equilibrio by the attraction of its particles in the inverse proportion of the square of the distance; all the level surfaces will be similar to the outer one: and any stratum of the fluid contained between two level surfaces will attract particles in the inside with equal force in opposite directions.

To this he devotes three pages of general reasoning, but I cannot allow that it is satisfactory. The proposition asserted is true in the case in which the fluid takes the form of an ellipsoid or of an oblatum, as we know from other sources; but we cannot affirm that it is necessarily true.

1428. On Ivory's page 125, combined with his page 98, we have a curious error.

He has shewn that if certain radii are in the same proportion then $K = K_1$; and he wants to shew conversely that if $K = K_1$ these radii are in the same proportion. Now he obtains in fact the equation

$$K - K_1 = \iint \log \frac{R'}{R_1'} C^{(2)} d\mu' d\varpi',$$

where R' and R_1' are the radii, and $C^{(2)}$ is a Laplace's coefficient of the second order; the integration is supposed to extend over the entire surface. To make this vanish it is not necessary that $\log \frac{R'}{R_1'}$ should be *constant*, as Ivory implies; it may be a Laplace's coefficient of any order except the second.

1429. If a homogeneous stratum be bounded by similar, similarly situated, and concentric ellipsoids, it exerts no attraction on an internal particle: this is well known. Conversely we might take this problem: Given that a homogeneous stratum bounded by similar, similarly situated, and concentric surfaces exerts no attraction on an internal particle, find the form of the surfaces from this condition. Ivory in fact discusses this, though he does not formally enunciate it in this way. By using the properties of Laplace's coefficients, he comes to the conclusion that the surfaces must be ellipsoids: see his pages 125...129. Ivory repeats this investigation in later memoirs; see page 512 of the memoir XXVII. and page 263 of the memoir XXX.

1430. Ivory says on his page 131:

> We are now to conclude that a homogeneous fluid mass cannot be in equilibrio by the attraction of its particles and a centrifugal force of rotation, unless it have the figure of an ellipsoid...

That is, Ivory claims to have solved the problem which I have called Legendre's in Art. 744, even without the limitation to surfaces of revolution. But it is almost needless to say that Ivory's process is unsatisfactory, for it is based on the principle which he unjustifiably assumed: see Art. 1422.

1491. In his pages 132...139 Ivory gives in effect a solution of the problem of the attraction of an ellipsoid on an internal particle. He exhibits the potential in the form of an expression which involves only single integrals.

1492. On his page 141 Ivory expresses in an interesting form the standard equation of Art. 581, namely

$$\frac{2q}{3} = \frac{(\lambda^2+3)\tan^{-1}\lambda - 3\lambda}{\lambda^3};$$

this may be written

$$\frac{2q}{9} = \frac{1}{3} - \left(\frac{1}{3} + \frac{1}{\lambda^2}\right)\left(1 - \frac{\tan^{-1}\lambda}{\lambda}\right).$$

Put $\tan^{-1}\lambda = \phi$; thus we get

$$\frac{2q}{9} = \frac{1}{3} - \left(\frac{1}{\sin^2\phi} - \frac{2}{3}\right)\left(1 - \frac{\phi}{\tan\phi}\right) \text{.............. (1).}$$

Now if $\frac{\phi}{\tan\phi}$ be expanded in powers of $\sin\phi$ it may be shewn that the expansion is of the form

$$1 - A_2\sin^2\phi - A_4\sin^4\phi - \dots - A_{2n}\sin^{2n}\phi - \dots,$$

where $A_2 = \frac{1}{3}$, and $A_{2n} = \frac{2.4.6\dots(2n-2)}{3.5\dots(2n+1)}$ if n is greater than 1;

see *Differential Calculus*, Art. 374.

Hence we find that (1) becomes

$$q = \frac{2}{5}\sin^2\phi + \frac{2}{5.7}\sin^4\phi - \frac{2.4.6}{5.7.9.11}\sin^6\phi$$

$$- 2\,\frac{2.4.6.8}{5.7.9.11.13}\sin^8\phi - \text{.......... (2).}$$

Ivory gives the terms so far with a slight misprint in the last. The general term on the right-hand side of (2) is

$$-\frac{2.4\dots(2n-2)}{5.7\dots(2n+3)}\cdot\frac{2n-6}{2}.\sin^{2n}\phi,$$

where n is supposed greater than 2.

The convergent series which forms the right-hand side of (2) vanishes when $\phi = 0$; it must also vanish when $\phi = \frac{\pi}{2}$, as we see by looking at the expression in (1), from which it was derived. It will be observed that the series presents *only one change of sign;* and if we differentiate with respect to ϕ we obtain the product of $\sin\phi\cos\phi$ into a series which has only *one change of sign.* Hence, by employing a principle which is explained in the Theory of Equations, we infer that as ϕ changes from 0 to $\frac{\pi}{2}$ the series is always positive, first increases continually from zero to its maximum value, and then decreases continually from its maximum value to zero. See *Theory of Equations,* Art. 22.

Thus from the form of the second side of (2) we have an evident demonstration of the result established in Art. 580.

1433. Ivory makes the following remarks on his pages 142 and 143:

When the rotatory velocity is greater than the maximum, the equilibrium cannot take place: for, on the one hand, the proposed rotation is inconsistent with the figure of an ellipsoid; and, on the other, it has been proved, that a homogeneous fluid cannot be in equilibrio unless it have that figure. In this case, therefore, the fluid would first extend itself, and flatten to a certain degree with a decreasing velocity of rotation, and then oscillate back with an increasing rotatory motion. But the tenacity of the particles would gradually diminish, and finally destroy, the oscillations of the fluid; which would therefore ultimately settle in one of the figures of equilibrium; that is, in an elliptical spheroid of revolution having the equatorial diameter more than 2·71... times the axis of revolution.

This for the most part is merely *assertion* on the part of Ivory, and it is obvious that difficult problems in hydrodynamics cannot be solved in this rapid manner.

1434. Ivory admits that the ordinary equation, which we denote by (1) in Art. 831, is *necessary* for equilibrium; this is his equation (A). But he asserts that it is not *sufficient* for equilibrium; one reason which he gives for this assertion, on which he seems to lay great stress, is quite unintelligible to me; he says on his page 144:

M'Laurin first proved synthetically that the ellipsoid, whatever be the degree of oblateness, fulfils all the conditions requisite for maintaining the equilibrium of a homogeneous fluid mass that revolves about an axis. If therefore the equation (*A*) were alone sufficient for the equilibrium, the ellipsoid must be deducible from it, not in particular suppositions and approximately, but generally, and by an accurate process of reasoning. But this has not been accomplished, nor even attempted, by any geometer.

See also his pages 145 and 150.

1435. We have next to notice the article *Attraction*, which Ivory wrote for the Supplement to the *Encyclopædia Britannica*; the article occupies pages 627...644 of the volume, published in 1824, and it forms a good elementary treatise, proceeding as far as a complete account of the attraction of homogeneous ellipsoids. The following points may be noticed:

Ivory deduces the attraction of a sphere on an external particle from the attraction of a sphere on a particle at its surface, by an elementary process of the same kind as he used in establishing the theorem on the attraction of ellipsoids, which is called by his name.

Ivory investigates Laplace's theorem which we have given in Art. 1046. Ivory adopts the method of expansion which we have noticed at the end of the Article. He says that "Laplace has arrived at the same conclusion by a different process": but Ivory's process is rather a modification of Laplace's than essentially different.

Towards the end of his article Ivory says:

In the preceding investigations, we have followed the method of Maclaurin for points situated in the surface of a spheroid, or within the solid. This method has always been justly admired; but neither its inventor, nor, as far as we know, any other Geometer, has applied it, excepting to spheroids of revolution; and it is here, for the first time, extended to ellipsoids.

But it must be observed that the extension of Maclaurin's method to ellipsoids in general is so obvious that it does not require any formal explanation; D'Alembert for instance, as we have seen in Art. 615, took this view.

1436. An article entitled *Remarks on the Theory of the Figure of the Earth*, occurs on pages 339...348 of the *Philosophical Magazine* for May, 1824.

This article gives first a good sketch of the history of the subject, and then a brief account of Ivory's peculiar views on fluid equilibrium, with a reference to the memoir V. for proofs.

The following passage occurs on the first page:

To whatever branch of the philosophical system of the universe we turn our attention, we are immediately led to the immortal author of the true theory founded on the law of universal gravitation. Newton not only laid down the principles: he, in a great measure, reared the superstructure; or, at least, he sketched out so accurately the proper view to be taken of every part of the subject, that his followers have done little else but fill up his original outlines. The modern theory of the figure of the planets, still imperfect in some respects, coincides in the main with the physical ideas of Newton, which the progress of the mathematical sciences has enabled the philosophers of the present day to develop and extend.

1437. An article entitled *On the Theory of the Figure of the Earth* occurs on pages 241...249 of the *Philosophical Magazine* for April, 1825.

This consists mainly of the two Propositions which we have noticed in Arts. 1426 and 1427.

1438. An article entitled *On the Variation of Density and Pressure in the interior Parts of the Earth* occurs on pages 321...329 of the *Philosophical Magazine* for November, 1825.

This is substantially coincident with the matter contained in the *Mécanique Céleste*, Livre XI. § 6, to which Ivory refers: see Art. 1325.

1439. We have next to notice an article entitled *On the Theory of the Figure of the Planets contained in the Third Book of the Mécanique Céleste.* This is published in the *Philosophical Magazine* in three parts, which occur respectively on pages 429...439 of the number for December, 1825, on pages 31...37 of the number for January, 1826, and on pages 81...88 of the number for February, 1826.

The first two parts repeat the objections against Laplace's favourite equation, and the consequence which he drew from it: we have sufficiently considered the matter in Chapter XXX.

The third part is devoted to Ivory's peculiar views on fluid equilibrium, as developed by him in the memoir numbered V.

The following passage occurs at the end of the second part of this series of papers:

An attentive reader who considers the foregoing observations must allow that some material inadvertencies and inaccuracies have originally slipt into the analysis of Laplace. But the theory having been published, it has been deemed advisable to repel all objections, and to defend it *to the utterance*.

1440. An article entitled *Notice relating to the Theory of the Equilibrium of Fluids* occurs on pages 439...442 of the *Philosophical Magazine* for June, 1826.

Ivory repeats the statement of his peculiar opinions on fluid equilibrium.

1441. An article entitled *On the equilibrium of a Fluid attracted to a fixt Centre* occurs on pages 10 and 11 of the *Philosophical Magazine* for July, 1826.

The equation (2) of Art. 57 represents an ellipse approximately when $a\omega^2 + \frac{\mu}{a^2}$ is small. Ivory proceeds to interpret the equation to a closer order of approximation, which he does accurately. We shall notice the matter hereafter in connexion with a paper published by Dr Thomas Young in 1826.

1442. The six papers which we have brought together under the number XIV. are not very closely connected with our subject. The first just touches on our theories. After having stated his peculiar opinions on fluid equilibrium, Ivory says on his page 5:

The theory we have been explaining has been opposed, and has been rejected superciliously without examination. But it is founded on truth, and will ultimately be adopted. No other way but by investigating the physical properties of equilibrium, can be successful in simplifying a very difficult subject, and in rendering it completely satisfactory.

Ivory then states that he has carried a certain process of approximation so far as to include the squares of the ellipticities; and accordingly he gives without demonstration, an equation which corresponds to Clairaut's primary equation, extended so as to include small quantities of the second order.

I have not been encouraged to attempt to verify Ivory's equation; indeed it is not quite intelligible, for it contains a symbol A which is described as an "unknown function" of the polar axis of a stratum of equal density. We are told however on page 6, that A vanishes if the density is constant; and thus we can test one of Ivory's formulæ, and indeed his main result. He gives an expression for the value of gravity which is meant to be true to the second order of small quantities. According to this expression the value of *Clairaut's fraction* is

$$\frac{5}{2}\beta - \alpha + \frac{\alpha^2}{2} - \frac{17}{14}\alpha\beta,$$

where α and β have the meaning assigned in Art. 978.

Now the fluid being homogeneous, we know that Clairaut's fraction is exactly equal to α; see Art. 922.

Hence we must have, true to the second order

$$\alpha = \frac{5}{2}\beta - \alpha + \frac{\alpha^2}{2} - \frac{17}{14}\alpha\beta,$$

therefore
$$\alpha = \frac{5}{4}\beta + \frac{\alpha^2}{4} - \frac{17}{28}\alpha\beta,$$

therefore
$$\alpha = \frac{5}{4}\beta - \frac{165}{448}\beta^2.$$

But this does not agree with the last result given in Art. 978; so that we may infer the incorrectness of Ivory's formula.

1443. The paper which we have numbered XV. occurs on pages 170...172 of the *Philosophical Magazine* for March, 1827.

Ivory draws attention to the discrepancy between an observation made by Lieutenant Foster at Port Bowen, and an observation made by Captain Sabine at Greenland.

1444. An article entitled *Some Remarks on a Memoir by M. Poisson, read to the Academy of Sciences at Paris, Nov. 20, 1826, and inserted in the Conn. des Tems, 1829*, occurs on pages 324...331 of the *Philosophical Magazine* for May, 1827.

This article relates to the subject of the expansion of functions in a series of Laplace's functions; I do not see anything of importance in the paper in addition to what Ivory had already given. Poisson replied to the criticism: see Art. 1384.

Ivory adverts to the paper by Professor Airy read to the *Cambridge Philosophical Society* in May, 1826: see Art. 1227. The tone which Ivory adopts in controversy is so confident that it may be fairly called arrogant.

The following passage of interest occurs on the last page of the article:

The theory of the figure of the planets originated with Newton and Huygens: it has been the subject of incessant discussion for a century: it has been attended with greater difficulty, and has occasioned a greater number of memoirs, than any other branch of the system of the world.

1445. A brief notice will suffice of the six papers which we have numbered XVII. The titles are the following:

A letter to Professor Airy, in reply to his Remarks on some Passages in a Paper by Mr Ivory; this occurs on pages 16...20 of the volume.

Letter to G. B. Airy, Esq., Lucasian Professor of the Mathematics in the University of Cambridge; this occurs on pages 88...92.

Letter from Mr Ivory to the Editors of the Philosophical Magazine and Annals of Philosophy; this occurs on pages 93 and 94.

On the Figure of Equilibrium of a Homogeneous Planet in a Fluid State; in reply to the *Observations of M. Poisson....* This is in three parts, which occur respectively on pages 161...168, 241...247, and 321...326 of the volume.

There is nothing to call for special remark in these papers, as they merely repeat Ivory's known opinions. But it may be of interest to observe some acknowledgement, however slight, of fallibility. We have on page 17 the words: "...I find that I have drawn a wrong inference from my analysis..." and on page 90 the words: "...I have expressed myself rather unguardedly with respect to M. Poisson's theorem:..."

1446. The papers which we have numbered XVIII. consist of numerical application. The three which relate to pendulum experiments occur respectively on pages 165...173, 206...210, and 241...243 of the volume. The two which relate to measured arcs occur respectively on pages 343...349 and 431...436 of the volume.

Ivory considers that the arcs measured in Peru, India, France and England give $\frac{1}{309}$ for the ellipticity; and that Svanberg's Swedish arc is consistent with this.

1447. An article entitled *Some Remarks on an Article in the Bulletin des Sciences Mathématiques Physiques et Chimiques, for March,* 1828, occurs on pages 245...248 of the *Philosophical Magazine* for October, 1828.

Ivory asserts the accuracy of his peculiar views on fluid equilibrium; and says he will address a short work on the subject to the Royal Society.

1448. The titles of the first four of the set of papers which we have numbered XX. are the following:

On the Latitudes and Difference of Longitude of Beachy Head and Dunnose in the Isle of Wight...; this occurs on pages 6...11 of the volume.

On Measurements on the Earth's Surface perpendicular to the Meridian; this occurs on pages 189...194 of the volume.

On the Method employed in the Trigonometrical Survey for finding the Length of a Degree perpendicular to the Meridian; this occurs on pages 241...245 of the volume.

On the Method in the Trigonometrical Survey for finding the Difference of Longitude of two Stations very little different in Latitude; this occurs on pages 432...435 of the volume.

The most interesting matter considered in these papers is a theorem to which we alluded in Art. 1037. The original investigation of the theorem was obscure and unsatisfactory; and Ivory was led to the erroneous conclusion that the theorem was inaccurate: see page 244 of the volume. He speaks of the method of calculation based on the theorem as "the greatest delusion that has ever prevailed in practical mathematics"; and he pronounces an unfavourable opinion on a demonstration of the theory published by Dr Tiarks: see page 435 of the volume.

1449. The titles of the last two of the set of papers which we have numbered XX. are the following:

On the Method of deducing the Difference of Longitude from the Latitudes and Azimuths of two Stations on the Earth's Surface; this occupies pages 24...28 of the volume.

On the Method of deducing the Difference of Longitude from the Azimuths and Latitudes of two Stations; this occupies pages 106...109 of the volume.

Ivory shews in the first paper that the theorem to which we have just alluded, is really very approximately true for an oblatum which is nearly spherical; and in the second paper he extends the range of the theorem to the case of any figure of revolution which is nearly spherical. He makes no reference to the contrary opinion, which, as we have observed in the preceding Article, he had formerly held. See also a paper by Dr Tiarks on pages 52 and 53 of the volume.

1450. An article entitled *Some Arguments tending to prove that the Earth is a Solid of Revolution,* occurs on pages 205...209 of the *Philosophical Magazine* for March, 1829. Ivory arrives at the conclusion that certain measurements, transverse to the meridian, agreed well with the hypothesis that the Earth is an ellipsoid of revolution.

1451. An article entitled *Some Remarks on an Article in the "Bulletin des Sciences Mathématiques" for June,* 1829, § 269, occurs on pages 272...275 of the *Philosophical Magazine* for October, 1829. Ivory states briefly and obscurely some of his peculiar opinions on fluid equilibrium, and on the expansion of a function in a series of Laplace's functions. Ivory asserts that in Clairaut's theory of the equilibrium of fluids some of the forces which act are omitted; but it is needless to say that this assertion is contrary to the fact.

1452. An article entitled *Letter relating to the Figure of the Earth* occurs on pages 241...244 of the *Philosophical Magazine* for April, 1830. Ivory merely states in a controversial tone his peculiar opinions on fluid equilibrium. He says in his first paragraph:

It is not my intention to add anything new on this subject, but merely to state briefly what I have contributed to the theory, and to assert my claim to my own proper notions.

1453. An article entitled *On the Figure of the Earth* occurs on pages 412...416 of the *Philosophical Magazine* for June, 1830.

Biot had inferred from pendulum experiments that the lengths of the pendulum in different latitudes "are not accurately represented by the formula usually employed"; he found that "the coefficient of the term proportional to the square of the sine of the latitude, is not an invariable quantity, as usually assumed, but a quantity decreasing gradually from the pole to the equator." Ivory does not agree with Biot's opinion.

1454. The titles of the two papers which we have connected in number XXV. are the following:

A direct Method of finding the shortest Distance between two Points on the Earth's Surface when their Geographical Position is given; this occurs on pages 30...34 of the volume. *On the Shortest Distance between two Points on the Earth's Surface*; this occurs on pages 114...117 of the volume. These two papers belong rather to Solid Geometry than to our subject.

1455. A memoir entitled *On the Equilibrium of Fluids, and the Figure of a Homogeneous Planet in a Fluid State*, is contained in the *Philosophical Transactions* for 1831, published in that year. The memoir occupies pages 109...145 of the volume; it was read on January 13 and 20, 1831.

1456. The memoir seems to me quite destitute of value; it contains nothing that is new, and repeats the errors which Ivory had already published in his memoir of 1824.

It will be sufficient to give a few specimens of the statements which Ivory makes, and for which there is no foundation. He says on page 121:

In a homogeneous planet in a fluid state, there are forces which prevail in the interior parts and vanish at the surface; and, as Clairaut's theory notices no forces except those in action at the surface, it leaves out some of the causes tending to change the figure of the fluid, and therefore it cannot lead to an exact determination of the equilibrium.

Let ϕ be such a function that $\frac{d\phi}{dx}$, $\frac{d\phi}{dy}$, and $\frac{d\phi}{dz}$ denote the accelerating forces parallel to the corresponding axes; then Ivory says on his page 124:

...for ϕ must be always positive, and it must increase continually from the centre of gravity to the surface of the fluid.

Ivory asserts on his page 133 that an ellipsoid cannot be a form of fluid equilibrium unless two of the axes are equal.

The language is often vague and scarcely intelligible. Thus on page 111 we have:

...*fdm* is the motive force of the cylinder or prism, or the effort it makes to move in the direction...

On page 126 we have:

It may be proper to add that the mass of fluid has no tendency to turn upon an axis. For no motion of this kind can be produced by the pressures propagated inward from the surface, the directions of which pass through the centre of gravity. Neither can the accelerating forces urging the particles, cause any such motion, these being wholly employed in counteracting the inequality of pressure.

1457. A memoir entitled *On the Equilibrium of a Mass of Homogeneous Fluid at liberty* is contained in the *Philosophical Transactions* for 1834, published in that year. The memoir occupies pages 491...530 of the volume; it was read May 29, 1834.

1458. This memoir also seems to me quite destitute of value; the old errors are repeated, and statements made without any foundation. Thus Ivory asserts on his pages 494 and 498 that the forces must be such as to vanish at the centre of gravity; "for without this condition the equilibrium of the mass of fluid would be impossible."

On his page 501 he has two functions $\phi(x, y, z)$ and $\phi'(x, y, z)$; he says that $\phi'(x, y, z)$ must not contain such terms as Ax, By, Cz; and that as $\phi(x, y, z)$ coincides with $\phi'(x, y, z)$ at the surface, $\phi(x, y, z)$ can contain no such term. But this is untenable. For suppose $u = 1$ to be the equation to the surface; and let

$$\phi'(x, y, z) = f(x, y, z) + Ax + By + Cz + (u - 1)(Ax + By + Cz),$$

and

$$\phi(x, y, z) = f(x, y, z) + Ax + By + Cz.$$

Then $\phi'(x, y, z)$ does not contain such terms as Ax, By, Cz; while $\phi(x, y, z)$ does contain them; and yet the two coincide at the surface.

On page 513 Ivory undertakes to demonstrate that fluid in the form of an ellipsoid with three unequal axes cannot be in equilibrium: but we know that his result is untrue.

1459. We have stated in Art. 1422 the new principle of fluid equilibrium which Ivory assumed. In his memoir of 1831 he modified his statement; see page 133 of that memoir. In the present memoir he calls attention to the circumstance that there was something exceptionable in the memoir of 1824, but that the memoir of 1831 is not liable to the same reproach: see pages 528 and 529 of the present memoir. The difference between his two opinions may be thus expressed in modern language; at first he assumed that the potential of the fluid bounded by the level surfaces would be constant *throughout* the space enclosed by the interior surface, but afterwards he assumed that it would be constant for all points *on* the interior surface. However we now know that there is really no difference between the two opinions; for, by a theorem due to Gauss, if the potential is constant for all points of the interior surface, it will also be constant for all points of the space bounded by that surface; see Gauss's memoir, *Allgemeine Lehrsätze...* 1840; or the *Cambridge and Dublin Mathematical Journal*, Vol. IV. page 200.

1460. A memoir entitled *Of such Ellipsoids consisting of Homogeneous Matter as are capable of having the Resultant of the Attraction of the Mass upon a Particle in the Surface, and a Centrifugal Force caused by revolving about one of the axes, made perpendicular to the surface*, is contained in the *Philosophical Transactions* for 1838, published in that year. The memoir occupies pages 57...66 of the volume; it was read December 11, 1837. There is a note connected with the memoir on pages 265 and 266 of the succeeding volume of the *Philosophical Transactions*.

The memoir discusses Jacobi's theorem; it contains numerous important errors, which I have corrected in a paper published in the *Proceedings of the Royal Society*, Vol. XIX. 1871.

1461. The titles of the papers which we have numbered XXIX. are the following:

On the Conditions of Equilibrium of a Homogeneous Planet in a Fluid State; this occupies pages 81 and 82 of the volume.

A Remark on an Article of M. Poisson's Traité de Mécanique (Edition 2nd. No. 593); this occupies pages 274...270 of the volume.

On a Principle laid down by Clairaut for determining the Figure of Equilibrium of a Fluid, the Particles of which are urged by accelerating forces ...; this occupies pages 321...324 of the volume.

These papers are merely repetitions of the peculiar views which Ivory had already frequently published; perhaps they are expressed even with more than the usual confidence. The following sentences from page 62 may serve as an illustration of the style:

Now the least attention to the nature of this equation will show that the attraction of the matter without the level surface is entirely independent of the rest of the equation...

Now these two equations are the same with those given in a paper in the Philosophical Transactions for 1824, and in two subsequent papers written for the purpose of obviating some objections (I had almost said, frivolous objections) of M. Poisson.

1462. A memoir entitled *On the Conditions of Equilibrium of an Incompressible Fluid, the Particles of which are acted upon by Accelerating Forces*, is contained in the *Philosophical Transactions* for 1839, published in that year. The memoir occupies pages 243...264 of the volume; it was read June 20, 1839.

1463. This memoir is only a reproduction of the same unsatisfactory matter as Ivory had already often published; there does not seem to be any improvement, nor even any novelty. Ivory still holds to his assumption that it is necessary for equilibrium that the fluid between two level surfaces should exert no tangential action on a particle placed on the inner surface; that is, in modern language he assumes that the potential of the stratum is constant all over the inner surface. A reference is given on page 257 to Poisson, who had recorded his dissent from Ivory's opinion; and it is plainly suggested that this dissent arises from the want of a "little patience."

On page 253 we are told that the true principle of the equilibrium of a fluid is that "the level surfaces at all depths must have determinate figures": it may be safely said that no valuable result could be deduced from such an obvious truism.

1464. The writings of Ivory on our subject, disregarding those which were published in the *Philosophical Magazine*, occupy about 300 quarto pages of the *Philosophical Transactions*; probably all which is valuable in them could be compressed into a tenth of that space. I consider these meritorious investigations to consist of three parts; the demonstration in the first memoir of the theorem which is usually called Ivory's, the indication in the second memoir of a weakness in one of Laplace's demonstrations, and some analytical results in the memoir of 1824, relating to Laplace's coefficients: in the last part however Rodrigues had anticipated Ivory; see Art. 1187.

But the discussions on fluid equilibrium are unworthy of Ivory, and their publication reflects little credit on the state of English mathematics, or on the administration of the Royal Society, at the epoch. It might perhaps have been permitted to a writer of Ivory's reputation to expound once his peculiar opinions; though even this is doubtful, since these opinions were opposed to the principles received by every scientific authority of the period: but even if the appearance of the memoir of 1824 is thus excused, there can be no justification for the repetition of the same unsatisfactory matter in the memoirs of 1831, 1834, and 1839.

CHAPTER XXXVII.

PLANA.

1465. THE writings of Plana, arranged chronologically, which belong to our subject are the following:

I. Sulla teoria dell'attrazione degli sferoidi elittici. I have already noticed this: see Art. 1147.

II. Mémoire sur l'attraction des Sphéroïdes Elliptiques Homogènes. Gergonne's *Annales de Mathématiques*, 1812 and 1813.

III. A Letter relating to Saturn's Ring occurs in De Zach's *Correspondance Astronomique*, Vol. I. 1818. We have already noticed this letter in Art. 867.

IV. Solution de différens Problèmes relatifs à la loi de la résultante de l'attraction... Turin *Memorie*, Vol. XXIV. 1820.

V. Note sur la Densité et la Pression des Couches du Sphéroïde Terrestre. De Zach's *Correspondance Astronomique*, Vol. V. 1821.

VI. Mémoire sur Différens Procédés d'intégration, par lesquels on obtient l'attraction d'un Ellipsoïde Homogène... Crelle's *Journal für...Mathematik*, 1840.

VII. Note sur l'intégrale $\int \frac{dM}{r} = V$.... This occurs in the same volume as number VI.

VIII. Appendix to the memoir number VI. Crelle's *Journal für...Mathematik*, 1843.

IX. Two Notes relating to propositions in Newton's Principia occur in the Turin *Memorie*, Vol. XI. 1851.

X. Note sur la densité moyenne de l'écorce superficielle de la Terre. *Astronomische Nachrichten*, Vol. XXXV.

XI. Note sur la Figure de la Terre et la loi de la Pesanteur à sa surface d'après l'hypothèse d'Huygens, publiée en 1690. *Astronomische Nachrichten*, Vol. XXXV.

XII. Sur la Théorie mathématique de la Figure de la Terre, publiée par Newton en 1687. Et sur l'état d'équilibre de l'ellipsoïde fluide à trois axes inégaux. *Astronomische Nachrichten*, Vol. XXXVI.

XIII. Sur la loi des Pressions, et la loi des Ellipticités des couches terrestres, ... *Astronomische Nachrichten*, Vol. XXXVI.

XIV. Sur la loi de la Pesanteur à la Surface de la mer, dans son état d'Equilibre. *Astronomische Nachrichten*, Vol. XXXVIII.

I proceed to give an account of such of these writings as have not been already noticed; the first of these is that numbered II.

1466. A memoir entitled *Mémoire sur l'attraction des Sphéroïdes Elliptiques Homogènes* is contained in the third volume of Gergonne's *Annales de Mathématiques*, which is dated 1812 and 1813. The memoir occupies pages 273...279 of the volume.

1467. The memoir may be described as a commentary on a passage of Lagrange's *Mécanique Analytique*, which occurs on pages 113 and 114 of the second edition, and on pages 106...108 of the third edition.

Plana investigates the general result which is quoted in Art. 1004; this he does by transforming the variables in the manner of Ivory, to whom he refers. The transformation is the same as was also used by Gauss: see Art. 1173.

Plana illustrates the advantage of Lagrange's result in a subsequent memoir: see Crelle's *Journal für... Mathematik*, Vol. XX. page 279.

1468. We now come to a memoir entitled *Solution de différens problèmes relatifs à la loi de la résultante de l'attraction exercée sur un point matériel par le cercle, les couches cylindriques, et quelques autres corps qui en dépendent par la forme de leurs élémens.*

This memoir is contained in Vol. XXIV. of the Turin *Memorie* which was published in 1820. The memoir occupies pages 389...450 of the volume. The memoir was read on the 28th of February, 1819.

1469. The first problem considered is the attraction exerted by the perimeter of a circle on a given point which is not necessarily in the plane of the circle: this occupies pages 394...408. Plana gives expressions for the two components into which the attraction may be resolved: the expressions involve complete elliptic integrals.

1470. In the particular case in which the given point is situated on the straight line drawn through the centre of the circle at right angles to its plane, the resultant attraction is $\frac{2\pi k s}{(s^2+k^2)^{\frac{3}{2}}}$; where k is the radius of the given circle, and s is the distance of the given point from the centre of the given circle. Plana says on his pages 403 and 404:

Cette expression est remarquable par sa simplicité, et en ce qu'elle nous fait voir, que la masse de la périphérie du cercle agit comme si elle était toute concentrée dans un quelconque de ses points.

This remark seems to me unnecessary: it is of course obvious, without any calculations, that every element of the circumference of the circle is at the same distance from the given point, and also exerts the same attraction along the direction of the resultant.

1471. Plana's formulæ may be applied to the case in which the given point is in the plane of the circle. Likewise he obtains immediately the attraction which a sphere exerts on a ring which is outside it in a plane passing through the centre of the sphere: this leads to the remark as to the instability of such a system made by Laplace, to whom Plana refers. See Art. 872.

1472. Plana passes naturally to consider the attraction of a shell, supposed to be of uniform infinitesimal thickness, and in the form of a surface of revolution, at a given point in the axis. He states the definite result for the case in which the shell is in the form of an oblongum, with the given point at the focus: see his page 407. But his result is wrong. Let e be the excentricity; then adopting his mode of expression, the result should be

$$\frac{4\pi e}{3}-\frac{4\pi}{e}-\frac{2\pi^3\sqrt{(1-e^2)}}{e^2}+\frac{8\pi\sqrt{(1-e^2)}}{e^2}\tan^{-1}\frac{\sqrt{(1+e)}}{\sqrt{(1-e)}}.$$

But instead of the first two terms Plana has

$$-\frac{4\pi e}{3}-\frac{12\pi}{e}.$$

Since $\tan^{-1}\dfrac{\sqrt{(1+e)}}{\sqrt{(1-e)}} = \dfrac{\pi}{4} + \dfrac{1}{2}\sin^{-1}e$, the result may be expressed thus:

$$\frac{4\pi e}{3} - \frac{4\pi}{e} + \frac{4\pi\sqrt{(1-e^2)}}{e^2}\sin^{-1}e.$$

This is the attraction *from* the centre.

It is shewn in *Differential Calculus*, Art. 374, that

$$\sqrt{(1-e^2)}\sin^{-1}e = e - \frac{1}{3}\left\{e^3 + \frac{2}{5}e^5 + \frac{2.4}{5.7}e^7 + \frac{2.4.6}{5.7.9}e^9 + \dots\right\}.$$

Hence the attraction *towards* the centre is

$$\frac{4\pi}{3}\left\{\frac{2}{5}e^3 + \frac{2.4}{5.7}e^5 + \frac{2.4.6}{5.7.9}e^7 + \dots\right\}.$$

1473. Plana concludes this section of his memoir thus, referring to the particular case just considered:

Ce cas particulier satisfait pour faire voir qu'un point matériel ne saurait demeurer en équilibre dans l'intérieur d'une couche elliptique d'épaisseur constante: il faut pour cela, que l'épaisseur soit variable comme l'intervalle compris entre deux ellipses dont le rapport des axes est le même.

He leaves his readers to establish this statement for themselves. The argument may take the following shape, which holds whether the surface be of revolution or not.

The inner surface of the shell is assumed to be that of an ellipsoid; if possible let the outer surface be of some other form, and not a similar and similarly situated and concentric ellipsoid. Take a similar, similarly situated, and concentric ellipsoidal surface *just* big enough to include the supposed outer surface, and therefore touching it at least at one point, say P. Then we have two shells which exert no attraction on an internal particle; namely one shell by hypothesis, and another by a known demonstration. Hence the difference of these two shells exerts no attraction. This difference is a shell which is of zero thickness at P. Suppose a particle *very* near to P inside the shell; draw a plane through the particle parallel to the tangent plane at P. Then of the two parts into which the shell is thus divided, that round P ultimately exercises no attraction: the attraction being in fact $2\pi\rho$ where ρ is ultimately zero: see Art. 993. Hence the attraction of the other part is unbalanced and the particle cannot be in equilibrium.

Thus the outer boundary of the shell can be nothing but an ellipsoid homothetical with the inner boundary.

1474. I will for an example solve one problem suggested by this section of Plana's memoir.

To find the attraction exerted by the circumference of a circle at an external point in the plane of the circle.

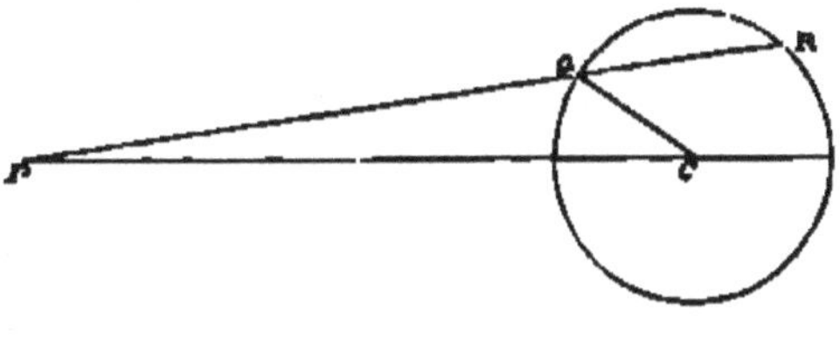

Let C be the centre, P the external particle; let k denote the radius, and p the distance CP.

Draw from P any straight line PQR to cut the circle. Let $CPQ = \theta$, and $CQR = \phi$; let $PQ = r_1$, and $PR = r_2$.

The attraction of the element of the circumference of the circle at Q may be denoted by $\frac{r_1\, d\theta \sec\phi}{r_1^2}$, and the attraction of the element of the circumference at R by $\frac{r_2\, d\theta \sec\phi}{r_2^2}$.

Hence the attraction of the two elements resolved along PC

$$= \frac{(r_1 + r_2)\sec\phi \cos\theta\, d\theta}{r_1 r_2} = \frac{2p\cos^2\theta \sec\phi\, d\theta}{p^2 - k^2}.$$

Now $k\sin\phi = p\sin\theta$; therefore $k\cos\phi d\phi = p\cos\theta d\theta$. Hence the whole attraction of the circle

$$= \frac{4k}{p^2 - k^2}\int_0^{\frac{\pi}{2}} \cos\theta d\phi = \frac{4k}{p^2 - k^2}\int_0^{\frac{\pi}{2}} \sqrt{\left(1 - \frac{k^2}{p^2}\sin^2\phi\right)} d\phi.$$

Thus the whole attraction is expressed as a complete elliptic integral of the second order.

Plana's result involves two complete elliptic integrals; one of the first order and one of the second order; see his page 401 at the top. He rather has in view a point within the circle, but the requisite slight modification is easily made. His result can be made to agree with that just given by means of the known properties of complete elliptic integrals.

We may put the above expression thus

$$\frac{4k}{p\sqrt{(p^2-k^2)}}\int_0^{\frac{\pi}{2}}\sqrt{\left(\frac{p^2-k^2\sin^2\phi}{p^2-k^2}\right)}\,d\phi.$$

As p diminishes, the factor outside the integral sign increases; and so also does the expression under the integral sign. Thus the attraction continually increases as P approaches the circumference; this might probably have been anticipated, though the demonstration is not immediately obvious.

1475. The second problem considered by Plana is the attraction exerted by a circular lamina on a particle which is not necessarily in the plane of the circle.

This occupies pages 408...421. Here, as in the first problem, the expressions obtained for the components of the attraction involve elliptic integrals.

1476. On his page 410, Plana omits p in the second term of his expression for an attraction. The error is obvious because it makes the two terms of his expression of *different dimensions*. Nevertheless the error is continued on pages 417, 418, and 421. There are several instances of this kind of error or misprint in the memoir. It is strange that such an elementary consideration as the necessity of having the various terms of an expression of the same dimension should apparently have been quite disregarded in writing the memoir, or in correcting the press.

1477. On his page 410, Plana says:

...il me semble qu'il conviendrait d'employer ici les formules données par Cotes pour avoir des valeurs approchées des intégrales.

I suppose that Plana here refers to pages 30...33 of the Chapter *De Methodo Differentiali Newtoniana* by Cotes. They relate to what we should now speak of as the approximate calculation of the area of a curve by the method of equidistant ordinates.

1478. On his page 417, Plana says:

Je présume, que à l'aide des formules précédentes on peut démontrer qu'un point placé entre les centres de deux cercles qui ne se coupent pas doit prendre un mouvement oscillatoire sur la ligne qui joint les centres des cercles sans pouvoir jamais atteindre la circonférence du plus petit cercle.

It is not quite clear to me what Plana means. But I presume that one circle is supposed to fall entirely *within* the other; or we may for facility of conception suppose the two circles parallel, but indefinitely close, and the particle to move between them.

Of course he cannot mean the two circles to be in the same plane, and one quite without the other. For in such a case there is indeed a position of equilibrium for a particle on the line joining the centres and between the two circumferences: but the equilibrium is *unstable*, and if the particle is moved towards either circle, it will move up to contact with that circle. This follows from the fact that if a particle in the plane of a circle, outside the circle, move *towards* the circle, the attraction *continually increases*. This fact is established by supposing the circle decomposed into thin strips at right angles to the straight line on which the particle is supposed to move; for the attraction of a strip varies inversely as the distance from the particle, and directly as the sine of half the angle subtended by the strip at the particle: hence the attraction of every strip increases as the particle approaches the circle.

1479. On his pages 418...420 Plana gives some numerical calculations as to the attraction of a thin ring, like that of Saturn, on a particle near the inner or outer boundary. He uses his results to throw doubts on some remarks made by Laplace in his sixth memoir: see Art. 871.

1480. I will give here a simple investigation of a problem connected with this section of Plana's memoir.

Find the attraction exerted by a circular cylinder on a particle placed in contact with the curved surface at a point equally distant from the ends of the cylinder.

Let $2h$ be the height of the cylinder, and a the diameter of the cylinder.

Take the origin at the point considered; let the axis of x be the normal at this point, and the axis of z the generating line of the cylinder.

Then the resultant attraction is obviously directed along the axis of x, and is equal to

$$\iiint \frac{x\,dx\,dy\,dz}{(x^2+y^2+z^2)^{\frac{3}{2}}},$$

the integration extending over the whole cylinder. Integrate with respect to z; the limits are $-h$ and h: thus we obtain

$$2h \iint \frac{x\,dx\,dy}{(h^2+r^2)^{\frac{1}{2}}\,r^2},$$

where r^2 stands for x^2+y^2.

Transform in the usual way to polar coordinates; thus we obtain

$$2h \iint \frac{\cos\theta\,d\theta\,dr}{(h^2+r^2)^{\frac{1}{2}}}.$$

Integrate with respect to r; the limits are 0 and $a\cos\theta$; thus we obtain

$$2h \int \cos\theta \log \frac{\sqrt{(h^2+a^2\cos^2\theta)}+a\cos\theta}{h}\,d\theta.$$

The limits for θ are $-\frac{\pi}{2}$ and $\frac{\pi}{2}$. Integrate by parts, and the integral reduces to

$$4ah \int_0^{\frac{\pi}{2}} \frac{\sin^2\theta\,d\theta}{\sqrt{(h^2+a^2\cos^2\theta)}}.$$

We may express this as

$$\frac{4h\sqrt{(h^2+a^2)}}{a}(F-E),$$

where F and E denote complete elliptic integrals of the first and second order respectively; the modulus being $\frac{a}{\sqrt{(h^2+a^2)}}$.

This is exact. If we suppose h very small compared with a, we have approximately

$$F = \log \frac{4\sqrt{(a^2+h^2)}}{h}, \text{ and } E = 1;$$

so that finally the attraction approximately

$$= 4h\left(\log \frac{4a}{h} - 1\right).$$

It will be found that this agrees with a result given by Plana on his page 418, after we correct the mistake noticed in Art. 1476. We must in his formula suppose $p = k$, and double the result to get the attraction of the whole cylinder.

For a numerical example suppose $h = \frac{1}{30}$ and $a = 2$; we get $\frac{4}{30}(\log 240 - 1)$, that is $\frac{4}{30}$ of 4·48064, that is ·59742. This agrees with the result given by Plana on his page 419 in the form $2 \times$ ·29871.

For another example suppose $h = \frac{1}{30}$ and $a = \frac{14}{5}$; we get $\frac{4}{30}(\log 336 - 1)$, that is $\frac{4}{30}$ of 4·81711, that is ·64228. This does not agree with the result given by Plana on his page 419 in the form $2 \times$ ·342933.

1481. The third problem considered by Plana is the attraction of a right cylindrical surface with a circular base on a particle in the plane of the base. This occupies pages 422...445. As before, the expressions obtained involve elliptic integrals.

1482. In the particular case in which the height of the cylinder is infinite, the attraction on an internal particle resolved along the plane of the base is zero. This may be easily verified. For suppose the cylinder to extend to infinity both above and below the plane in which the particle is situated; then the attraction vanishes, as may be shewn by a process like that of Newton for an ellipsoidal shell. Hence, as the attractions of the parts above and below the plane resolved along the plane are obviously equal, each must vanish. In exactly the same manner, if a particle be placed inside an ellipsoidal shell, at any point of a principal plane, the attraction of each half into which the principal plane divides the shell resolved along the plane is zero.

1483. There are numerous misprints in this section. For instance, on page 427 in Plana's formulæ (e) and (i), for $2\lambda - 1$ read λ. This misprint is obvious from the principle of dimensions to which I have referred in Art. 1476; the misprint extends its influence over many of the subsequent formulæ.

1484. On his pages 439...443 Plana investigates the values of the following definite integrals:

$$\int_{-a}^{a} \tan^{-1} \frac{\beta \sqrt{(a^2 - x^2)}}{a (q - x)} dx \text{ and } \int_{-a}^{a} x \tan^{-1} \frac{\beta \sqrt{(a^2 - x^2)}}{a (q - x)} dx.$$

His process may be much improved.

Assume with him $x = \frac{(y^2 - 1)\, a}{y^2 + 1}$; then the definite integrals become

$$4a \int_0^\infty \frac{y\, dy}{(1 + y^2)^2} \tan^{-1} \frac{2\beta y}{q + a + y^2 (q - a)},$$

and

$$4a^2 \int_0^\infty \frac{y(y^2 - 1) dy}{(1 + y^2)^3} \tan^{-1} \frac{2\beta y}{q + a + y^2 (q - a)}.$$

After this his process becomes very laborious. A better way will be to assume

$$\tan^{-1} \frac{2\beta y}{q + a + y^2 (q - a)} = \tan^{-1} My - \tan^{-1} Ny;$$

this gives $$M - N = \frac{2\beta}{q + a}, \quad MN = \frac{q - a}{q + a};$$

thus $$M + N = \frac{2}{q + a} \sqrt{(\beta^2 + q^2 - a^2)}.$$

Hence M and N are known.

Consider the first integral, which Plana denotes by X, so that

$$X = 4a \int_0^\infty (\tan^{-1} My - \tan^{-1} Ny) \frac{y dy}{(1 + y^2)^2}.$$

Integrate by parts; thus

$$X = 2a \int_0^\infty \left(\frac{M}{1 + M^2 y^2} - \frac{N}{1 + N^2 y^2} \right) \frac{dy}{1 + y^2} \quad \text{............ (1).}$$

But $\displaystyle\int_0^\infty \frac{dy}{(1+h^2y^2)(1+k^2y^2)} = \frac{1}{h^2-k^2}\int_0^\infty\left(\frac{h^2}{1+h^2y^2}-\frac{k^2}{1+k^2y^2}\right)dy$

$$= \frac{1}{h^2-k^2}(h-k)\frac{\pi}{2} = \frac{\pi}{2(h+k)} \quad\ldots\ldots\ldots\ldots\ldots (2).$$

From (1) and (2) we see that

$$X = \pi\alpha\left(\frac{M}{M+1}-\frac{N}{N+1}\right) = \frac{(M-N)\pi\alpha}{1+M+N+MN}$$

$$= \frac{\alpha\beta\pi}{q+\sqrt{(q^2+\beta^2-\alpha^2)}} = \frac{\pi\alpha\beta\left[q-\sqrt{(q^2+\beta^2-\alpha^2)}\right]}{\alpha^2-\beta^2}.$$

Now consider the second integral, which Plana denotes by X'. Integrate by parts; thus

$$X' = 2\alpha^2\int_0^\infty\left(\frac{M}{1+M^2y^2}-\frac{N}{1+N^2y^2}\right)\frac{y^2dy}{(1+y^2)^2} \quad\ldots\ldots\ldots\ldots (3).$$

If we differentiate (2) with respect to k we find that

$$\int_0^\infty \frac{y^2dy}{(1+h^2y^2)(1+k^2y^2)^2} = \frac{\pi}{4k(h+k)^2} \quad\ldots\ldots\ldots\ldots\ldots\ldots (4).$$

From (3) and (4) we see that

$$X' = \frac{\pi\alpha^2}{2}\left\{\frac{M}{(M+1)^2}-\frac{N}{(N+1)^2}\right\}$$

$$= \frac{\pi\alpha^2}{2}\,\frac{M(N+1)^2-N(M+1)^2}{(M+1)^2(N+1)^2}$$

$$= \frac{\pi\alpha^2}{2}\,\frac{(M-N)(1-MN)}{(M+1)^2(N+1)^2}$$

$$= \frac{\pi\alpha^2}{2}\,\frac{\beta}{\left[q+\sqrt{(q^2+\beta^2-\alpha^2)}\right]^2}$$

$$= \frac{\pi\alpha^2\beta}{2}\left\{\frac{q-\sqrt{(q^2+\beta^2-\alpha^2)}}{\alpha^2-\beta^2}\right\}^2.$$

Thus, as Plana says on his page 443, we have

$$X' = \frac{\alpha}{2\pi\beta}X^2;$$

he adds: "ce qui constitue un théorème assez remarquable."

Let there be an infinite right elliptic cylinder; let 2α be the major axis and 2β the minor axis of the ellipse. Then $4X$ expresses the attraction of the cylinder on a particle situated on the axis major produced, and at a distance q from the centre. This is easily shewn by cutting up the cylinder into infinitesimal rods parallel to the generating lines, and summing the attractions of the rods which form a slice at right angles to the major axis of the ellipse.

1485. The last problem considered by Plana is the attraction of an infinite right elliptic cylinder on an external particle. This occupies pages 445...450.

Let f and g be the coordinates of the attracted particle, estimated from a fixed point in a plane at right angles to the generating lines of the cylinder. Let V denote the potential of the cylinder. Then we know that V will be a function of f and g determined by

$$\frac{d^2V}{df^2} + \frac{d^2V}{dg^2} = 0.$$

Hence $$V = \phi\,[f + g\sqrt{(-1)}] + \psi\,[f - g\sqrt{(-1)}],$$
where ϕ and ψ denote functions at present undetermined.

The resolved attractions we know are equal to $\frac{dV}{df}$ and $\frac{dV}{dg}$ respectively. Hence we obtain for these resolved attractions expressions like that given for V.

Now Plana obtains these expressions for the resolved attractions without any use of the potential; and this is the only novelty in his solution: see his page 449.

1486. In the fifth volume of De Zach's *Correspondance Astronomique...*, published in 1821, is a note by Plana entitled, *Note sur la densité et la pression des couches du sphéroïde terrestre;* it occupies pages 68...79 of the volume.

1487. The note relates to the law of density and the law of pressure discussed by Laplace in the *Connaissance des Tems* for 1822: see Arts. 1283 and 1325.

Plana calculates the values of the density and of the pressure at various depths below the surface of the Earth by Laplace's formulæ; but he adds nothing to the theoretical investigations. He also makes some comparison of the theory with observations.

I will notice two or three points which present themselves.

1488. His formula (2) on page 70 involves some strange mistake or misprint; the term $\left(\frac{5\pi}{6\sqrt{6}}+\frac{1}{3\sqrt{2}}\right)^2$ is wrong.

1489. Plana works out in detail the comparison of theory with observation, to which Laplace himself gave some attention: see Art. 1329. By this comparison Plana arrives at the result that the ratio of the Moon's mass to the Earth's mass is ·0122651, which he says lies between $\frac{1}{82}$ and $\frac{1}{83}$; but it should be between $\frac{1}{82}$ and $\frac{1}{81}$.

Plana says:

La masse de la lune ·0122651 diffère sensiblement de la fraction $\frac{1}{68\cdot5}$ que l'on obtient autrement comme l'on sait. Mais malgré cela on doit, ce me semble, admettre, que cette loi de la densité des couches du sphéroïde terrestre s'accorde assez bien avec l'ensemble des phénomènes connus. Cela posé, il me paraît certain, que la valeur de

$$\frac{\int \rho a^4 da}{\int \rho a^2 da} = \cdot 27216,$$

trouvée par M. le Baron de *Lindenau* (voyez Éphémérides de Berlin pour 1820, page 211) doit etre trop éloignée de la véritable, puisque la théorie précédente donne ·485967 avec un degré d'approximation plus plausible.

Au reste, il est essentiel d'observer, que le rapport de ces deux intégrales définies a été calculé par M. de *Lindenau* à l'aide de la formule

$$\frac{\int \rho a^4 da}{\int \rho a^2 da} = \frac{3}{5}\left(1 - \frac{j}{2e}\right),$$

qui lui a été communiquée par M. *Gauss*.

J'ignore dans ce moment le juste degré d'approximation de cette formule ainsi que le moyen de la dériver de la condition de l'équilibre de l'océan...

I have expressed the formula given by Gauss in my own notation; ϵ is the Earth's ellipticity and j is, as usual, the ratio of the centrifugal force at the equator to the attraction there.

I am, like Plana, quite ignorant as to the origin of the formula. I have no faith in it. It looks to me as if Gauss had intended to suppose the Earth to be composed of similar elliptic strata, surrounded by a film of fluid; then the $\frac{3}{5}$ must be cancelled. See equation (2) of Art. 323; and there suppose ϵ constant.

1490. Plana brings evidence from a comparison of the lunar theory with observation to confirm the value of the moon's mass he had obtained, which, as we have seen, was rather smaller than the value found otherwise. The present received value seems still smaller than that adopted by Plana.

1491. Plana concludes thus:

Telles sont toutes les principales conséquences qui dérivent immédiatement de la loi supposée pour la densité des couches du sphéroïde terrestre. En considérant l'ensemble des phénomènes, l'accord est tellement satisfaisant, que l'on parait autorisé à regarder cette loi comme celle de la nature.

Perhaps this is expressed rather too strongly. Let us consider what is meant by *l'ensemble des phénomènes*.

Taking with Legendre and Laplace, as in Art. 1326, for the density the expression $A\frac{\sin na}{a}$, it is found that if n is supposed equal to $\frac{5\pi}{6}$, the ellipticity of the Earth agrees well with observation. But we can hardly say that this gives any evidence in favour of the supposed law of density; the coincidence has in fact been secured by a proper selection of a certain arbitrary constant.

Thus all that seems to remain is that the value of $\frac{\int \rho a^4 da}{\int \rho a^2 da}$, as found by our expression for ρ, agrees reasonably well with the phenomena of precession and nutation.

1492. The next memoir by Plana which we have to consider is entitled *Mémoire sur différens procédés d'intégration, par lesquels on obtient l'attraction d'un ellipsoïde homogène dont les trois axes sont inégaux, sur un point extérieur.* This memoir occupies pages 189...270 of Crelle's *Journal für...Mathematik*, Vol. XX., which was published in 1840.

1493. Plana begins by alluding to some discussions in the Academy of Paris, which we have seen took place towards the close of Poisson's career: see Art. 1404. He says that he drew up the memoir principally for his own instruction, after reading that by Poisson in Vol. XIII. of the Memoirs of the Academy of Paris. On reflection his memoir seemed to him to gain a greater degree of importance by the novelty of the methods employed, either to obtain the known results or to illustrate them.

1494. The memoir is divided into four sections; each section might be considered in certain respects as an independent memoir: but in uniting them they afforded mutual assistance.

1495. The following sentences from the introduction may be noticed as giving Plana's opinion on a point discussed between Poisson and Poinsot:

On verra dans le quatrième paragraphe quelle est mon opinion et ma manière de considérer la solution donnée par *Legendre* en 1788. C'est un chef-d'œuvre d'analyse; si par le mot *analyse*, on veut bien entendre une suite de transformations des formules primitives dans lesquelles le raisonnement est en partie remplacé par le mécanisme du calcul. Abstraction faite de la longueur des calculs c'est, à mon avis, la solution la plus directe qu'on ait donné de ce problème jusqu'à ce jour.

Plana proceeds then to explain in what sense he takes the word *direct*, so as to justify the high commendation he thus pronounces on Legendre's solution: but the remarks do not altogether commend themselves to my judgement.

1496. The memoir is useful as bringing together various investigations which were originally published in other places, and supplying some explanatory comment, yet it cannot be said to contain anything essentially new and important.

1497. Suppose we have a given ellipsoid, and we wish to determine a confocal ellipsoid which shall pass through a given point: this problem we know occurs in Ivory's mode of treating the attraction of an ellipsoid. The problem leads to a cubic equation. Plana's first section is devoted to the discussion of this problem: it occupies his pages 193...206.

1498. The following passage may be noticed:

...*Legendre* avait dit (et Mr. *De Pontécoulant* a répété d'après lui, voyez p. 354 du second volume de sa Théorie analytique du système du monde) que cette équation n'a qu'une seule racine réelle (voyez p. 512 du tome 1er de son Traité des fonctions elliptiques); mais la réalité de ses trois racines est maintenant hors de doute.

Plana is wrong in his charge so far as relates to Pontécoulant. The equation which Pontécoulant takes is

$$\frac{a^2}{k^2} + \frac{b^2}{k^2 + e^2} + \frac{c^2}{k^2 + e'^2} = 1;$$

this equation has doubtless three roots for k^2, namely, one positive and two negative, but the latter do not make k^2 real: so there is but one *real* value of k^2.

Legendre indeed is incautious; his equation is

$$\frac{f^2}{\alpha^2 + \xi} + \frac{g^2}{\beta^2 + \xi} + \frac{h^2}{\gamma^2 + \xi} = 1;$$

and he says that there is only one real value of ξ: but the context suggests that he means only one real *positive* value.

1499. Plana's second section relates to the equation of the cone which has its vertex at a given point and circumscribes a given ellipsoid; it occupies pages 206...216. He shews how to put this equation in its simplest form by changing the axes of coordinates. The last twelve lines of the section contain some troublesome misprints in formulæ which relate to the circular sections of the cone.

1500. Plana's third section purports to treat of the formulæ for determining the attraction of an indefinitely thin ellipsoidal shell bounded by similar surfaces, on an external particle; it occupies pages 210...240. But the title is inadequate, as there are many other formulæ in the section besides those which relate to the attraction of the film bounded by homothetical ellipsoids.

1501. In the beginning of this section Plana draws various useful deductions from the known formulæ for the attraction of an ellipsoid on an external point. We will reproduce some of these; but it will be sufficient for us to confine ourselves to one of the three components of the attraction.

Denote by X the component considered in Art. 885. Then with the notation there employed we have

$$X = \frac{4\pi abcf}{k} \int_0^1 \frac{x^2\,dx}{\sqrt{k^2 + (b^2 - a^2)x^2}\,\sqrt{k^2 + (c^2 - a^2)x^2}},$$

where k^2 has to be found from an equation there given.

Put $m = \frac{a^2}{b^2}$, $n = \frac{a^2}{c^2}$, $k^2 = a^2(1 + \nu)$.

Then the equation for determining ν and k^2 becomes

$$\frac{f^2}{1+\nu} + \frac{mg^2}{1+m\nu} + \frac{nh^2}{1+n\nu} = a^2 \text{..............} (1),$$

and we have

$$X = \frac{4\pi f}{\sqrt{(1+\nu)}} \int_0^1 \frac{x^2 dx}{\sqrt{m(1+\nu) + (1-m)x^2}\,\sqrt{n(1+\nu) + (1-n)x^2}}.$$

Now if we form $\frac{dX}{d\nu} d\nu$, we shall obtain the expression for the component attraction of a film bounded by homothetical ellipsoids; because by the change of a into $a + da$ and of ν into $\nu + d\nu$, we pass from one ellipsoid to a similar, similarly situated, and concentric ellipsoid infinitesimally different from the former.

In this way we can verify Poisson's theorem as to the attraction of a certain film: see Art. 1304. Poisson himself says that he did this: his words are quoted in Art. 1405. If however we differentiate with respect to ν the value of X in the form in which we have left it, the integral sign does not obviously disappear. It is convenient to change the variable. Put

$$x^2 = \frac{1+\nu}{1+u};$$

then we obtain

$$X = 2\pi f \int_0^\infty \frac{(1+u)^{-\frac{3}{2}}\, du}{\sqrt{(1+mu)(1+nu)}} \text{ ....(2).}$$

Therefore
$$\frac{dX}{d\nu} = -\frac{2\pi f (1+\nu)^{-\frac{3}{2}}}{\sqrt{(1+m\nu)(1+n\nu)}}.$$

Hence the component $\frac{dX}{d\nu} d\nu$ is expressed free from the integral sign.

Similarly the other components can be expressed free from the integral sign.

Plana's formulæ give the direction of the resultant attraction; but we do not delay on this point.

1502. Put

$$(1+u)\,a^2 = v_1^2, \quad (1+mu)\,b^2 = v_2^2, \quad (1+nu)c^2 = v_3^2.$$

Then (1) becomes

$$X = 4\pi fabc \int_k^\infty \frac{dv_1}{v_1^2 v_2 v_3} \text{ ....(3).}$$

This form agrees with that given by Rodrigues; see Art. 1396.

Thus, as Plana observes, Rodrigues had only one step to take in order to obtain from his formulæ the attraction of the film. He had in fact only to pass from (3) to (2), and then differentiate with respect to ν. However Rodrigues did not take this step; and the result was first given by Poisson.

1503. Plana finishes the section with remarks which demand some criticism.

By a certain transformation of the expression already given for X he obtains a result of the form

$$X = 4\pi \int f(\omega)\, d\omega,$$

where the limits of the variable ω are 0 and $\sqrt{A'}$.

Then he says he will interpret this geometrically. Now a certain series of cones has presented itself in his investigations, determined by a parameter which varies between 0 and $\sqrt{A'}$.

Hence he infers that the resolved attraction of that part of the ellipsoid which is contained between the cones corresponding to the values ω and $\omega + d\omega$ is $4\pi f(\omega)\, d\omega$. This is quite unsound. We might on the same ground take the expression for X given in the beginning of Art. 1501, change x into $\frac{\omega}{\sqrt{A}}$, so that X becomes say $\int \psi(\omega)\, d\omega$, and then assert that $\psi(\omega)\, d\omega$ represents the attraction of the conical element. In other words Plana proves fairly that $4\pi \int f(\omega)\, d\omega$ between the specified limits represents the resolved attraction of the ellipsoid; but he has *no ground whatever* for his assertion as to the geometrical meaning of $4\pi f(\omega)\, d\omega$. The assertion is indeed true, for Legendre demonstrated it; but Plana gives no demonstration of his assertion.

1504. The fourth section of Plana's memoir is devoted to the demonstration of a certain formula of Legendre's; this section occupies pages 240...270. The formula is that to which we alluded in the preceding Article.

In our account of Legendre's third memoir we have stated that Legendre supposes a certain series of cones having their common vertex at the attracted point. Two consecutive conical surfaces will determine a conical shell; Legendre finds an expression for the attraction at the vertex of that part of such a shell which is bounded by the attracting ellipsoid. This expression is that of his formula (g') in page 479: it is

$$\frac{2\pi\omega' d\omega}{F}\left\{\frac{G}{\sqrt{H}} + \frac{G'}{\sqrt{H'}}\right\},$$

where ω is a parameter which determines the conical shell, and the capital letters denote certain complicated functions of the parameter, the coordinates of the attracted point, and the semi-axes of the ellipsoid.

Legendre devotes his pages 470...479 to the investigation of the formula. As I have said however his process is rather indicated than worked out.

1505. Now Plana gives the operations in detail; and this constitutes the business of his thirty pages. He says himself on his page 240:

Les artifices de calcul par lesquels *Legendre* est parvenu à une démonstration directe de la formule...sont fort ingénieux. Mais, il est impossible d'apprécier au juste le degré de la complication et le mérite de la difficulté vaincue, sans suivre pas à pas les transformations et les réductions à travers lesquelles on doit passer pour mettre en évidence la propriété caractéristique du résultat obtenu par le procède de son intégration. Malgré les indications laissées par *Legendre* il n'est pas fort aisé (du moins pour moi) de retrouver, ni les résultats intermédiaires ni le résultat final. Je pense qu'il ne sera pas tout-à-fait inutile d'exposer ici avec détail la marche que j'ai suivie par y parvenir.

1506. I had myself gone over Legendre's work in his own order and with his own notation, before I had seen Plana's memoir. I find that the additional developments thus required will fill about seven quarto pages like Plana's. To a patient student such a course would not be more laborious than the study of Plana's pages.

1507. Legendre effects a certain transformation which at first sight does not appear to have produced any simplification. Plana observes that perhaps it was only after trial justified by the subsequent investigations that he found the advantage of his transformation. Plana adds on his page 248:

Un récit naïf des tâtonnements de ce genre serait fort instructif; mais, par des motifs difficiles à deviner, il est rare de rencontrer dans les écrits des grands géomètres des exemples comparables à ceux qui nous frappent en lisant les ouvrages d'*Euler*.

1508. Legendre's formula (g') presents a very complicated appearance; Legendre simplifies it by a peculiar process, not by direct calculation: see what has been said with respect to this point in Arts. 888 and 889. Plana appears to allude to this in the last paragraph of his memoir; he says:

Le calcul dont nous venons de parler (fort pénible même dans le cas particulier où le point attiré serait placé dans le plan d'une des trois sections principales de l'ellipsoïde) est sans doute celui que Mr. Poisson qualifie d'*inextricable*: mais la démonstration de *Legendre* ne réclame point, ni l'exécution effective de ce calcul, ni l'appui d'aucun théorème dérivé d'une autre source. Le théorème qu'il avait en vue, et son expression analytique sous la forme la plus simple, découlent des principes les plus rigides du Calcul Intégral. L'état progressif de l'analyse fait espérer que cette espèce de lacune laissée par *Legendre* sera un jour heureusement remplie. Il est même probable qu'elle le sera par un de ces traits de génie qui attestent la force des principes, et la difficulté de saisir le véritable mode d'en faire l'application.

I do not feel quite certain as to Plana's meaning; but I presume by *le calcul* he means a simplification of Legendre's formula (g') by actual algebraical work. When he says that Legendre's demonstration does not require this work, I suppose he means that the peculiar process adopted by Legendre is sound; in this I agree with him. Thus Plana, according to my view, is correct in the sense he assigns to Poisson's word *inextricable*; the meaning being that at a certain stage Legendre could not obtain his desired result by actual algebraical work, but adopted a peculiar process. This actual algebraical work Plana describes on his last page as "effrayant pour le plus intrépide algèbriste."

1509. The next memoir by Plana is entitled *Note sur l'intégrale* $\int \frac{dM}{r} = V$, *qui exprime la somme des élémens de la masse d'un ellipsoïde, divisés respectivement par leur distance à un point attiré.* This occurs in the same volume as the preceding memoir: it occupies pages 271...282.

1510. The memoir investigates a certain expression for the potential of an ellipsoid which agrees with that given in Art. 1184. Let the notation be as in Art. 1501; then the potential V is determined by

$$V = -\pi \int_\nu^\infty \left\{\frac{f^2}{1+u} + \frac{mg^2}{1+mu} + \frac{nh^2}{1+nu} - a^2\right\} \frac{du}{\sqrt{U}} \ldots\ldots(4),$$

where $U = (1+u)(1+mu)(1+nu)$.

For an external point ν must be found by the equation (1) of Art. 1501. For an internal point we must put zero for ν.

Plana considered the potential for an internal point again in a subsequent memoir. See the *Astronomische Nachrichten*, Vol. XXXVI. page 172, and Art. 1543.

To justify his formula we observe that if we differentiate with respect to f, g, or h, we obtain a correct result. Thus, for example, it gives

$$\frac{dV}{df} = -2\pi f \int_{\nu}^{\infty} \frac{du}{(1+u)\sqrt{U}},$$

for no term arises from the variation of ν, since the expression under the integral sign vanishes when $U = \nu$, by equation (1) of Art. 1501.

Now this value of $\frac{dV}{df}$ is correct, because $-\frac{dV}{df}$ we know is equal to the X of Art. 1501; and so we see that our result is true by (1) of that Article.

In like manner, if we differentiate (4) with respect to g or to h, we obtain a correct result. Hence (4) must be true, provided it is true for any special values of f, g, h; that is, we have only to shew that no arbitrary constant is required in (4).

Plana effects this comparison by calculating directly the value of V for an external particle, and shewing that the expression agrees with (4).

A simple mode is by shewing that when f, g, and h are infinite, the expression for V in (4) vanishes as it should.

When f, g, h are infinite, we have ν infinite by (1) of Art. 1501.

But $\frac{f^2}{1+u} + \frac{mg^2}{1+mu} + \frac{nh^2}{1+nu}$ is numerically less than when $u = \nu$; so that it is finite. And $\sqrt{U}$ is greater than $u^{\frac{3}{2}}\sqrt{(mn)}$; hence $\int_{\nu}^{\infty} \frac{du}{\sqrt{U}}$ vanishes when ν is infinite; and so also does $\int_{\nu}^{\infty} \frac{vdu}{\sqrt{U}}$ if v be any finite quantity.

1511. Plana shews that his value of V involves a demonstration of a result originally obtained by Jacobi and investigated by Poisson: see Art. 1401.

1512. If the value of V is required in a series proceeding according to inverse powers of the distance of the attracted point from the centre of the ellipsoid, Plana considers that the best way would be to combine the methods of Laplace and Lagrange. Laplace's method is that which is contained in the *Mécanique Céleste*, Livre III. Chapitre II. Lagrange's method is that to which we allude in Art 1467. Plana applies both methods.

1513. The next memoir by Plana is entitled *Appendice au Mémoire sur l'attraction de l'ellipsoïde homogène imprimé dans le Tome* XX. *de ce journal.* This memoir occupies pages 132...146 of Crelle's *Journal für...Mathematik*, Vol. XXVI., which was published in 1843.

1514. In this memoir Plana gives in detail the operations required in the third Section of Legendre's memoir: see Art. 860.

Legendre himself left much of the work to be done by the reader. A patient student would find it perhaps as easy to fill up the steps for himself in Legendre's memoir as to read Plana's.

Plana concludes thus:

On doit comprendre maintenant que l'analyse de *Legendre* avoit besoin de ce long développement, pour pouvoir saisir plusieurs des motifs secrets qui ont guidés la marche de son calcul. J'ignore s'il est permis de soutenir, que de tels motifs peuvent être aisément devinés d'après le texte de *Legendre*.

1515. Two notes by Plana were published in the Turin *Memorie*, Vol. XI. 1851. The first note purports to relate to Newton's Proposition LXXI., and the second note to Newton's Propositions LXXX. and LXXXIV. The notes occupy pages 391...406 of the volume.

1516. The notes may be said to consist of translations of Newton's investigations respecting attractions into analytical language. Thus, take Newton's Proposition LXXI. We have seen in Art. 4, that Newton's investigation leads to this result: the attraction $= \frac{4\pi k\rho a}{c^2}\int\frac{pdp}{\sqrt{(a^2-p^2)}}$, the integration being taken between proper limits. Thus we have $\frac{4\pi k\rho a}{c^3}\int d.\sqrt{(a^2-p^2)}$ between proper limits. Now $\sqrt{(a^2-p^2)}$ is equal to half a certain chord in Newton's figure, which he calls HK. Plana arrives at this expression by a method somewhat different from that which we have used.

1517. Plana gives what he considers an exposition of Newton's mode of establishing his Proposition LXX; we may however state briefly that the exposition is inaccurate. Plana asserts that Newton uses the property about the intersecting chords of a circle; Euclid III. 35 is meant. Newton does not use this property: he seems to use the equality of angles in the same segment.

1518. There is nothing important or remarkable in Plana's analytical translations of Newton's processes; they are substantially the same as would occur to any person acquainted with the elements of the Differential and Integral Calculus, and had in fact been given by Maupertuis in his memoir of 1732.

1519. Plana seems to think that Newton adopted geometrical forms in order to conceal his peculiar methods. Thus he says:

Mais, Newton, qui, probablement, ne voulait pas dévoiler à ses lecteurs toutes les ressources de ce calcul, alors naissant, a imaginé une démonstration, où, le procédé de l'intégration est déguisé...

Again:

On ne peut éviter cette traduction algébrique sans nuire à la clarté, et sans renforcer l'opinion que Newton ne voulait pas exposer ses découvertes avec cette ingénuité qui les aurait rendues accessibles aux hommes doués d'une médiocre intelligence.

1520. The next memoir by Plana is contained in the *Astronomische Nachrichten*, No. 828, which was published in September 1852: this number forms part of Vol. XXXV. of the Journal.

The memoir consists of three parts, namely a *Note* and two *Additions*. The memoir is entitled *Note sur la densité moyenne de l'écorce superficielle de la Terre*.

1521. Plana begins thus:

C'est un fait incontestable, que la densité des couches elliptiques du sphéroïde terrestre est croissante depuis la surface jusqu'au centre. La loi de cet accroissement est inconnue; mais comme elle est intimement liée à tous les phénomènes qui dépendent du mouvement de rotation de la Terre autour de son centre de gravité, on peut faire des hypothèses sur son expression algébrique, et comparer ensuite les résultats qu'elles fournissent à ceux que l'observation a fait connaître avec un précision suffisante. Je vais examiner, sous ce point de vue, la plus simple de toutes les hypothèses, proposée par *Laplace* à la page 46 du cinquième volume de la *Mécanique Céleste*.

1522. Thus Plana's Note relates to the hypothesis proposed by Laplace in the page just cited. The hypothesis is that the density increases from the surface to the centre in Arithmetical Progression. The hypothesis is expressed symbolically in the following manner. Let (ρ) denote the density at the surface, ρ the density at the distance a from the centre, the mean radius of the Earth being taken as unity. Then $\rho = (\rho)\,[1 + e - ea]$, where e is some constant to be determined.

Taking the density of the sea for unity, Laplace assumed that $(\rho) = 3$, which is about the density of granite: then he obtained $e = 2{\cdot}349$.

1523. Plana considers that 3 is too large a value for (ρ). He does not assume a value for (ρ), but considers that it has been well established by the experiments of Reich that the *mean* density of the Earth is 5·44. Taking this for granted, and employing certain formulæ from other parts of the *Mécanique Céleste*, Plana finds that e is about 7·8907.

1524. The Note contains nothing new in theory; it consists entirely of numerical calculation. I have not verified this. Indeed there are some difficulties, which though not very serious discourage any attempt to go over the work. I will state them, as some other student may be fortunate enough to explain them.

(1) I do not see how the coefficient of μ in Plana's equation (6) is obtained; apparently it should be the product of ·12065 and ·0163; but it is not equal to this.

(2) In comparing his equations (6) and (7) the sign of the coefficient of μ seems arbitrarily changed from $-$ to $+$.

(3) Having arrived at his equation (17) Plana seems to assert that the density at the centre is more likely to be greater than

16·301 than less. I cannot conjecture how he obtained any knowledge as to the numerical value of the density at the centre.

(4) Towards the beginning of his Note, Plana says that 16·27 is the density at the centre; but from his equation (17) it would appear that this density is greater than 16·301. I cannot reconcile these statements.

(5) Plana says that Legendre's law of density leads to less satisfactory results. According to this law, see Art. 1326,

$$\rho = \frac{(\rho)\sin an}{a\sin n}.$$

If $n = \frac{5\pi}{6}$, Plana obtains for the density at the centre 0·4235; and if $n = \frac{7\pi}{8}$, he obtains 12·89. Then I gather that these results are unsatisfactory to him, because they do not make the density at the centre great enough. But, as I have already said, I cannot conjecture how Plana obtained any knowledge as to what the density at the centre ought to be.

(6) Suppose that in the last expression for ρ we give to n a small increment δn; then Plana says that ρ becomes

$$\frac{(\rho)\sin an}{a\sin n} + \frac{(\rho)\,\delta n\sin(n - an)}{\sin^2 n}.$$

It seems to me that the second term should be

$$(\rho)\,\delta n\,\frac{a\cos an\sin n - \sin an\cos n}{a\sin^2 n}.$$

1525. Laplace obtained for the mean density of the Earth 4·76. Plana finds that the same value of the mean density would follow from supposing $(\rho) = 1·6$, instead of $(\rho) = 3$; and remarks that this is rather singular.

1526. It appears from the conclusion of the Note that Plana undertook the discussion at the request of Humboldt. Towards the beginning of the Note Plana refers to what Humboldt had said on the subject in his *Kosmos*, Vol. I. page 177, of the German edition. See Vol. I. page 159, of Sabine's English translation, fifth edition.

1527. In the first Addition to his Note, Plana adverts to the determination of the density of a mountain by the aid of pendulum experiments. Laplace had touched on this subject in the *Mécanique Céleste*, Vol. v. pages 55 and 50. Plana adds nothing of importance.

Plana gives in this Addition the calculation necessary to solve the following problem. Suppose a sphere to have the same mass as the mountain Schehallien, find at what distance it must be placed from a pendulum, to produce the same amount of deviation as the mountain Schehallien produced, according to Maskelyne's observations.

1528. Plana's second Addition relates to the expression of the force of gravity at various places of the Earth's surface.

Let P denote the value of gravity at the latitude θ, and P' that for the latitude $\sin^{-1}\frac{1}{\sqrt{3}}$; then Plana says that

$$P = P' + P'\left(\sin^2\theta - \frac{1}{3}\right)\left\{2j - \epsilon\frac{\int_0^1 \rho a^4\,da}{\int_0^1 \rho a^2\,da}\right\}.$$

This formula is obtained on the supposition that all the strata are similar, and that ϵ is the ellipticity. It may be easily verified from results given by Clairaut: see Arts. 336 and 323. With the law of density which is under examination Plana obtains a value of P, which he says agrees well with observation.

Plana proceeds to a calculation as to the value of a certain term in the expression for the depth of the sea. But the formula which he quotes is wrong. It amounts in fact to taking the expression for h' given in the *Mécanique Celeste*, Vol. v. page 30, and *omitting the denominator.*

Plana himself in a subsequent memoir gives the same formula as Laplace: see *Astronomische Nachrichten*, No. 861, equation (38).

1529. The next memoir by Plana is contained in the *Astronomische Nachrichten*, No. 830, which was published in December 1852; this number forms part of Vol. xxxv. of the Journal. The memoir is entitled *Note sur la figure de la terre et la loi de la pesanteur à sa surface, d'après l'hypothèse d'Huygens, publiée en* 1690.

1530. Plana begins thus:

Quoique cette hypothèse soit démentie par l'ensemble des faits observés, il est curieux de l'exhumer, et de la présenter développée avec le langage de l'analyse moderne, afin de pouvoir juger si *Huygens*, dans l'appendice à sa dissertation sur la cause de la gravité, a réellement été capable de démontrer que son hypothèse conduit aux deux résultats que *Laplace* lui attribue dans sa notice historique, exposée vers le commencement du $\text{XI}^{\text{ème}}$ livre de sa Mécanique Céleste. Comme je pense que, *Huygens* n'était pas en possession de principes suffisants pour comprendre dans son analyse la double hypothèse de la force constante et celle de la force variable, j'ai voulu mettre en évidence par quelle espèce d'heureuse divination, il a pu, à travers des calculs bornés, en conclure le véritable rapport entre la pesanteur au pôle et la pesanteur à l'équateur, dans le cas, où, sans admettre l'attraction de molécule à molécule, on suppose, que chaque molécule d'une masse fluide homogène, tournant sur un axe, tend vers le centre de gravité de cette masse, en raison inverse du carré de sa distance à ce point. On verra que cela tient à une extension hasardée que *Huygens* donnait au résultat obtenu pour le cas de la force constante.

Plana maintains that in two points rather more has been ascribed to Huygens than is justly due to him.

1531. Let us take the first point. Laplace says with respect to Huygens in the *Mécanique Céleste*, Vol. v. page 5:

Il n'admet point l'attraction de molécule à molécule, et il suppose que chaque molécule d'une masse fluide homogène, tournant sur un axe, tend vers le centre de gravité de cette masse, en raison inverse du carré de sa distance à ce point.

But this is not correct, for Huygens assumes that the central force is constant. It is true that the result as to the ellipticity, supposed small, is the same as if the force varies according to any law of the distance; but this was not shewn by Huygens, though he does indeed make such a statement: see Art. 56.

There is however another matter connected with this which has given occasion to error: see Art. 64.

1532. Now let us take the second point. Laplace says on the page already cited:

Il trouve ensuite que la pesanteur croît de l'équateur aux pôles, proportionnellement au carré du sinus de la latitude, et de manière que la pesanteur étant supposée 1 à l'équateur, elle est $1 + 2\phi$ aux pôles.

Laplace's ϕ is our j.

Plana, however, holds that Huygens did not obtain this definite result; but that Laplace has himself given his own extension to the theory of Huygens. I do not agree with Plana; I hold that Laplace's words fairly represent what Huygens obtained. The passage to be examined is page 160 of the *Discours de la Cause de la Pesanteur*.

1533. The next memoir by Plana is contained in the *Astronomische Nachrichten*, Numbers 850 and 851, which were published in March, 1853; these numbers form part of Vol. XXXVI. of the Journal. The memoir is entitled *Sur la Théorie mathématique de la Figure de la Terre, publiée par Newton en* 1687. *Et sur l'état d'équilibre de l'ellipsoïde fluide à trois axes inégaux.*

Thus it is seen that the memoir consists of two parts.

1534. The first part of the memoir consists of the ordinary analytical theory of the relative equilibrium of a rotating oblatum of fluid: it adds nothing substantially to what we find in the *Mécanique Céleste*, Livre III. Chapitre III. The object of Plana in reproducing this known theory seems to have been to point out what he considers an important error in Newton's process; in Plana's words, *un vice inhérent à son analyse.* But the error exists only in Plana's imagination.

Newton required the value of the attraction of a nearly spherical oblatum on a particle at the equator. He determined this indirectly, by the theorem that this attraction may be considered to be a mean proportional between the attraction of a certain sphere on a particle at its surface, and the attraction of a certain oblongum on a particle at its pole. Plana then holds that Newton considered this theorem to be *absolutely* true; there can be no doubt that Newton only considered it to be *approximately true:* which it is. I think that Plana is alone in this untenable opinion as to Newton's meaning. See Art. 22.

1535. Plana says with reference to his supposed correction of Newton:

...Si je ne me trompe, cette remarque n'avait pas encore été faite. Elle a échappé au Commentateur *Calandrini*, au point, qu'il a entrepris de démontrer, que le théorême dont je parle étoit vrai sans restriction (voyez les pages 83 et 84 du troisième volume des Principia. Edition de Genève).

This assertion with regard to Calandrini is in direct contradiction to the fact. The commentator certainly considered the theorem as an approximate truth: the words *quam proxime* occur four times in the demonstration.

1536. Thus it is sufficiently obvious that the first part of Plana's memoir was quite unnecessary, being founded on a misconception of his own. Some incidental points may be noticed.

1537. Plana begins thus:

Laplace, en imitant les raisonnemens, que *Clairaut* avait développés dans la première partie de son Introduction à l'ouvrage des Principes, a exposé sans l'emploi des formules algébriques, cette Théorie de *Newton* au commencement du XI Livre de la Mécanique Céleste.

This rather vague reference to Clairaut is, I presume, intended to apply to the translation of the Principia into French by Madame du Chastellet, who was assisted by Clairaut. See Art. 560.

1538. The equation which connects the excentricity of the oblatum with the angular velocity is in Laplace's symbols

$$\tan^{-1}\lambda = \frac{9\lambda + 2q\lambda^3}{9 + 3\lambda^2}.$$

Plana gives the approximate solution of the equation for the case when λ is a small quantity, carried further than Laplace. Plana's results are

$$\frac{5}{4}q = \frac{\lambda^2}{2} - 15\left\{\frac{2.2}{5.7}\left(\frac{\lambda^2}{2}\right)^2 - \frac{3.2^2}{7.9}\left(\frac{\lambda^2}{2}\right)^3 + \frac{4.2^3}{9.11}\left(\frac{\lambda^2}{2}\right)^4 - \dots\right\},$$

from which by reversion of series

$$\frac{\lambda^2}{2} = \frac{5}{4}q + \frac{12}{7}\left(\frac{5}{4}q\right)^2 + \frac{148}{49}\left(\frac{5}{4}q\right)^3 + \frac{21673}{11319}\left(\frac{5}{4}q\right)^4 + \dots$$

I have not verified the last line.

It is known that for a given angular velocity we have in general two solutions: for example, if the angular velocity is small besides the solution in which λ is a small quantity there is a solution in which λ is great. Plana says that this second solution was first indicated by Laplace; this is wrong, for, as we have remarked, Thomas Simpson and D'Alembert preceded him: see Art. 580.

1539. Plana puts in a convenient form the equation which shews that the *oblongum* is not a possible form of relative equilibrium.

Suppose the equation to the oblongum to be

$$\frac{x^2+y^2}{a^2}+\frac{z^2}{a^2(1+\gamma^2)}=1.$$

Let the attractions of the oblongum resolved parallel to the axes at a point (x, y, z) within it or on its surface be denoted respectively by

$$Ax, \quad Ay, \quad Cz.$$

Then the values of A and C are well known; and Plana puts them in the form

$$A=\frac{4\pi\rho\cos^3\psi}{\sin^3\psi}\int_0^\psi\frac{\sin^2 x}{\cos^3 x}\,dx,$$

$$C=\frac{4\pi\rho\cos^3\psi}{\sin^3\psi}\int_0^\psi\frac{\sin^2 x}{\cos x}\,dx,$$

where ψ is such that $\tan\psi=\gamma$; and ρ is the density.

The condition for relative equilibrium is

$$A-\omega^2=C(1+\gamma^2),$$

where ω denotes the angular velocity.

Put q for $\frac{3\omega^2}{4\pi\rho}$; then the condition becomes

$$q=\frac{3}{\sin^3\psi}\int_0^\psi\frac{\sin^2 x}{\cos^3 x}(\sin^2 x-\sin^2\psi)\,dx.$$

This is impossible: for the left-hand member is positive and the right-hand member is negative.

1540. We now pass to the second part of the memoir. In this Plana demonstrates Jacobi's theorem, that a fluid ellipsoid rotating round its least axis may be a form of relative equilibrium. The demonstration presents nothing novel. Plana however employs the expressions for the attraction of an ellipsoid by means of elliptic integrals, which had been used by Legendre and Poisson: see Art. 1398. Hence he calculates certain numerical results which give him the dimensions of various ellipsoids which satisfy the condition of relative equilibrium. These results are collected in the following table. The smallest semiaxis is taken for unity, this is that about which the body is supposed to rotate. The largest semiaxis is $\sqrt{(1+\gamma')}$; the values of this are given in the first column. The other semiaxis is $\sqrt{(1+\gamma'')}$; the values of this are given in the second column. The third column contains the values of $\frac{3\omega^2}{4\pi\rho}$, where ω is the angular velocity and ρ is the density.

1·41545	1·41448	·21159
1·41836	1·41564	·22128
1·42672	1·42120	·21545
1·43524	1·42405	·20591
1·52425	1·47261	·225875
1·06164	1·52340	·260562
1·83008	1·54643	·27235
2·09574	1·54976	·280143
2·13005	1·52015	·27612
2·20269	1·50573	·27543
2·28117	1·43210	·263034
2·45860	1·26100	·220114
2·70044	1·09756	·1042
5·24086	1·03561	·21524

These numbers rest on Plana's authority. The entry ·1042 in the third column looks suspiciously small; but he gives it in two places, namely his pages 151 and 170.

1541. There are some misprints in the memoir which will not cause serious trouble; except perhaps in Plana's equations (30). In the last of these equations for Δ read Δ'.

1542. On his page 170, Plana speaking with respect to the ellipsoid in Jacobi's theorem, says:

La grande inégalité entre les deux aplatissements d'un tel ellipsoïde, ne permet pas de le considérer comme un corps semblable à la Terre.

This is wrong. There is not necessarily such a great difference between the two ellipticities: it is sufficient for example to look at the first case given in the preceding table.

For information as to Jacobi's theorem I may refer to my paper cited in Art. 1460.

1543. Plana finds the value of the potential at any internal point of an ellipsoid; this subject he had formerly considered: see Art. 1510. He now definitely expresses the potential by means of elliptic integrals.

Suppose that the attractions at the point (x, y, z) parallel to the axes are respectively

$$Ax, \quad By, \quad Cz.$$

Let V denote the potential: then

$$V = \Pi - \frac{Ax^2}{2} - \frac{By^2}{2} - \frac{Cz^2}{2},$$

where Π is some constant to be determined.

It is obvious that Π must be equal to the potential for a point at the centre of the ellipsoid.

Take for the equation to the ellipsoid

$$\frac{x^2}{a^2} + \frac{y^2}{a^2(1+\gamma^2)} + \frac{z^2}{a^2(1+\gamma'^2)} = 1.$$

Then with the usual polar notation

$$\Pi = \rho \iiint r \sin\theta \, d\theta \, d\phi \, dr.$$

We have first to integrate with respect to r from $r = 0$ to

$$r^2 = \frac{a^2}{\cos^2\theta + \sin^2\theta\left(\frac{\cos^2\phi}{1+\gamma^2} + \frac{\sin^2\phi}{1+\gamma'^2}\right)}.$$

Thus $$H=\frac{\rho a^3}{2}\iint\frac{\sin\theta\,d\theta\,d\phi}{\cos^2\theta+\sin^2\theta\left(\dfrac{\cos^2\phi}{1+\gamma^2}+\dfrac{\sin^2\phi}{1+\gamma'^2}\right)}.$$

Then we integrate with respect to ϕ; the limits are 0 and 2π. We may take 0 and $\frac{\pi}{2}$ for the limits, and multiply by 4. Thus we get

$$H=\pi\rho a^3\int_0^\pi\frac{\sin\theta\,d\theta}{\sqrt{\left\{\left(\cos^2\theta+\dfrac{\sin^2\theta}{1+\gamma^2}\right)\left(\cos^2\theta+\dfrac{\sin^2\theta}{1+\gamma'^2}\right)\right\}}}$$

$$=2\pi\rho a^3\sqrt{[(1+\gamma^2)(1+\gamma'^2)]}\int_0^{\frac{\pi}{2}}\frac{\sin\theta\,d\theta}{\sqrt{[(1+\gamma^2\cos^2\theta)(1+\gamma'^2\cos^2\theta)]}}.$$

By assuming $\cos\theta=\dfrac{\tan\psi}{\gamma}$ we obtain

$$H=\frac{2\pi\rho a^3}{\gamma}\sqrt{[(1+\gamma^2)(1+\gamma'^2)]}\int_0^\beta\frac{d\psi}{\Delta},$$

where $\Delta^2=1-\dfrac{\gamma^2-\gamma'^2}{\gamma^2}\sin^2\psi$, and $\tan\beta=\gamma$.

Thus H is expressed by an elliptic function. And A, B, C can also be similarly expressed: see Art. 1540. Thus finally V can be so expressed.

1544. Poisson had maintained that for points *on the surface* of a body

$$\frac{d^2V}{dx^2}+\frac{d^2V}{dy^2}+\frac{d^2V}{dz^2}=-2\pi\rho;$$

but, as we have remarked, this cannot be considered satisfactory: see Art. 1253.

Plana here objects to Poisson's opinion; but Plana is himself equally unsatisfactory. Plana affirms that at the surface of a homogeneous ellipsoid the right-hand member of the equation must be $-4\pi\rho$. Plana's sole argument is, that such is the value for *any internal* point, and consequently such must be the value at the surface. It is astonishing that Plana did not see the unsoundness of the argument. For we know that for any external point the right-hand side of the equation is *zero*; and so we might just as readily assert that it must be *zero* for a point on the surface.

1545. On his page 171, Plana referring to Jacobi's theorem, speaks of what he calls *les formules primitives de Jacobi*. I do not know what this means. I am not aware that Jacobi gave any investigation of his theorem, or any formulæ.

1546. The next memoir by Plana is contained in the *Astronomische Nachrichten*, Numbers 860 and 861, which were published in May, 1853; these numbers form part of Vol. XXXVI. of the Journal. The memoir is entitled *Sur la loi des pressions, et la loi des ellipticités des couches terrestres, en supposant leur densité uniformément croissante depuis la surface de la Terre jusqu'à son centre.*

1547. Assuming that the earth consists of fluid elliptical strata, Clairaut obtained an equation connecting the law of the ellipticity with the law of the density: see Art. 341. Plana then assumes the law of density given in Art. 1522, and applies Clairaut's equation to determine the ellipticity. He integrates Clairaut's differential equation in the form of a convergent series. He obtains about $\frac{1}{320}$ for the ellipticity at the surface.

Again with the same law of density Plana now takes the hypothesis that the earth consists of a solid part surrounded by a layer of fluid. If this layer is treated as infinitesimal, and the solid strata be assumed to have all the same ellipticity, he finds that the ellipticity is about $\frac{1}{308}$. If the layer is supposed of finite but small depth, and the solid strata have as before the same ellipticity, he finds that the ellipticity of the solid part is about $\frac{1}{301}$; and that of the surface of the fluid about ·000044. The latter he makes to be $\frac{1}{74\cdot471}$ of the former; in his introductory part he speaks of this as about $\frac{1}{100}$.

The theory is again Clairaut's, being all involved in equation (2) of Art. 323.

1548. The memoir contains nothing new in theory; it may be regarded as a mathematical exercise. But it shews I think that the law of density which is assumed merits nearly as much attention as that which Laplace discussed after Legendre. The English elementary treatises of Airy, Pratt, and O'Brien, seem to me to attach implicitly a greater importance to this law than Laplace himself did; see his language on page 10 of his Volume v. where he explains his object in adopting this law.

Besides Legendre's law of density, and the present law, a third has been considered, namely that by Roche and Resal: see the *Traité Elémentaire de Mécanique Céleste*, page 232.

1549. In his introductory remarks Plana states his opinion on some points connected with the formation of the earth and with the temperature of its interior; they seem to me expressed too positively as if they involved known facts instead of hypotheses.

1550. Plana gives at the end of his memoir a theorem for the approximate calculation of an integral. He says it is new; but I do not see what is the novelty which is claimed. The main result seems to be this

$$\int_0^h y\,dx = h\,\frac{\Delta y}{\log(1+\Delta y)},$$

where the expression on the right hand is to be expanded in powers of Δy, and any power as $(\Delta y)^n$ changed to $\Delta^n y$.

But this cannot be called new. See for instance De Morgan's *Differential and Integral Calculus*, pages 262 and 265.

1551. The next memoir by Plana is contained in the *Astronomische Nachrichten*, Numbers 903, 904, and 905, which were published in May, 1854: these numbers form part of Vol. XXXVIII. of the Journal. The memoir is entitled *Sur la loi de la pesanteur à la surface de la mer, dans son état d'équilibre*.

1552. The memoir may be said to go over the same ground as Laplace in the *Mécanique Céleste*, Livre XI. Chapitre II. §§ 2...5. Plana thus describes what Laplace proposed to effect:

La loi de la pesanteur à la surface de la mer, a été donnée au No. 33 du 3ème livre de la Mécanique Céleste, en supposant le sphéroïde terrestre entièrement recouvert par une couche très-mince d'eau en équilibre. La petitesse de la profondeur que l'on attribue ainsi à la mer, et à sa masse totale, permet de négliger l'action qu'elle exerce sur ses propres molécules, soit comparativement à celle de la Terre, soit en comparaison de la force centrifuge, (beaucoup plus petite) née de sa rotation diurne. *Laplace*, considérant ensuite que sa théorie, fondée sur la double hypothèse d'une inondation générale, et de la nullité d'action à l'égard de la masse de la mer, ne pouvait pas représenter le cas de la nature, a repris la question dans le XIème livre de son ouvrage, pour la tracer avec plus de généralité.

Accordingly as I have said Plana goes over the same ground as Laplace did in the *Mécanique Céleste*. Plana's analysis is rather more elaborate; but substantially the process is the same as Laplace gave.

We will notice a few points of detail.

1553. In Art. 1306 I have given in my own notation the equation

$$\frac{1}{2} X + a \frac{dX}{dr} = 0 \text{ ....(1).}$$

This equation is much used by Laplace in Chapter II. of his Eleventh Book.

Plana begins by considering this equation. He regards Laplace's demonstration as insufficient; and substitutes another which rests on the theory of the expansion of functions in a series of Laplace's functions. In this way he arrives at a more general form of the theorem. He finds that

$$\frac{1}{2} X + a \frac{dX}{dr} = -2\pi\rho a s \text{ ....(2),}$$

where ρ denotes the density of the matter of which X is the potential, and s may be called the elevation above the level of the sea. Thus s is in fact discontinuous, being zero for any point on the surface of the sea, and for any point on a continent expressing the height above the level of the sea. This is the same result as Lagrange obtained: see Art. 1100.

The analysis is simple and interesting by which Plana proves his equation (2); and this may be considered an improvement on Laplace's process. Perhaps Plana overestimates the importance of the result. He says it is indispensable that equation (1) should be true exclusively for the part of the earth which is covered by the sea. It seems to me that for Laplace's purpose it is sufficient to know that the equation is true for this part, and it is of scarcely any interest to know whether the equation is or is not true for other parts of the earth.

1554. Plana quotes from Fourier's *Théorie de la Chaleur*, page 245, the statement that the series

$$2\varpi\left\{\frac{\sin\varpi\sin\theta}{\pi^2-\varpi^2}+\frac{\sin 2\varpi\sin 2\theta}{\pi^2-2^2\varpi^2}+\frac{\sin 3\varpi\sin 3\theta}{\pi^2-3^2\varpi^2}+\dots\right\},$$

is equal to $\sin\theta$ for values of θ comprised between 0 and ϖ: it should be equal to $\sin\frac{\pi}{\varpi}\theta$. See *Integral Calculus*, Art. 325.

1555. I have already called attention to a passage in this memoir: see Art. 1310.

1556. Plana deduces from his investigations the formulæ given by Clairaut on his pages 217 and 220 of his work; and takes the opportunity of stating that D'Alembert was wrong in his objections to Clairaut's formulæ. Plana says, with reference to Clairaut's equation on page 220,

...Ainsi, il est démontré que cette équation est conforme à l'hypothèse d'une profondeur constante de la mer, admise par *Clairaut*; ce qui fait tomber la critique publiée par *D'Alembert* en 1773, (après la mort de *Clairaut*) dans les pages 227...230 du sixième volume de ses Opuscules Mathématiques. Ce jugement de *D'Alembert* prouve qu'il n'avait pas senti toute la justesse de la Théorie de *Clairaut*.

I quite concur with the last sentence. I have pointed out the precise point at which D'Alembert went wrong in my Art. 634.

1557. In his equation (49) Plana gives an expression for the length of a degree of the meridian. But the signs and the numerical values of many of the terms in this expression seem to me wrong. In his equation (58) he puts the expression into a numerical form; but if I am right his expression will be quite unsatisfactory.

1558. Plana touches on the subject of the Tides in the course of his memoir; I do not however enter on this subject in the present work. But one remark must be made. In sections V. and XI. of his memoir Plana quotes and accepts a formula given by Laplace in the *Mécanique Céleste*, Vol. II. page 192. But in a note to section XI. Plana objects to the formula. He says:

Mais en examinant de plus près les calculs par lesquels cette formule a été déduite, je viens de reconnaître qu'elle n'est pas le véritable résultat fourni par l'intégration des trois équations différentielles qui déterminent les oscillations de l'Océan.

Plana then proceeds to state his own results which he says he has obtained by an analysis, too long to be exhibited here. He concludes his note thus:

C'est de quoi je donnerai ailleurs une démonstration, fondée sur l'intégration des équations qui renferment implicitement, l'explication de tous les phénomènes que présente le flux et le reflux de la mer.

I do not find that the intention thus expressed was carried into effect.

1559. In the last section of his memoir Plana applies his formulæ to the case in which the Earth is supposed to be homogeneous and entirely fluid. He obtains an expression for the radius vector, and an expression for the value of gravity; then from these it follows that the latter varies inversely as the former. Plana then says:

C'est en vertu de cette transformation de la valeur précédente de p, que *Newton* disait dans sa Proposition XX. du 3ème livre des Principes que "les poids des corps dans quelque région de la Terre que ce soit, sont réciproquement comme les distances des lieux du centre de la Terre."

I do not wish to lay undue stress on the words; but they seem to imply that Newton must have obtained his result in the way Plana verifies it; namely in virtue of a certain transformation. But it is probable that Newton adopted quite a different way; see Art. 33. The matter is not important; but there are other instances in which Plana seems to assume that the course which his own analysis takes in verifying Newton's results must have been the course by which Newton originally obtained them.

1560. In closing the survey of Plana's memoirs on our subject it will be obvious that although extensive in quantity they add but little to what had been already obtained. The most valuable of them are those which we introduce to notice in Arts. 1520, 1546, and 1551: these may be considered as forming extensions of parts of the second Chapter of Laplace's eleventh Book.

CHAPTER XXXVIII.

MISCELLANEOUS INVESTIGATIONS BETWEEN THE YEARS 1801 AND 1825.

1561. THE present Chapter will contain an account of various miscellaneous investigations between the years 1801 and 1825.

The works of La Lande and Reuss, to which allusion is made in Arts. 738 and 739, do not afford us guidance beyond the close of the eighteenth century; and thus it is possible that in the present Chapter some books and memoirs which ought to have been noticed, may have been omitted from ignorance of their existence.

1562. A memoir entitled *Observations on the Figure of the Earth*, by Joseph Clay, is published in the *Transactions of the American Philosophical Society, held at Philadelphia*... Vol. v, 1802. The memoir occupies pages 312...319 of the volume.

The memoir commences thus:

The subject of this paper was suggested to me by a perusal of the "Studies of Nature," by Bernardin de St. Pierre. The positive manner in which that author asserts that the earth is a prolate spheroid, the arrogance with which he challenges refutation, and above all the erroneous theories which he has built on this assertion, seem to require all doubts to be removed by a mathematical demonstration.

The error of St Pierre was that of Keill and Cassini: see Art. 972. It was scarcely necessary to correct this error at the beginning of the nineteenth century.

The mathematical investigation of the memoir amounts to establishing the following theorem: let A be the extremity of the axis major, and B of the axis minor of an ellipse, P the point on the arc AB where the tangent is equally inclined to the axes; then the arc BP is longer than the arc AP.

The process adopted is rather rude. If a denote the semiaxis major, and b the semiaxis minor, it is to be shewn that

$$\int_0^{\frac{a^2}{c}} \frac{\sqrt{(a^4 - a^2x^2 + b^2x^2)}}{a\sqrt{(a^2 - x^2)}}\, dx \text{ is greater than } \int_0^{\frac{b^2}{c}} \frac{\sqrt{(b^4 - b^2y^2 + a^2y^2)}}{b\sqrt{(b^2 - y^2)}}\, dy,$$

where c is put for $\sqrt{(a^2 + b^2)}$.

Clay expands the numerator and the denominator of the expression to be integrated, divides the former result by the latter, then integrates, and thus obtains an infinite series. We may arrive at the required result more simply by assuming $x = az$ in the first integral, and $y = bz$ in the second; then the integrals become respectively

$$\int_0^{\frac{a}{c}} \frac{\sqrt{(a^2 - a^2z^2 + b^2z^2)}}{\sqrt{(1 - z^2)}}\, dz \text{ and } \int_0^{\frac{b}{c}} \frac{\sqrt{(b^2 - b^2z^2 + a^2z^2)}}{\sqrt{(1 - z^2)}}\, dz.$$

Each element of the first integral is greater than the corresponding element of the second, for the values of z within the range of the second integral; moreover the upper limit of the first integral is greater than the upper limit of the second: therefore the first integral is the greater.

But the result follows *immediately* from the fact that the radius of curvature increases continually from A to B, so that it is greater through PB than through AP.

1563. We may just refer to the work entitled *Dr. Benzenberg's Versuche über die Umdrehung der Erde*, which was published in 1804 at Dortmund. This is an octavo volume, containing xii + 542 pages, and a page of errata; there are seven plates besides the frontispiece and the engraved title-page.

Benzenberg made experiments with the view of determining the deviation from the vertical of a body falling through a considerable space. Some of the experiments were made from a high church-tower, that of St Michael, in Hamburg; and others down the shaft of a coal-mine. In both cases the mean of the results gave a deviation towards the east; in the former case they gave also a deviation towards the south: but there were considerable discrepancies in the experiments among themselves. The church-tower still, I believe, adorns the city of Hamburg.

There is also a work which may be regarded as a supplement to the preceding, entitled *Versuche über die Umdrehung der Erde. Aufs Neue berechnet von Dr. Benzenberg.* Düsseldorf, 1845.

Benzenberg's two publications are very interesting and contain much historical information connected with the subject; which is however beyond our limit. The student who wishes to pursue it should consult the *Cambridge and Dublin Mathematical Journal*, Vol. IV. page 97; and the collected edition of the works of Gauss, Vol. V. page 495.

1564. The first two volumes of the *Mécanique Céleste* were translated into German, and published at Berlin under the title *Mechanik des Himmels von P. S. Laplace, ... Aus dem Französischen übersetzt und mit erläuternden Anmerkungen versehen von J. C. Burckhardt.* The first volume is dated 1800, and the second 1802.

The notes are neither very numerous nor very important; they supply the detail of some of the analytical processes which Laplace himself treated rather briefly. The most useful note in the part of the work with which we are concerned is that to which we have already alluded in Art. 1060.

1565. It may be convenient here to notice other publications of the nature of Burckhardt's; by some strange accident the like character of incompleteness seems to attach to them all.

A translation of the first book of the *Mécanique Céleste* with notes was published at Nottingham in 1814, by the Rev. John Toplis; this forms one octavo volume.

A translation of the first volume of the *Mécanique Céleste* with notes was published by the Rev. Henry Harte at Dublin, in two quarto parts, the first in 1822 and the second in 1827.

The translation by Bowditch with notes, to which I have frequently referred, extends to the first four volumes of the original; the translation is in four quarto volumes, published at Boston in America, between 1829 and 1839.

Pontécoulant's *Théorie Analytique du Système du Monde* consists of four octavo volumes, published between 1829 and 1840; the subjects with which we are concerned are discussed in the second volume and an Appendix. The work is still unfinished. A new edition of the first and second volumes appeared in 1850.

The first and only volume of a work entitled *Elementi di Meccanica Celeste di Francesco Bertelli* was published at Bologna in 1841. The volume is in quarto. It does not treat on our subject.

1566. We now come to a memoir by Playfair, entitled *Investigation of certain Theorems relating to the Figure of the Earth.* This memoir appears in Volume v. of the *Transactions of the Royal Society of Edinburgh;* it occupies pages 1...30 of the volume; the date of publication of the volume is 1805: the memoir was read on the 5th of February, 1798.

1567. The memoir relates to the geometry of the subject; investigations are given with respect to the lengths of arcs of the meridian, of arcs perpendicular to the meridian, and of arcs parallel to the equator.

1568. Suppose $2a$ and $2b$ the major and minor axes of an ellipse; then the radius of curvature at the point where the normal is inclined at an angle ϕ to the major axis is known to be $\frac{a^2b^2}{(a^2\cos^2\phi + b^2\sin^2\phi)^{\frac{3}{2}}}$. By integrating this with respect to ϕ we obtain the length of an arc of the meridian. Let $b = a(1-\epsilon)$; then neglecting powers of ϵ above the second we shall find that the length of the arc measured from the equator to the latitude ϕ is

$$a\phi - \frac{a\epsilon}{2}\left(\phi + \frac{3}{2}\sin 2\phi\right) + \frac{a\epsilon^2}{16}\left(\phi + \frac{15}{4}\sin 4\phi\right);$$

this will be found to be consistent with Art. 1014, observing that $\epsilon = \frac{e^2}{2} + \frac{e^4}{8}$.

Playfair applies this formula, neglecting ϵ^2, to the Peruvian arc of 3° 7′ 1″ and the French arc of 8° 20′ 2″; and he obtains for ϵ the value $\frac{1}{300}$ nearly.

1569. Playfair in like manner gives a formula for the case of an arc supposed of small extent, measured perpendicular to the meridian; the arc in this case may be considered to belong to a circle which has for its radius the length of the part of the normal intercepted between the point at which the arc begins and the minor axis of the generating ellipse.

Also finally Playfair gives a formula for the case of an arc of a parallel of latitude.

1570. Thus there are three kinds of arcs considered. The elements of the Earth's figure can be determined either from two arcs of different kinds, or from two arcs of the same kind in different latitudes. Playfair gives various combinations and discusses the merits of each. But these discussions are of small importance, because they relate only to the forms of trigonometrical expressions, and take no account of the relative accuracy with which the necessary astronomical and geodetical operations can be performed.

1571. It appears from his pages 28 and 29, that Playfair intended to pursue his investigations on what may be called spheroidal trigonometry; but the intention does not seem to have ever been carried into execution.

1572. Playfair gives an example on his page 17, taken from the degrees of the meridian and perpendicular measured in the South of England. He refers to the *Philosophical Transactions*, 1795, page 537. In the separate account of this survey, the corresponding place is Vol. I. page 309: see Art. 984. The ellipticity deduced is about $\frac{1}{148}$, which is of course far too great. But it seems "that all the other comparisons of the degrees of the meridian, with those of the curve perpendicular to it, made from the observations in the South of England, agree nearly in giving the same oblateness to the terrestrial spheroid." "The authors of the *Trigonometrical Survey* seem willing, therefore, to give up the elliptic figure of the earth." Here Playfair refers to page 527 of the above volume of the *Philosophical Transactions*; the passage is in Vol. I. page 302 of the separate work.

In the *Philosophical Transactions* the sentence runs thus: "Now this comparison between the measured and computed degrees, sufficiently proves that the Earth is not an ellipsoid,..." But in the separate work instead of "sufficiently proves" we have "seems to prove."

Playfair thinks that the anomaly may arise from the fact that in the part of England where these measures were taken, "the strata are of chalk, and though of great extent, are bordered on all the sides that we have access to examine by strata much denser and more compact." See his pages 6, 18, and 19.

1573. The following passage occurs on page 29:

...In the mean time, I think it is material to observe, that the principle laid down by Mr Dalby, viz. that in a spheroidal triangle, of which the angle at the pole and the two sides are given, the sum of the angles at the base is the same as in a spherical triangle, having the same sides, and the same vertical angle, is not strictly true, unless the excentricity of the spheroid be infinitely small, or the triangle be very nearly isosceles."

The pages seem to be 524 and 529 which Playfair has in view; in the separate work the corresponding places are, Vol. I. pages 298 and 302; perhaps Mr Dalby intended to limit the principle to the case of a triangle nearly isosceles. See Art. 1037.

1574. Playfair offers a brief criticism on a passage in the *Philosophical Transactions*, page 529, which corresponds to Vol. I. page 303 of the separate work. He says:

"This shews, that the method of ascertaining the figure of the earth, proposed by the authors of the *Trigonometrical Survey* as a subject of future inquiry, is less exact than that which is founded on their own observations."

1575. A work was published at Stockholm in 1805, entitled *Exposition des opérations faites en Lapponie, pour la détermination d'un arc du méridien en* 1801, 1802 *et* 1803;... *redigée par Jons Svanberg.*

This is an octavo volume containing xxxi + 196 pages, besides the Title and three Plates. The work gives an account of the remeasurement of the Lapland arc, as stated in Art. 197; for some further information respecting it, I may refer to the memoir named in Art. 199.

1576. In De Zach's *Monatliche Correspondenz*, Vol. XIII. 1806, we have a memoir entitled *Gedanken über die Figur der Erde von dem...Anton Freyherrn von Zach.* The memoir occupies pages 221...235 of the volume.

The writer was apparently a brother of the editor of the periodical. The memoir is not mathematical, and belongs rather to Physical Geography than to our subject; there are indeed some remarks which depend on the principles of mechanics, but they exhibit inaccuracy of knowledge or at least of expression.

1577. We will now advert briefly to the work which gives an account of the great French measurement commenced towards the end of the eighteenth century.

There are three quarto volumes entitled *Base du système métrique décimal, ou mesure de l'arc du méridien compris entre les parallèles de Dunkerque et Barcelone, exécutée en 1792 et années suivantes, par MM. Méchain et Delambre. Rédigée par M. Delambre...*

The first volume was published in 1806, the second in 1807, and the third in 1810.

A fourth volume, also in quarto and connected with these, was published in 1821: it is entitled *Recueil d'observations géodésiques, astronomiques et physiques,...rédigé par MM. Biot et Arago.*

It would appear from page XXX of the Introduction to this volume, that Arago intended to publish in another volume an account of some operations which he carried on alone; but this intention does not seem to have been realised.

There are, however, two memoirs subsequently published by Biot, which relate to pendulum observations, and may be considered as connected with the present work.

One is entitled *Mémoire sur la Figure de la Terre*; this was read to the French Academy, on December 5th, 1827, and is published in Vol. VIII. of the *Mémoires de l'Académie Royale...de France.*

The other is entitled *Mémoire sur la latitude de l'extrémité australe de l'arc méridien de France et d'Espagne;* this was read to the French Academy on May 15th, 1843, and is published in Vol. XIX. of the *Mémoires.*

The Introduction to the volume of 1821, and the pages 521...541, are reprinted in Biot's *Mélanges Scientifiques et Littéraires,* Vol. I. 1858; the reprint is followed by two papers, the first of which gives a popular account of other operations of the author, bearing on the determination of the Figure of the Earth, and the second offers suggestions as to future labours. The whole series is extremely interesting, and well worthy of Biot's great literary and scientific reputation.

I may observe that on page 119 of Vol. II. of the *Base du système métrique,* we have one of the four formulæ which are usually, but improperly, called *Gauss's Theorems,* namely

$$\sin \frac{1}{2} c \cos \frac{1}{2}(A-B) = \sin \frac{1}{2} C \sin \frac{1}{2}(a+b).$$

As this volume is dated July, 1807, the formula may have been now *printed* for the first time; Delambre refers to this place in the *Connaissance des Tems* for 1809, which is dated April, 1807: here the four formulæ were first *published.* I have vindicated Delambre's claim to the formulæ in an article in the *Philosophical Magazine* for February, 1873.

On page 306 of Vol. III. of the *Base du système métrique,* we have the unpretending name of the English mathematician Dalby refined into d'Alby.

The metre, as is well known, was intended to be such that 10,000,000 metres should be the length of a quadrant of the Earth's meridian. On page 158 of the work cited in Art. 100, we have a suggestion by J. Cassini, that the unit of length might be such that 10,000,000 units should be the length of the Earth's radius.

1578. Two papers bearing on our subject by Dr Young, were published in 1808 in *Nicholson's Journal,* and are reprinted on pages 120...128 of the second volume of the *Miscellaneous Works* *of the late Thomas Young....*The first paper is entitled *A concise Method of determining the Figure of a Gravitating Body revolving round another.* The second paper is entitled *Calculation of the Direct Attraction of a Spheroid, and Demonstration of Clairaut's Theorem.*

These papers are of no importance. The conciseness which is claimed for them is obtained by *stating* results in words instead of *demonstrating* them by the aid of symbols. The process would not be intelligible to a reader, unless he could supply the usual mathematical investigation; and would be superfluous if he could.

At the end of the second paper there is an absurd misprint, both in the original and in the reprint; we have *upper real diminution* instead of *apparent diminution*: the misprint is corrected in a volume in my possession, containing a copy of the original paper, which seems to have formerly belonged to Dr Young himself.

1579. In De Zach's *Monatliche Correspondenz*, Vol. XXI. 1810, we have a memoir entitled *Über Densität der Erde und deren Einfluss auf geographische Ortsbestimmungen;* the memoir occupies pages 293...310 of the volume.

No name is mentioned; but we may safely ascribe the memoir to the editor of the periodical himself. We have here a brief popular account of the subject, especially of the operations of Bouguer at Chimborazo, and of Maskelyne at Schehallien. The possibility and actual existence of local attractions are said to be put beyond doubt also by more recent observations, due to Schiegg, to Méchain, and to Mudge. The writer urges the advantage which will follow from further investigations on the subject, and points out suitable localities in Germany and elsewhere.

The following two points may be noticed.

On page 297, some numerical statements are taken from Bouguer's *Figure de la Terre*, but not quite accurately; for instance it is said to be possible to approach to within 18 toises of the centre of Chimborazo, but it should be 1800 toises.

After giving an account of the operations at Schehallien, the writer points out that the result obtained from them agreed with Newton's conjecture that the ratio of the density of water to the mean density of the earth might lie between the ratios of 1 to 5 and 1 to 6. He then adds on his page 307:

...Die unter verschiedenen Breiten beobachteten Längen des einfachen Secunden-Pendels, gaben nach gehöriger Rechnung jenes Verhältniss 1 : 3·7.

I do not know what observations and calculation the writer here has in view.

1580. The second edition of the *Mécanique Analytique* consists of two volumes, the first of which was published in 1811, and the second in 1815. A third edition with notes by Bertrand was published in 1853...1855.

A few remarks relating to Attraction occur in the first volume, on pages 111...115 of the second edition, and 105...108 of the third edition. These remarks treat very briefly on the value of the potential of an ellipsoid at any external point; they are connected with the memoir by Lagrange in the Berlin *Mémoires* of 1792 and 1793: see Art. 1004. The precise relation of these remarks to the memoir is however not quite obvious.

In that memoir Lagrange treated of the value of the potential, and shewed that certain terms depending on P_2, P_4, and P_6 could be expressed as functions of b^2-a^2, and c^2-a^2: see Art. 1011. Now in the present work he substantially asserts that such a result will hold universally, that is for the term depending on P_n, where n is any positive integer. He gives however no demonstration of this statement, except what may be derived from the following words:

M. Laplace a donné, dans sa *Théorie des attractions des sphéroïdes*, une très-belle formule par laquelle on peut former successivement tous les termes de la série,...

J'ai trouvé qu'en partant de ce résultat et faisant usage du théorème que j'ai donné dans les Mémoires de Berlin de 1792—3, on pouvait construire tout d'un coup la série dont il s'agit...

The *très-belle formule* must doubtless be one of those contained in the *Mécanique Céleste*, Livre III., Chapter I.; but I am not certain which is meant: nor am I certain to which theorem in his own memoir of 1792—3 Lagrange alludes. Moreover

Lagrange's words would seem to suggest that the *très-belle formule* had been given subsequent to 1792—3, and that by combining this with a theorem of his own he had been able to arrive at the general result. But this is not the case, for all that Laplace published on the subject is to be found substantially in his memoir of 1782.

Thus finally it seems that if the entire series could be constructed *tout d'un coup* in 1811, it might have been also in 1792; and Lagrange ought to have explained more fully the statement he made in 1811.

There is a memoir by Plana in the third volume of Gergonne's *Annales de Mathématiques*, which forms a commentary on this passage of the *Mécanique Analytique;* but it does not touch on the point I have noticed. See Art. 1406.

On the whole it seems to me that the case may be stated thus: In the memoir of 1792—3, Lagrange attempted to put the series for the potential in such a form as to furnish a proof of the theorem due to Laplace, usually called by the name of Ivory; but the attempt was attended with only slight success. In the book Lagrange gives up this attempt, and assuming the truth of Laplace's theorem, deduces the constitution of the series.

1581. A few remarks relating to our subject occur also in the first volume of the *Mécanique Analytique*, on pages 199...204 of the second edition, and 188...193 of the third edition.

Lagrange had previously investigated the conditions of fluid equilibrium; and he now applies them to the case of fluid surrounding a solid nucleus, when the nucleus is an ellipsoid, and the outer surface of the fluid that of another ellipsoid. The two ellipsoids have the same centre, and the same directions for their axes, and differ but little from spheres.

The first thing required is the potential of a homogeneous ellipsoid at an external particle. Let a, b, c be the semiaxes of an ellipsoid; and suppose that

$$b^2 = a^2 + \beta^2, \quad c^2 = a^2 + \gamma^2;$$

let M denote the mass of the ellipsoid. Then if we confine ourselves to the first powers of β^2 and γ^2 we have for the potential at the point (x, y, z)

$$\frac{M}{r}\left\{1-\frac{\beta^2+\gamma^2}{10r^2}+\frac{3(\beta^2y^2+\gamma^2z^2)}{10r^4}\right\},$$

where $r^2 = x^2 + y^2 + z^2$.

Lagrange had given this formula in the investigations which we have just noticed; it may be easily verified.

For we have with the usual notation,

$$V=\iiint\frac{r'^2d\mu'd\psi'dr'}{\sqrt{(r^2-2rr'\lambda+r'^2)}},$$

where λ stands for

$$\mu\mu'+\sqrt{(1-\mu^2)}\sqrt{(1-\mu'^2)}\cos(\psi-\psi').$$

Now if r_1' denote the value of r' at the surface of the ellipsoid we have

$$r_1'^2\left\{\frac{\mu'^2}{a^2}+\frac{1-\mu'^2}{a^2+\beta^2}\cos^2\psi'+\frac{1-\mu'^2}{a^2+\gamma^2}\sin^2\psi'\right\}=1;$$

thus to our order of approximation

$$r_1'^2\left\{1-\frac{\beta^2\cos^2\psi'+\gamma^2\sin^2\psi'}{a^2}(1-\mu'^2)\right\}=a^2,$$

and
$$r_1'=a\left\{1+\frac{\beta^2\cos^2\psi'+\gamma^2\sin^2\psi'}{2a^2}(1-\mu'^2)\right\}.$$

Then in the usual way we obtain

$$V=\iiint\frac{1}{r}\left\{1+P_1\frac{r'}{r}+P_2\frac{r'^2}{r^2}+\dots\right\}r'^2d\mu'd\psi'dr'.$$

The first term of the series here exhibited is $\frac{M}{r}$.

The second term which depends on P_1 vanishes by the property of Laplace's coefficients.

Next $\int r'^4dr'=\frac{r_1'^5}{5}$; and to our order this

$$=\frac{a^5}{5}\left\{1+\frac{5(1-\mu'^2)}{2a^2}(\beta^2\cos^2\psi'+\gamma^2\sin^2\psi')\right\};$$

and to facilitate the integrations with respect to ψ' and μ' we arrange this as

$$\frac{a^5}{5}+\frac{a^5}{2}\left\{\frac{1}{3}(\beta^2+\gamma^2)+\frac{\beta^2+\gamma^2}{2}\left(\frac{1}{3}-\mu'^2\right)+\frac{\beta^2-\gamma^2}{2}(1-\mu'^2)\cos 2\psi'\right\}.$$

Moreover $P_2 =$

$$\frac{9}{4}\left(\mu^2-\frac{1}{3}\right)\left(\mu'^2-\frac{1}{3}\right)+3\mu\mu'\sqrt{(1-\mu^2)}\sqrt{(1-\mu'^2)}\cos(\psi-\psi')$$
$$+\frac{3}{4}(1-\mu^2)(1-\mu'^2)\cos 2(\psi-\psi').$$

Thus the term in Vr^3 which depends on P_2 reduces to

$$\frac{3a^5}{16}(\beta^2-\gamma^2)(1-\mu^2)\iint(1-\mu'^2)^2\cos 2(\psi-\psi')\cos 2\psi'\,d\mu'\,d\psi'$$
$$+\frac{9a^5}{16}(\beta^2+\gamma^2)\left(\frac{1}{3}-\mu^2\right)\iint\left(\frac{1}{3}-\mu'^2\right)^2 d\mu'\,d\psi'.$$

The limits for ψ' are 0 and 2π, and the limits for μ' are -1 and 1. Hence our result

$$=\frac{3a^5}{16}(\beta^2-\gamma^2)(1-\mu^2)\frac{16}{15}\pi\cos 2\psi+\frac{9a^5}{16}(\beta^2+\gamma^2)\left(\frac{1}{3}-\mu^2\right)\frac{16}{45}\pi$$

$$=\frac{a^5\pi}{5}\left\{(\beta^2-\gamma^2)(1-\mu^2)\cos 2\psi+(\beta^2+\gamma^2)\left(\frac{1}{3}-\mu^2\right)\right\}$$

$$=\frac{a^5\pi}{5r^2}\left\{(\beta^2-\gamma^2)(y^2-z^2)+(\beta^2+\gamma^2)\left(\frac{r^2}{3}-x^2\right)\right\}$$

$$=\frac{a^5\pi}{5r^2}\left\{2\beta^2y^2+2\gamma^2z^2-\frac{2}{3}r^2(\beta^2+\gamma^2)\right\}$$

$$=-\frac{M}{10}(\beta^2+\gamma^2)+\frac{3M}{10r^2}(\beta^2y^2+\gamma^2z^2)$$

to our order of approximation.

Thus the proposed formula is verified.

Now let σ be the density of the solid, and ρ the density of the fluid. Then we may consider that we have an ellipsoid of density ρ, and another of the density $\sigma-\rho$, as in Art. 383. Let a_1, b_1, c_1 be the semiaxes of the fluid surface; and let $\beta_1^2=b_1^2-a_1^2$

and $\gamma_1^2 = c_1^2 - a_1^2$. Then the whole potential at the point (x, y, z) will be

$$\frac{4\pi a_1 b_1 c_1 \rho}{3r}\left\{1 - \frac{\beta_1^2 + \gamma_1^2}{10r^2} + \frac{3(\beta_1^2 y^2 + \gamma_1^2 z^2)}{10r^4}\right\}$$
$$+ \frac{4\pi abc(\sigma - \rho)}{3r}\left\{1 - \frac{\beta^2 + \gamma^2}{10r^2} + \frac{3(\beta^2 y^2 + \gamma^2 z^2)}{10r^4}\right\}.$$

Let there also be at the point (x, y, z) the accelerations fx, gy, hz, parallel to the axes of x, y, z respectively, and directed outwards. Then if V denote the whole potential we must have for equilibrium

$$V + \frac{1}{2}(fx^2 + gy^2 + hz^2) = \text{constant} \quad \ldots\ldots\ldots\ldots\ldots (1).$$

But by hypothesis the surface is an ellipsoid determined by the equation

$$\frac{x^2}{a_1^2} + \frac{y^2}{b_1^2} + \frac{z^2}{c_1^2} = 1 \quad \ldots\ldots\ldots\ldots\ldots\ldots\ldots\ldots (2).$$

Hence by comparing (1) and (2) we arrive at the conditions which must hold.

To obtain these conditions we may substitute in (1) the approximate value for r, namely

$$r = a_1\left\{1 + \frac{\beta_1^2 y^2 + \gamma_1^2 z^2}{2a_1^4}\right\}.$$

Then (1) reduces to

$$\frac{fx^2}{2} + \left(B + \frac{g}{2}\right)y^2 + \left(C + \frac{h}{2}\right)z^2 = \text{constant};$$

where $B = \dfrac{4\pi}{3a_1^3}\left\{abc(\sigma - \rho)\left(\dfrac{3}{10}\beta^2 - \dfrac{1}{2}\beta_1^2\right) - \dfrac{1}{5}a_1 b_1 c_1 \rho \beta_1^2\right\}$,

and $C = \dfrac{4\pi}{3a_1^3}\left\{abc(\sigma - \rho)\left(\dfrac{3}{10}\gamma^2 - \dfrac{1}{2}\gamma_1^2\right) - \dfrac{1}{5}a_1 b_1 c_1 \rho \gamma_1^2\right\}$.

And to make this agree with (2) we must have

$$B + \frac{g}{2} = \frac{fa_1^2}{2b_1^2}, \quad C + \frac{h}{2} = \frac{fa_1^2}{2c_1^2}.$$

that is,

$$\frac{g}{2}+\frac{4\pi}{3a_1^3}\left[\frac{3}{10}abc(\sigma-\rho)\beta^2-\left\{\frac{abc(\sigma-\rho)}{2}+\frac{a_1b_1c_1\rho}{5}\right\}\beta_1^2\right]=\frac{fa_1^2}{2b_1^2},$$

$$\text{and } \frac{h}{2}+\frac{4\pi}{3a_1^3}\left[\frac{3}{10}abc(\sigma-\rho)\gamma^2-\left\{\frac{abc(\sigma-\rho)}{2}+\frac{a_1b_1c_1\rho}{5}\right\}\gamma_1^2\right]=\frac{fa_1^2}{2c_1^2}.$$

Suppose that the stratum of fluid is very thin; then our equations may be written

$$\frac{g}{2}+\frac{4\pi bc}{3a^2}\left\{\frac{3}{10}(\sigma-\rho)\beta^2-\left(\frac{\sigma-\rho}{2}+\frac{\rho}{5}\right)\beta_1^2\right\}=\frac{fa_1^2}{2b_1^2},$$

$$\frac{h}{2}+\frac{4\pi bc}{3a^2}\left\{\frac{3}{10}(\sigma-\rho)\gamma^2-\left(\frac{\sigma-\rho}{2}+\frac{\rho}{5}\right)\gamma_1^2\right\}=\frac{fa_1^2}{2c_1^2}.$$

As an example suppose that $f=0$, and that $g=h=\omega^2$; then we obtain

$$\beta^2(\sigma-\rho)=\beta_1^2\left\{\frac{5\sigma}{3}-\rho\right\}-\frac{5\omega^2}{4\pi}\cdot\frac{a^2}{bc},$$

$$\gamma^2(\sigma-\rho)=\gamma_1^2\left\{\frac{5\sigma}{3}-\rho\right\}-\frac{5\omega^2}{4\pi}\cdot\frac{a^2}{bc}.$$

These results, allowing for difference of notation, agree with those in Art. 383.

1582. Lagrange proceeds to the case in which the mass is entirely composed of homogeneous fluid rotating with uniform angular velocity; and here some points require to be noticed.

Let a, b, c be the semiaxes of an ellipsoid; x, y, z the corresponding coordinates of any point at the surface. Then it is known that the attractions at (x, y, z) parallel to the axes are of the form $M\lambda x$, $M\mu y$, $M\nu z$ respectively, where M is the mass of the ellipsoid, and λ, μ, ν are certain constants in the form of definite integrals. Lagrange says that hence

$$V=\frac{M}{2}(\lambda x^2+\mu y^2+\nu z^2).$$

This is however inaccurate. In the first place this result could not be obtained from the fact that the attractions *at the surface* take the specified form, but from the fact that they do so *throughout the body.* In the second place there should be a *constant* added to the value of V; although for the object in view it is not necessary to determine this constant.

Lagrange then considers whether the rotating ellipsoid can be in relative equilibrium. If so the equation $V + \frac{\omega^2}{2}(x^2 + y^2) = \text{constant}$ must agree with the equation $\frac{x^2}{a^2} + \frac{y^2}{b^2} + \frac{z^2}{c^2} = 1$. This leads to the conditions

$$\frac{M\lambda + \omega^2}{M\nu} = \frac{c^2}{a^2}, \qquad \frac{M\mu + \omega^2}{M\nu} = \frac{c^2}{b^2}.$$

Lagrange says that these give $a = b$, because λ and μ are like functions of a, b and b, a. We now know that this inference is inaccurate; it is not *necessary* that $a = b$: the discovery is due to Jacobi. This inaccuracy is corrected by Bertrand in the third edition of the *Mécanique Analytique.*

1583. We come next to a memoir by Professor Playfair entitled *Of the Solids of Greatest Attraction, or those which, among all the Solids that have certain Properties, Attract with the greatest Force in a given Direction.* This is published in the *Transactions of the Royal Society of Edinburgh*, Vol. VI. 1812; it occupies pages 187...243 of the volume: it was read on January 5th, 1807.

1584. Playfair first discusses Silvabelle's problem: see Arts. 531 and 682. Playfair does not use the Calculus of Variations, but the easier method which amounts to making the attraction, resolved in the given direction, constant at all points of the bounding surface of the body. Playfair solves various simple exercises connected with the result which he obtains; thus for instance he finds the area of the generating curve, and the volume of the solid which it generates by revolution: see his pages 187...205.

Playfair does not refer to Silvabelle; but he says that the problem had been treated of by Boscovich. But Playfair had never been able to procure a sight of the memoir by Boscovich; nor have I been more fortunate.

1585. Playfair solves various problems respecting attractions, which are examples of the ordinary methods of maxima and minima explained in the Differential Calculus.

Thus on pages 206...209 he determines the form of a right circular cone of given volume, so that the attraction at the vertex may be the greatest possible.

On pages 209...214 he discusses the attraction which a right circular cylinder exerts at the centre of one of the circular ends; and he determines the ratio of the radius of the base to the height, so that when the volume is given the attraction may be the greatest possible. Let u denote the ratio of the radius of the base to the height; then to determine the value of u Playfair obtains the equation

$$(2-u)\sqrt{(1+u^2)} = 2-u^2 \text{ .... } (1).$$

By squaring we have

$$(2-u)^2(1+u^2) = (2-u^2)^2,$$

this reduces to $$4u^2 - 9u + 4 = 0 \text{ .... } (2).$$

From (2) we obtain

$$u = \frac{9 \pm \sqrt{17}}{8}.$$

Playfair has some trouble in convincing himself and his readers that we must take the lower sign in this expression for u. But the fact is very simple: although both expressions satisfy (2), yet it is only the expression with the lower sign which satisfies (1); and (1) is the equation which really must be satisfied.

1586. On his pages 215 and 216 Playfair determines the attraction of a rod of infinitesimal section and of finite length, on an external point, resolved in the direction perpendicular to the rod. Then on his pages 216...218 he applies this to demonstrate a result which he gives elsewhere without demonstration: see Art. 731. And on pages 218...220 he proposes "to find the figure of a semi-cylinder, given in magnitude, which shall attract a particle situated in the centre of its base, with the greatest force possible, in the direction of a line bisecting the base:" but the determination of the maximum leads to an equation of considerable difficulty, and he contents himself with an approximate solution obtained by trial.

1587. On his pages 220...225, Playfair considers the following problem: to determine the oblate spheroid of a given solidity which shall attract a particle at its pole with the greatest force. Here he makes some curious mistakes.

Let a be the major semiaxis, and b the minor semiaxis of the generating ellipse. Suppose the given volume to be denoted by $\frac{4\pi}{3} n^3$, so that $a^2 b = n^3$. Then the attraction of the oblatum at the pole is

$$\frac{4\pi n}{e^3} (1 - e^2)^{\frac{5}{6}} \left\{ \frac{e}{\sqrt{(1 - e^2)}} - \sin^{-1} e \right\};$$

this may be easily deduced from Art. 261.

Put $\sin^{-1} e = \phi$; thus we obtain

$$\frac{4\pi n}{(\cos \phi)^{\frac{4}{3}}} \frac{\tan \phi - \phi}{\tan^3 \phi} \text{.........................} (1).$$

When ϕ is very small this becomes approximately

$$4\pi n \left(1 + \frac{2}{3} \tan^2 \phi\right) \left(\frac{1}{3} - \frac{1}{5} \tan^2 \phi\right),$$

that is,

$$\frac{4\pi n}{3} \left(1 + \frac{1}{15} \tan^2 \phi\right).$$

Thus when e, and therefore ϕ, increases from zero, the attraction begins by *increasing*; but from the expression (1) it is obvious that the attraction vanishes when $\phi = \frac{\pi}{2}$: hence there must be a maximum for some value of ϕ between 0 and $\frac{\pi}{2}$.

But Playfair, on the contrary, implicitly denies the existence of this maximum, and asserts that there is a maximum when $\phi = 0$. This is the more curious, because he obtains correctly the equation which determines when (1) is a maximum, namely

$$\phi = \frac{\tan \phi \, (9 + 2 \tan^2 \phi)}{9 + 5 \tan^2 \phi} \text{....................} (2).$$

He seems to have believed that (2) has no solution except $\phi = 0$; but it is clear from what we have said that there must be a solution between $\phi = 0$ and $\phi = \frac{\pi}{2}$; and it is easy to establish this statement from considering the equation itself.

Playfair finishes this section of his memoir with the following paragraph:

> If the oblateness of a spheroid diminish, while its quantity of matter remains the same, its attraction will increase till the oblateness vanish, and the spheroid becomes a sphere, when the attraction at its poles, as we have seen, becomes a maximum. If the polar axis continue to increase, the spheroid becomes oblong, and the attraction at the poles again diminishes. This we may safely conclude from the law of continuity, though the oblong spheroid has not been immediately considered.

But the statements are inaccurate; the attraction will really *decrease* till the oblateness vanishes; and there is *no maximum* when the spheroid becomes a sphere: while the axis of revolution *continually increases*, as here supposed, and does not deviate sensibly from the other axis, the attraction *continually decreases*. This is in fact quite as consistent with the law of continuity as Playfair's erroneous result.

1588. Playfair on his pages 225...228 finds the attraction of a rectangular lamina at a point which is on the straight line drawn at right angles to the plane of the lamina through one corner, resolved along the direction of this straight line: this is probably the first appearance of the result in *finite terms*. We have seen in Art. 1017 that Cavendish failed to obtain it; and Playfair on his page 237 adverts to this circumstance.

Playfair makes an easy application of his result to determine the attraction which a right pyramid on a square base exerts at the vertex; and he finds the form of the pyramid so that the attraction may be the greatest possible when the volume is given: see his pages 228...231.

1589. Playfair arrives on his page 233 at a general result, which we may enunciate thus: suppose a lamina of any shape to attract an external particle; then the resolved attraction in the direction perpendicular to the lamina is measured by the product of the thickness of the lamina into the solid angle subtended by the lamina at the particle. The solid angle is to be measured in the usual way, by the portion of the surface of a sphere of radius unity having its centre at the particle, which is determined by a straight line from this centre which describes the boundary of the lamina.

Playfair demonstrates this by employing the expression which he had obtained for the resolved attraction of a rectangular lamina. But it may be obtained more simply by considering the action of an infinitesimal element. Let δS denote an infinitesimal element of the lamina, r its distance from the attracted particle, θ the angle between the direction of r and the perpendicular from the particle on the lamina, κ the thickness of the lamina. Then the resolved attraction of the element is $\frac{\kappa\delta S}{r^2}\cos\theta$; and it is obvious that $\frac{\delta S\cos\theta}{r^2}$ is equal to the element of the spherical surface which corresponds to δS.

1590. On his pages 235...237, Playfair applies the general result of the preceding Article to establish a proposition which is now given in our elementary books, namely, that "whatever be the figure of any body, its attraction will decrease in a ratio that approaches continually nearer to the inverse ratio of the squares of the distances, as the distances themselves are greater." He considers that this proposition is usually taken for granted without any other proof than "some indistinct perception of what is required by the law of continuity."

1591. On his pages 239...243 Playfair investigates the attraction of a rectangular parallelepiped resolved parallel to an edge at a point on the edge produced. This is an easy deduction from the result he had obtained as to a rectangular lamina: see Art. 1588. Playfair's formula on his page 242 must have its sign changed if the attraction is to be a positive quantity.

1592. It will be seen from our account that the main contributions of the memoir to our subject are the resolved attraction of a rectangular lamina given in Art. 1588, and the general result of Art. 1589.

1593. The next memoir which we have to notice is entitled *Of the Attraction of such Solids as are terminated by Planes; and of Solids of greatest Attraction.* By Thomas Knight.

This memoir is contained in the *Philosophical Transactions* for 1812, published in that year; it occupies pages 247...309 of the volume. The memoir was read on March 19th, 1812.

1594. The memoir begins thus:

Mathematicians, in treating of the attraction of bodies, have confined their attention, almost entirely, to those solids which are bounded by continuous curve surfaces; and Mr. Playfair, if I do not mistake, is the only writer, who has given any example of that kind of inquiry, which is the chief object of the present paper. This learned mathematician has found expressions for the action of a parallelopiped; and of an isosceles pyramid, with a rectangular base, on a point at its vertex; and observes, on occasion of the first mentioned problem, that what he has there done, "gives some hopes of being able to determine generally the attraction of solids bounded by any planes whatever."

It is this general problem, that I venture to attempt the solution of, in what follows:...

1595. Thus it appears that Knight's memoir was suggested by Playfair's; but, as we shall soon see, proceeded somewhat farther.

Let OPQ be a right-angled triangle, having the right angle at P; through O draw a straight line at right angles to the plane of the triangle; then Knight determines the components of the attraction which a lamina of infinitesimal thickness in the shape of the triangle exerts at any point of the straight line. Playfair took a rectangle instead of a triangle, and confined himself to estimating the value of the component which is along the straight line: thus his investigations are more restricted than Knight's.

It may be said that the problem thus enunciated is the basis of nearly the whole of Knight's memoir.

1596. Let N denote the point in the straight line; let $NO = a$, $OP = b$, $PQ = c$.

Let X denote the component of the attraction along NO, let Y denote the component parallel to OP, and Z the component parallel to PQ. Take O as the origin, OP as the axis of y, and a straight line through O parallel to PQ as the axis of z. Let y and z be the coordinates of any point of the triangle.

Then

$$X = \mu a \iint \frac{dydz}{r^3}, \quad Y = \mu \iint \frac{ydydz}{r^3}, \quad Z = \mu \iint \frac{zdydz}{r^3};$$

where μ represents the infinitesimal thickness of the lamina, and r stands for $\sqrt{(a^2 + y^2 + z^2)}$. The integrations must extend over the whole area of the triangle.

1597. It is easy to effect the integrations; we will not follow Knight extremely closely.

$$\int \frac{dz}{r^3} = \frac{z}{(a^2 + y^2)\sqrt{(a^2 + y^2 + z^2)}};$$

the limits of z are 0 and ty, where t denotes the tangent of POQ; so that $t = \frac{c}{b}$. Thus

$$X = \mu a t \int \frac{ydy}{(a^2 + y^2)\sqrt{(a^2 + y^2 + t^2y^2)}}.$$

Assume $a^2 + y^2(1 + t^2) = v^2$; then we find that

$$X = \mu a t \int \frac{dv}{v^2 + t^2a^2} = \mu \tan^{-1} \frac{v}{at}.$$

And taking this between the appropriate limits we obtain finally

$$X = \mu \left\{ \tan^{-1} \frac{\sqrt{[a^2 + b^2(1 + t^2)]}}{at} - \tan^{-1} \frac{1}{t} \right\}.$$

Similarly by effecting the integration with respect to z we obtain

$$Y = \mu t \int \frac{y^2dy}{(a^2 + y^2)\sqrt{(a^2 + y^2 + t^2y^2)}}$$

$$= \mu t \int \frac{dy}{\sqrt{(a^2 + y^2 + t^2y^2)}} - \mu t a^2 \int \frac{dy}{(a^2 + y^2)\sqrt{(a^2 + y^2 + t^2y^2)}}.$$

The first term is immediately integrable; and to integrate the second assume $v = \frac{y}{\sqrt{(a^2 + y^2)}}$, so that

$$\int \frac{a^2 dy}{(a^2 + y^2)\sqrt{(a^2 + y^2 + t^2 y^2)}} = \int \frac{dv}{\sqrt{(1 + t^2 v^2)}}.$$

Hence finally by taking the integrals between the appropriate limits we get

$$Y = \frac{\mu t}{\sqrt{(1+t^2)}} \log \frac{b\sqrt{(1+t^2)} + \sqrt{\{a^2 + b^2(1+t^2)\}}}{a} - \mu \log \frac{bt + \sqrt{\{a^2 + b^2(1+t^2)\}}}{\sqrt{(a^2 + b^2)}}.$$

And $$\int \frac{z dz}{(a^2 + y^2 + z^2)^{\frac{3}{2}}} = -\frac{1}{(a^2 + y^2 + z^2)^{\frac{1}{2}}}.$$

Thus $$Z = \mu \int \left\{ \frac{1}{\sqrt{(a^2 + y^2)}} - \frac{1}{\sqrt{(a^2 + y^2 + t^2 y^2)}} \right\} dy;$$

and integrating between the appropriate limits we get

$$Z = \mu \log \frac{b + \sqrt{(a^2 + b^2)}}{a} - \frac{\mu}{\sqrt{(1 + t^2)}} \log \frac{b\sqrt{(1+t^2)} + \sqrt{\{a^2 + b^2(1+t^2)\}}}{a}.$$

1598. By decomposing any rectilinear lamina into triangles, Knight can estimate the component attractions which it exerts at any point. Then for any solid which can be decomposed into such laminæ the component attractions may always be reduced to the form of single integrals; and for various examples he actually works out the integration.

Four out of the five sections of the memoir are devoted to these subjects; and the last section to the problem of Solids of greatest Attraction.

The mathematical processes are sound and satisfactory, though sometimes the results might be obtained with greater ease and elegance by special methods instead of the general process which Knight uniformly employs. I will offer a few remarks on some miscellaneous points.

1599. The fifth section commences thus:

The subject of this section has occupied the attention of Mr. Playfair, in the same paper I have before noticed; it had previously been treated of by Silvabelle. Frisi also, in the third volume of his works, gives a solution of the same problem as that which is first considered by Mr. Playfair, but his result is an erroneous one. None of these writers have pursued the matter any further than what relates to the figure of a homogeneous solid of revolution.

It does not appear to me that the solution given by Frisi is wrong: see Art. 682.

The extension which Knight undertakes to supply to the problem is twofold. Instead of confining himself to the case of a body which can be cut up into circular slices, he considers also various bodies which can be cut up into rectilinear slices. And instead of confining himself to a homogeneous body he considers some cases of varying density.

1600. In the case of a solid of revolution Knight shews that the result obtained by Silvabelle and Playfair for the homogeneous body is also true when the density is any function of two assigned variables, namely the distance from the axis of revolution, and the distance of the plane of the circular slice from the origin. Knight uses the formal Calculus of Variations, and not the simple principle adopted by Playfair after Silvabelle, that the bounding surface must be one of equal resolved attraction. The extension which Knight obtains can be immediately deduced also by Playfair's principle. It will be observed that under such a law of density as Knight supposes the resultant attraction is along the axis of revolution.

1601. Suppose however that we modify the problem, and allow the density to be any function of the three coordinates of a point; then if we require, not the maximum resultant attraction but, the maximum value of the component along the axis of revolution, we shall still obtain the form assigned by Silvabelle and Playfair. This is also an immediate deduction from the principle adopted by Playfair; but is less clearly obvious according to Knight's method. Knight may have seen it but he does not make any mention of it. Let us apply his method.

1602. Take the attracted particle as the origin of coordinates, and the axis of x as that of revolution. Let s stand for

$\sqrt{(x^2+y^2+z^2)}$, and $\phi(s)$ for the law of attraction; then if ρ be the density at the point (x, y, z) the resolved attraction of an element is $\frac{\rho\, dx\, dy\, dz\, x\phi(s)}{s}$. Transform by putting $r\cos\theta$ for y, and $r\sin\theta$ for z; then the resolved attraction of an element becomes $\frac{\rho\, dx\, r dr d\theta\, x\, \phi(s)}{s}$, where ρ is some function of x, r, and θ. Suppose we integrate with respect to θ from 0 to 2π; the result will be some function of r and x which we may denote by $f(r)$, for it is not necessary to allude explicitly to x. Then integrate $f(r)$ with respect to r from 0 to y, where y now denotes the extreme value of r, that is the ordinate to the generating curve of the solid; denote the result by $\psi(y)$. Thus finally the resolved attraction is $\int \psi(y)\, dx$.

Then this is to be a maximum while the mass is constant. The mass may be denoted by $\int \chi(y)\, dx$, where $\chi(y)$ stands for

$$\int_0^y \int_0^{2\pi} \rho r dr\, d\theta.$$

By the usual principles we must make the expression

$$\int [\psi(y) + C\chi(y)]\, dx$$

a maximum, where C is some constant.

This leads in the usual way to

$$\psi'(y) + C\chi'(y) = 0,$$

that is to

$$\int_0^{2\pi} \frac{\rho x \phi(s)}{s} y d\theta + C\int_0^{2\pi} \rho y d\theta = 0,$$

that is to

$$\int_0^{2\pi} \left\{\frac{x\phi(s)}{s} + C\right\} y\rho d\theta = 0,$$

that is to

$$\left\{\frac{x\phi(s)}{s} + C\right\} y \int_0^{2\pi} \rho d\theta = 0.$$

Hence $\frac{\alpha\phi(s)}{s} + C = 0$. In this equation we have $s = \sqrt{(x^2 + y^2)}$; and thus the equation expresses the fact that the resolved attraction is to be constant over the surface of the solid. Thus we have the same form as we should obtain when the body is homogeneous.

1603. A formula in the Integral Calculus occurs on page 292, which may deserve notice, namely

$$\int \frac{dx}{(a^2+x^2)^{\frac{n+1}{2}}} = \left\{\frac{x}{a^2} + \frac{(n-2)\,x^3}{3a^4} + \frac{(n-2)(n-4)\,x^5}{3\,.\,5a^6} + \dots\right\} \frac{1}{(a^2+x^2)^{\frac{n-1}{2}}}.$$

The mode of demonstration will indicate more distinctly the form of the last term, which must be supplied when n is not an even positive integer.

$$\int \frac{dx}{(a^2+x^2)^{\frac{n+1}{2}}} = \int \frac{1}{a^2(a^2+x^2)^{\frac{n-2}{2}}} \frac{d}{dx} \frac{x}{\sqrt{(a^2+x^2)}}\, dx$$

$$= \frac{x}{a^2(a^2+x^2)^{\frac{n-1}{2}}} + \frac{n-2}{a^2} \int \frac{x^2}{(a^2+x^2)^{\frac{n+1}{2}}}\, dx \;\dots\dots\dots (1);$$

$$\int \frac{x^2\,dx}{(a^2+x^2)^{\frac{n+1}{2}}} = \frac{1}{3a^2} \int \frac{1}{(a^2+x^2)^{\frac{n-4}{2}}} \frac{d}{dx} \frac{x^3}{(a^2+x^2)^{\frac{3}{2}}}\, dx$$

$$= \frac{1}{3a^2} \frac{x^3}{(a^2+x^2)^{\frac{n-1}{2}}} + \frac{n-4}{3a^2} \int \frac{x^4}{(a^2+x^2)^{\frac{n+1}{2}}}\, dx \dots\dots\dots (2).$$

Substitute from (2) in (1), thus

$$\int \frac{dx}{(a^2+x^2)^{\frac{n+1}{2}}} = \frac{x}{a^2(a^2+x^2)^{\frac{n-1}{2}}} + \frac{n-2}{3a^4} \cdot \frac{x^3}{(a^2+x^2)^{\frac{n-1}{2}}}$$

$$+ \frac{(n-2)(n-4)}{3a^4} \int \frac{x^4\,dx}{(a^2+x^2)^{\frac{n+1}{2}}} \dots\dots\dots\dots\dots\dots (3).$$

The process may be continued by putting the last integral in (3) in the following form:

$$\int \frac{x^4\,dx}{(a^2+x^2)^{\frac{n+1}{2}}} = \frac{1}{5a^2} \int \frac{1}{(a^2+x^2)^{\frac{n-6}{2}}} \frac{d}{dx} \cdot \frac{x^5}{(a^2+x^2)^{\frac{5}{2}}}\, dx.$$

1604. One of Knight's examples may be of sufficient interest to be reproduced here; we shall however adopt a method which is simpler than his.

Suppose that the law of attraction is that of the inverse n^{th} power of the distance: find an expression for the attraction of a prism of infinitesimal section, but of infinite length both ways, at an external point.

Let x denote the perpendicular distance of the point from the prism; let any other straight line drawn from the point to the prism make an angle θ with the perpendicular distance; let μ be the area of a section of the prism. Then the volume of an element of the prism will be $\mu d.x\tan\theta$, that is $\frac{\mu x\, d\theta}{\cos^2\theta}$. Hence the resultant attraction is $\int \frac{\cos\theta}{(x\sec\theta)^n}\frac{\mu x\, d\theta}{\cos^2\theta}$, that is $\frac{\mu}{x^{n-1}}\int \cos^{n-1}\theta\, d\theta$. The limits of θ are $-\frac{\pi}{2}$ and $\frac{\pi}{2}$; so that the attraction becomes $\frac{A\mu}{x^{n-1}}$, where A is a function of n alone. The value of A can be determined immediately if n is a positive integer; but we do not require this value for the application we have in view.

Required the form of an infinitely long cylinder so that the attraction may be a maximum at an external point.

It will follow by the use of the principle which Playfair adopted that the resolved attraction must be constant throughout the curve formed by a section of the cylinder by a plane at right angles to the generating lines and passing through the external point. Let r be the distance from the external point to an element of the curve formed by this section of the cylinder. Let θ be the angle between the direction of r and that of the resultant attraction. Then we must have $\frac{A\cos\theta}{r^{n-1}}$ constant; therefore $\frac{\cos\theta}{r^{n-1}}$ must be constant. The result is considered remarkable by Knight: see his page 301.

If $n = 2$ the equation is that of a circle, which passes through the attracted point.

1605. A treatise was published in 1814 by De Zach entitled *L'Attraction des Montagnes et ses effets.* The work is in octavo; it contains xix + 715 pages, and three plates. De Zach made observations on a mountain a few miles to the north-west of Marseilles, and also on an island a few miles to the south-east. He found on the whole that the mountain produced a deviation of very nearly two seconds in the direction of the plumb-line. But it has been doubted by the most competent judges whether the small repeating circle which De Zach used was adequate to such a delicate operation. See Arago's *Œuvres Complètes,* Vol. XI. pages 149...164, and the article *Figure of the Earth* in the *Encyclopædia Metropolitana,* page 173. There are no theoretical investigations to engage our attention: I have alluded to the work in Art. 727, and will merely notice a few points here.

1606. A preliminary discourse which occupies pages 1...28 of De Zach's work gives a history of the attempts made to ascertain the attraction of mountains.

De Zach observes that it is not necessary to have great mountains in order to cause a deviation in the direction of the plumb-line; for a defect of homogeneity in the internal strata of the earth near the point of observation would produce the same effect, *comme Newton l'a prouvé.* To justify these words a reference is given to Lib. III. Prop. 20 of the Principia. Newton makes indeed such a remark in this place, but cannot be said to *prove* anything.

De Zach refers to the Chimborazo operations; see Art. 363. Here it was not possible to make observations both on the north and south sides of the mountain; so that one observation was made at the foot of the mountain at the south side, and another at a second station about a league and a half to the east of the first. Then in a note De Zach says:

Un Auteur très-illustre, en rapportant cette expérience, s'est trompé; il a cru et supposé que ces Académiciens avoient observé au Nord et au Sud de la montagne, ce qui n'étoit pas le cas, comme on voit.

I do not know who is meant by this passage; it might have been supposed perhaps that De Zach was alluding to some recent or contemporary author, but the note had really been published about 60 years previously. It occurs on page 149 of the work which we have designated as XVIII. in Art. 332; and De Zach ought to have given a reference.

1607. Maskelyne makes this remark in the *Philosophical Transactions* for 1775, pages 502 and 503:

> Fortunately, however, Perthshire afforded us a remarkable hill, nearly in the centre of Scotland, of sufficient height, tolerably detached from other hills, and considerably larger from East to West than from North to South, called by the people of the low country Maiden-pap, but by the neighbouring inhabitants Schehallien; which, I have since been informed, signifies in the Erse language, Constant Storm: a name well adapted to the appearance which it so frequently exhibits to those who live near it, by the clouds and mists which usually crown its summit.

This must I presume be the place from which De Zach obtained the philological information which he thus curiously distorts on his page 21:

> ... il trouva toutes les conditions requises réunies dans le *Schehallien*, montagne appelée dans le pays, en langue Erse, *Maiden-Pap*, qui veut dire *orage perpétuel*.

A similar remark occurs on page 304 of the memoir which we have noticed in Art. 1579.

1608. A work entitled *Quotidiana Terræ conversio devio corporum casu demonstrata. Auctore A. Tadino* was published at Milan; the date given is *Anno 1° ab exacto Bonaparte*, which I presume is about 1814.

The work consists of 125 pages in octavo, with very large margins.

The author refers to some theoretical investigation of the deviation of falling bodies, which he had published in 1786, *Ticinensibus Ephemeridibus*. He gives an account of the experiments he made from a tower at Bergomi, about 100 feet high. The mean result of 143 trials was an easterly deviation, agreeing closely with what had been calculated from theory.

The work seems to be little known; it is not referred to by the authorities cited in Art. 1563.

1609. On pages 53...56 of the *Bulletin des Sciences par la Société Philomatique de Paris*, 1815, we have an article by Cauchy entitled *De la différence entre les attractions exercées par une couche infiniment mince sur deux points très-rapprochés l'un de l'autre, situés l'un à l'intérieur, l'autre à l'extérieur de cette même couche; par A. L. Cauchy, ingénieur des ponts et chaussées.*

The object of the article is to deduce from the general formulæ of attraction the theorem given in Poisson's first memoir on electricity: see Arts. 1357 and 1380. There is nothing in Cauchy's analysis which is specially interesting; it does not even seem so convincing as the synthetical investigation contained in Poisson's memoir on electricity, which we know is due to Laplace.

1610. An academical dissertation entitled *Dissertatio Academica de Figura Telluris ope Pendulorum determinanda* now presents itself to our notice. This seems to have consisted of various parts; but I have seen only Part 5 and Part 6. Part 5 is by Johannes Magnus a Tengstrom, and is dated 27th May, 1815. Part 6 is by Johannes Gabriel Bonsdorff, and is dated 27th June, 1815. Both parts were published at Abo.

I presume the entire dissertation contained a full account of the observations which had been made in various places with pendulums.

Parts 5 and 6 each consist of 10 pages.

Part 5 begins by adverting to some observations made by a Spanish navigator named Ciscar; a reference is given to an article by Oltmanns in De Zach's *Monatliche Correspondenz*, 1812, page 468, &c.

Let E denote the length of the seconds pendulum at the equator, p the length at the latitude l; then we learn from theory that $p = E + x \sin^2 l$, where x is some quantity which does not vary with the latitude. Hence if we know the length of the seconds pendulum at two different latitudes we can determine E and x. Let $P = E + x$, so that P is the length of the seconds pendulum at the pole. The values of P obtained from a large number of binary combinations of observations are given; and as an average of these combinations P is found to be 441·4933, expressed in Paris lines.

In Part 6 the values of E and x are calculated from a large number of observations by the method of least squares. This gives $x = 2·29695$ and $E = 439·2303$.

Then by Clairaut's theorem the ellipticity of the Earth is $\frac{5}{2} \cdot \frac{1}{289} - \frac{x}{E}$; and with the above values of x and E this becomes $\frac{1}{292·3}$.

But from theoretical grounds, for which reference is made to the work of Svanberg noticed in Art. 1575, it is considered that $\frac{1}{305}$ is the correct value of the ellipticity. Then it is stated that by omitting some of the pendulum observations, which appear to differ too much from the rest, a result can be obtained from the rest which does not deviate much from the fraction $\frac{1}{305}$. Thus if the observations at Kola, Mulgrave, Melita, Megasaki, Umatog, Rio Janeiro, and St Helena are omitted, the values found are $x = 2·32941$, $E = 439·20943$, and the ellipticity is $\frac{1}{298·5}$.

Then the author says:

Hac ratione plures instituimus comparationes, aliis aliisve omissis observationibus, quarum fides minor visa est, et præbuit nobis hic calculus valores ellipticitatis $\frac{1}{312·6}$, $\frac{1}{309·8}$, $\frac{1}{303·7}$, $\frac{1}{301·4}$, qui omnes aperte ostendunt, verum valorem ellipticitatis terræ ex observationibus penduli derivatum, utpote intra allatos hos limites medium, valori aliunde invento non modo non repugnare, sed potius optime ita convenire, ut, si ex diversissimis similiter sitis locis haberentur observationes penduli æque certæ, nullum esse videatur dubium, quin hæ etiam ellipticitatem indicent $= \frac{1}{305}$ uti maxime probabilem.

Finally the author assumes $\frac{1}{305}$ for the ellipticity; and he considers the length of the pendulum at Paris accurately known; thus he obtains the formula $p = 439·2221 + 2·3590 \sin^2 L$

1611. We may briefly advert to the *Essays on the Theory of the Tides, the Figure of the Earth, the Atomical Philosophy, and the Moon's Orbit.* *By Joseph Luckcock.*

This work was published at London in 1817; it is in a small quarto size, and consists of a Title-page, a Preface on iv pages, the Text on 96 pages, and five Plates. The Essay on the Figure of the Earth occupies pages 23...47 of the volume; it is a foolish production by an ignorant writer: he rejects what is usually called the centrifugal force, and denies that the Earth is elevated at the equator.

We may give a specimen of the work. The writer finds correctly, that if we take the Earth as a sphere of 8000 miles diameter, we have corresponding to a distance of one mile on the surface, a deviation of about 8 inches from a straight line. Then he proceeds thus on his page 41:

...But suppose a canal to be dug upon a meridian, from the pole to the equator; the correction between the telescopic and the true level would be grossly erroneous: the engineer who should have the temerity to work according to the rule, would find the banks of his canal at the equator 18½ miles deep! But the engineer happens to be right; and the rule will serve him in cutting his canal east, west, north or south; no matter what direction it may take; consequently the meridians are circles equally with the parallels of latitude, and here is a demonstration that the equatorial regions are not elevated above the natural level, otherwise there must be one rule for working east and west, and another rule for working north and south; but which rule has never yet been a desideratum, and which has never yet been heard of.

1612. We next notice a memoir entitled *Investigation of the Figure of the Earth, and of the Gravity in different Latitudes.* By Robert Adrain. This is published in the *Transactions of the American Philosophical Society*...Vol. I. New Series. Philadelphia, 1818. The memoir occupies pages 119...135 of the volume; it was read October 7th, 1817.

We have seen in Art. 1108 that Laplace deduced, by two methods, a general expression for the length of the seconds pendulum from fifteen observed lengths. Adrain takes the same fifteen observations, and treats them by the method of Least Squares instead of by either of Laplace's methods.

In Laplace's deduction of the most probable ellipse from the pendulum observations he made two mistakes of calculation; Adrain points them out and gives the correct work: the mistakes are the last two out of the four to which we allude in Art. 1110, and it is curious that Bowditch makes no reference to Adrain. But the brief account of the origin of Laplace's fourth mistake, which Adrain gives on his page 131, is not intelligible.

On his page 127 Adrain proposes an expression for the force of gravity, when the place of observation is above the level of the sea. Thus if R is the mean radius of the Earth, h the height of the place above the level of the sea, and g the force of gravity at the level of the sea, he takes $\frac{gR^2}{(R+h)^2}$ for the force of gravity at the place of observation. But this makes no allowance for the attraction of the matter which is between the place of observation and the level of the sea.

It may be observed that Adrain claims the method of Least Squares as his own discovery; he begins thus: "Having in the year 1808 discovered a general method of resolving several useful problems, by ascertaining the highest degree of probability where certainty cannot be found;..." The principles on which he established the method are explained by him elsewhere; and they have been examined by Mr Glaisher in the *Memoirs of the Royal Astronomical Society*, Vol. XXXIX. pages 75...81.

1613. The volume which contains the preceding memoir by Adrain contains also another by him, entitled *Research concerning the Mean Diameter of the Earth*. This occupies pages 353...366; it was read Nov. 7th, 1817.

The memoir consists of simple investigations relating to an oblatum which is nearly spherical.

Let a and b denote the major and minor semiaxes of the generating ellipse; and suppose it required to find the radius of the sphere which has the same volume as the oblatum. Let r denote the radius of the sphere: then we must have $r^3 = a^2b$.

If the difference between a and b is small we obtain *approximately* $r = \frac{2a+b}{3}$.

Now Adrain shows that the *same* approximate value is also obtained from the solution of various other problems; as for instance if we require that the surface of the sphere shall be equal to the surface of the oblatum; or if we require that the curvature of the sphere shall be equal to the mean curvature of the oblatum, with a suitable definition of mean curvature.

1614. On pages 486...517 of the *Philosophical Transactions* for 1818, published in that year, there is a memoir entitled, *An abstract of the results deduced from the measurement of an arc on the meridian, extending from latitude* 8° 9′ 38″·4, *to latitude* 18° 3′ 23″·6, N. By Lieut. Colonel William Lambton.... The memoir was read on May 21st, 1818.

The memoir gives a short account of the operations on the great Indian arc, with references for details to volumes of the *Asiatic Researches*. Lambton, by comparing his results with the lengths of arcs in France and Sweden, arrives at an ellipticity of about $\frac{1}{310}$.

In some formulæ which occur on pages 497 and 499, Lambton gives values for radii of curvature which are halves of what they should be; but as he only uses the values in the form of ratios, this does not lead to any final error. For a correction as to another point, see the article on the *Figure of the Earth* in the *Encyclopædia Metropolitana*, page 210.

A note connected with the memoir will be found on pages 27...33 of the *Philosophical Transactions* for 1823.

For the later history of the progress of the measurement of the Indian arc the reader must consult the works published by the late Sir George Everest: see the *Proceedings of the Royal Society*, Vol. XVI. pages xi...xiv.

1615. We have next to notice a publication entitled *Sopra l'identità dell' attrazione molecolare coll' astronomica Opera del Cavaliere Leopoldo Nobili*...Modena, 1818.

This consists of 64 quarto pages, with 4 plates. It is divided into two parts. The first part is a memoir which had been published in the *Giornale di Fisica*...Pavia, 1817; and the second part is a supplement to enforce the doctrine of the memoir.

The author holds that the law of attraction according to the inverse square of the distance will suffice for the explanation of the phenomena of molecular action, as well as for the phenomena of astronomy. He treats of adhesion, cohesion, and capillary attraction; and has scarcely anything which falls within our subject.

He investigates the formula for the attraction of an indefinitely thin spherical shell on any particle; but he does not use any symbol to represent the thickness of the shell: thus for example, if the particle is just on the outside of the shell he obtains 4π for the resultant attraction. Then, as in Art. 993, this attraction may be divided into two equal parts, one arising from the part of the shell which is close to the particle, and the other from the rest of the shell. Thus he gets the *finite* value 2π for the attraction of a particle on an adjacent particle. If he had explicitly introduced the thickness of the shell this apparently finite result would have been really infinitesimal. This omission would be of no consequence for many purposes; but with regard to the special object which Nobili has in view it constitutes a fatal objection to almost the whole of the work.

Some illustration of the result obtained by considering attraction like an emanation from a centre is given on pages 27...30: it seems to me altogether unsatisfactory.

Two results, which are correct when we supply a factor to represent the thickness of the shell, are obtained which may be noticed.

Let r denote the radius of the shell, δr the thickness; suppose a particle inside the shell at the distance c from the centre; then if the shell be divided into two parts by a plane through the particle at right angles to the radius on which it is situated the resultant attraction of each part is

$$\frac{2\pi r^2\delta r}{c^2} - \frac{2\pi r\sqrt{(r^2-c^2)}\,\delta r}{c^2};$$

Suppose such a shell as before, but let the law of attraction be that of the inverse cube of the distance; then the resultant attraction on an external particle at the distance c from the centre is

$$\frac{2\pi r^2 \delta r}{c\,(c^2-r^2)} + \frac{\pi r \delta r}{c^2} \log \frac{c+r}{c-r}.$$

See pages 13 and 18 of the work.

1616. A memoir by Dr Young is next to be noticed, which occupies pages 70...95 of the *Philosophical Transactions* for 1819: it is entitled *Remarks on the Probabilities of Error in Physical Observations, and on the Density of the Earth.* The memoir is reprinted in the *Miscellaneous Works of the late Thomas Young.* Vol. II. pages 8...28.

We are concerned with only two sections of the memoir, namely one entitled *On the mean density of the earth*, and another entitled *On the irregularities of the earth's surface.*

1617. The section *On the mean density of the earth* is important. Laplace in the *Mécanique Céleste*, Livre XI. Chapitre II. discussed the hypothesis involved in the relation $\frac{d\Pi}{d\rho} = 2k\rho$; this discussion was apparently suggested by the remarks made by Dr Young in the present section: see Art. 1330. Young's hypothesis, however, is not the same as that which Laplace adopted, but the more simple one which belongs to elastic fluids, namely that involved in the relation $\frac{d\Pi}{d\rho} = k$.

If x denotes the distance from the centre of the Earth supposed spherical we have from the ordinary principles of Hydrostatics

$$\frac{d\Pi}{dx} = -\frac{4\pi\rho}{x^2}\int_0^x \rho x^2 dx;$$

thus

$$\frac{d\rho}{dx} = -\frac{4\pi\rho}{kx^2}\int_0^x \rho x^2 dx.$$

Young in fact proposes to obtain from this equation a value of ρ in the form of a series in powers of x; and he gives some numerical calculations. He considers the hypothesis adequate to meet the facts of the subject.

I do not know what he means by "the experiment on the sound of ice": the language is strange if he is referring to an experimental determination of the velocity of sound transmitted through ice.

1616. The section *On the irregularities of the earth's surface* treats of the effect produced on the pendulum by "the attraction of a circumscribed mass, situated at a moderate depth below the earth's surface." The word *circumscribed* seems here strange and unnecessary. Some correct results are obtained, but the process is neither clear nor interesting: we will reproduce one of these results.

The earth is supposed fluid and nearly spherical; take its radius for the unit of length, and its mass for the unit of mass. Suppose there is an additional mass a at the depth c below what would be the spherical surface if there were no irregularity. Let R denote the distance of a point in the surface of the fluid from the centre of the sphere, and r the distance of the point from the centre of the additional mass, then the surface will be determined by the equation

$$\frac{1}{R} + \frac{a}{r} = \text{a constant}.$$

Let s denote the elevation produced by the disturbing mass; then by applying the above general equation to the top of the elevation, and also to the point diametrically opposite, we obtain

$$\frac{1}{1+s} + \frac{a}{c+s} = \frac{1}{1} + \frac{a}{2-c}.$$

If c is supposed small we have from this approximately

$$s = \frac{a}{c}.$$

This result is the same as Dr Young's, but the process seems to me much more natural than his, which begins thus: "the fluxion of the elevation is as the fluxion of the arc and as the deviation...conjointly;..."

We have at the end of the section some remarks as to the value of the earth's attraction at the summit of a mountain; these embody what is now usually called *Dr Young's Rule*. This Rule coincides with the formula originally given by Bouguer, and reproduced by D'Alembert; see Art. 593: Dr Young does not refer to any preceding writer. He takes the density of the mountain as $\frac{2\frac{1}{2}}{5\frac{1}{2}}$ of the mean density of the earth. Then if x be the height of the mountain, r the radius of the earth, and g the value of the attraction at the surface of the earth, the value of the attraction at the summit of the mountain by Bouguer's formula is

$$g\left\{1 - \frac{2x}{r} + \frac{3}{2}\cdot\frac{5}{11}\cdot\frac{x}{r}\right\},$$

that is

$$g\left\{1 - \frac{29}{22}\cdot\frac{x}{r}\right\},$$

or

$$g\left\{1 - \frac{29}{44}\cdot\frac{2x}{r}\right\}.$$

Thus the correction would be $\frac{2x}{r}$ if we paid no regard to the attraction of the mountain; but becomes $\frac{29}{44}$ of $\frac{2x}{r}$, that is $\frac{66}{100}$ of $\frac{2x}{r}$, when we allow for this.

1619. A paper was published by Dr Young in *Brande's Quarterly Journal* for 1820, and reprinted in the *Miscellaneous Works of the late Thomas Young*, Vol. II. pages 78...83, entitled *Remarks on Laplace's latest Computation of the Density of the Earth.*

The paper begins with a very just remark:

It cannot but be highly flattering to any native of this country, to have his suggestions on an astronomical subject admitted and adopted by the Marquis de Laplace:...

As we have stated in Art. 1617, Dr Young proposed the hypothesis $\frac{d\Pi}{d\rho} = k$, while Laplace adopted the hypothesis $\frac{d\Pi}{d\rho} = 2k\rho$. In the present paper Dr Young states his objections against Laplace's hypothesis. There are no theoretical investigations, but Dr Young gives a table which assigns the value of the ellipticity corresponding to various values of the superficial density of the earth, the mean density being taken as 5·4. I am at a loss to understand how the table was calculated; Dr Young says:

In these calculations, it has not been necessary to have recourse to any foreign authority or assistance whatever...The geographical elements of the problem have been supplied by the experiments and observations of Maskelyne and Cavendish, compared with those of General Mudge, Colonel Lambton, and Captain Kater.

As Dr Young disclaims all foreign assistance he did not calculate the ellipticity by the theory of Clairaut and Laplace; and I cannot conjecture what he substituted. Nor do I know what are the *geographical elements* which he obtained from his five countrymen.

Dr Young finishes thus:

It is unnecessary to enter into any inquiry respecting the precession and nutation, as connected with the earth's density, since these effects are known to depend on the ellipticity of the spheroid and of its strata alone, without any regard to the manner in which the density is distributed among them.

I do not understand this; on the contrary it seems to me that the calculation of precession and nutation cannot be completed until the law of density is assumed.

1620. Two other papers by Dr Young may be conveniently noticed, although they fall beyond the date which we have fixed as the limit of our survey. To these we proceed.

1621. A contribution by Dr Young to our subject is entitled *Estimate of the Effect of the Terms involving the Square of the Disturbing Force on the Determination of the Figure of the Earth. In a Letter to G. B. Airy, Esq.* This was published in *Brande's Quarterly Journal* for 1820, and is reprinted in the *Miscellaneous Works of the late Thomas Young*, Vol. II. pages 87 and 88.

The paper discusses only a very simple case of the general problem implied in the title; namely the case "of a fluid supposed to be without weight, and surrounding a spherical nucleus." By the strange phrase "without weight" is meant I believe that the attraction of the fluid itself is to be neglected. The problem is thus purely speculative, and is treated in Young's usual obscure and repulsive manner; the result, naturally enough, is wrong. It may be useful to give a correct and intelligible solution.

Let ω be the angular velocity, r the radius vector of any point in the surface of the fluid, and θ the inclination of r to the plane of the equator. The attraction tends accurately to the centre of the spherical nucleus, and may be denoted by $\frac{\mu}{r^2}$. Then resolving along the tangent to the meridian, we must have for relative equilibrium

$$\frac{\mu}{r^2}\sin\psi = r\omega^2\cos\theta\sin(\theta+\psi),$$

where

$$\tan\psi = -\frac{1}{r}\frac{dr}{d\theta}.$$

Let a be the equatorial semiaxis; then we assume

$$r = a(1 - \epsilon\sin^2\theta + u),$$

where ϵ is a small quantity, and u is small compared with ϵ.

Put j for $a\omega^2 \div \frac{\mu}{a^2}$; then our fundamental equation becomes

$$j\cos\theta\sin\theta + j\cos^2\theta\tan\psi = (1 - \epsilon\sin^2\theta + u)^{-3}\tan\psi,$$

where

$$\tan\psi = \frac{2\epsilon\sin\theta\cos\theta - \frac{du}{d\theta}}{1 - \epsilon\sin^2\theta + u}.$$

Substitute the value of $\tan\psi$: thus

$$j\cos\theta\sin\theta + j\cos^2\theta\left(2\epsilon\sin\theta\cos\theta - \frac{du}{d\theta}\right)(1 - \epsilon\sin^2\theta + u)^{-1}$$
$$= \left(2\epsilon\sin\theta\cos\theta - \frac{du}{d\theta}\right)(1 - \epsilon\sin^2\theta + u)^{-4}.$$

By comparing the terms of the first order we obtain

$$j = 2\epsilon.$$

Then by comparing the terms of the second order we obtain

$$2\epsilon j\cos^3\theta\sin\theta = 8\epsilon^2\sin^3\theta\cos\theta - \frac{du}{d\theta};$$

therefore

$$u = 2\epsilon^2\sin^4\theta + \epsilon^2\cos^4\theta + \text{constant}.$$

The constant must be determined so as to leave $r=a$ when $\theta=0$; hence finally

$$u=2\epsilon^2\sin^4\theta+\epsilon^2\cos^4\theta-\epsilon^2.$$

Dr Young makes $u=\epsilon^2\sin^4\theta+\epsilon^2\cos^4\theta$.

We may also obtain the correct result by the aid of equation (2) of Art. 57.

Assume $r=a(1-\epsilon\sin^2\theta+u)$; thus the equation becomes

$$(1-\epsilon\sin^2\theta+u)^{-1}+\frac{j}{2}\cos^2\theta(1-\epsilon\sin^2\theta+u)^2=1+\frac{j}{2}.$$

By comparing the terms of the first order we obtain

$$\epsilon\sin^2\theta=\frac{j}{2}(1-\cos^2\theta),$$

so that
$$\frac{j}{2}=\epsilon.$$

Then by comparing the terms of the second order we obtain

$$\epsilon^2\sin^4\theta-u-2\epsilon^2\sin^2\theta\cos^2\theta=0,$$

therefore
$$u=\epsilon^2\sin^4\theta-2\epsilon^2\sin^2\theta\cos^2\theta$$
$$=2\epsilon^2\sin^4\theta+\epsilon^2\cos^4\theta-\epsilon^2.$$

It follows that the polar semi-axis is equal to $a(1-\epsilon+\epsilon^2)$ to the second order. Dr Young maintained in fact the erroneous opinion that the difference of the two semiaxes would not involve a term in ϵ^2. Ivory, as we have said, treated the problem correctly at about the same date: see Art. 1441.

1622. The other paper by Dr Young is entitled *Determination of the Figure of the Earth from a single tangent*: this was first published in the *Life of Thomas Young*, 1855, pages 511...514.

The title does not give any idea of the contents of the paper. The problem discussed is this: given the difference of latitude and the difference of longitude of two adjacent places, and also the azimuth of each as seen from the other, to determine the ellipticity of the earth.

The investigation is rather obscure, and there is a misprint of

$\cos\frac{1}{2}(\alpha+\alpha')$ for $\sin\frac{1}{2}(\alpha+\alpha')$ throughout the third paragraph. Also, α and α' denoting the two azimuths, $\frac{1}{2}(\alpha-\alpha')$ is very strangely called the mean azimuth on page 513.

It is stated in the note that, "The preliminary propositions are involved in the method proposed by Dalby for determining arcs of parallel." I do not know what this means: Dr Young's preliminary propositions involve only Plane Trigonometry.

1623. The great and deserved reputation of Dr Young renders it necessary to state that his mathematical writings are dangerous for students, and should not be consulted by them except under sound professional advice. Speaking generally the processes will be found unintelligible except to persons well acquainted with the subject discussed, and then they are superfluous.

One obscure and abstruse work bears the singularly inappropriate title of *Elementary Illustrations of the Celestial Mechanics of Laplace*; if this fell into the hands of a beginner, who had not been warned of its character, he might be alienated permanently from the study of Physical Astronomy.

The absurd opinions which Dr Young expresses in his life of Lagrange prove that he was quite incapable even of appreciating the highest mathematical genius; they have drawn forth a just protest from Dr Peacock: see the *Miscellaneous Works of the late Thomas Young*, Vol. II. page 579.

1624. We may briefly notice a strange work entitled *Address of M. Hoene Wronski, to the British Board of Longitude, upon the actual state of the Mathematics, their reform, and upon the new Celestial Mechanics, giving the definitive Solution of the Problem of Longitude. Translated from the original in French by W. Gardiner*, London, 1820.

This is a duodecimo volume containing xii + 127 pages.

Wronski came to England with the hope of inducing the Board of Longitude to reward his mathematical labours, on the ground of their scientific value, and especially of their use with respect to the practical problem of determining the longitude; but he seems to have received no encouragement. From page xii we learn that he submitted to the Board a manuscript consisting on the whole of 930 quarto pages, which contained a new theory of Celestial Mechanics, and a comparison of it with the old theory.

The work is almost unintelligible, and it is obvious that the translator was an incompetent person, but the original French was probably very obscure: the chief peculiarity which strikes a reader is the perpetual reference to *the Absolute*, without any adequate explanation of the mysterious term.

The pages 35...66 have some relation to the Figure of the Earth, or as Wronski styles it, "the problem of the formation of the celestial globes." We have a sketch of the history of the subject, and then a statement of the main results of the new theory of the author, which are five in number. I do not profess to understand them; but as the fifth is the shortest I will quote that as a specimen. It occurs on page 63:

At last, the fifth result of this theory of the construction of celestial globes is, that by this known form of the earth, regular or irregular, simple or complex, we can discover immediately the distribution itself of the masses in the interior of our globe, that is to say, the interior structure of the earth—Thus, we shall, with an astonishing facility, penetrate into these mysterious retreats, where, distant from light, the plaistic mother, in her chambers of silence and obscurity, prepares in great measure the generation of all that animates this our globe; and into which the most unbridled imagination has not dared to enter, but with fear and trembling to shadow her fanciful chimeras.

For an opinion on Wronski see the *Miscellaneous Works* of Dr Thomas Young, Vol. II. page 65.

1625. In the second volume of the *Transactions of the Cambridge Philosophical Society*, which is dated 1827, we have a memoir entitled *On the Figure assumed by a Fluid Homogeneous Mass, whose Particles are acted on by their mutual Attraction, and by small extraneous Forces.* By G. B. Airy ...

The memoir occupies pages 203...216 of the volume; it was read March 15, 1824.

The memoir commences thus:

The principal difficulty in the solution of this problem, consists in the investigation of the attraction of any spheroid (differing little from a sphere) upon a point in its surface. This has been found by Laplace, in a manner so general, and by an analysis so powerful, that any new investigations might seem entirely unnecessary. But the abstruse nature of that analysis, it must be acknowledged, is such as to make a more simple investigation desirable: and the obscurities which have led Laplace himself into error, serve to shew the value of a process which involves nothing more difficult than the common applications of the differential calculus.

I am not certain what error of Laplace's is here alluded to; but perhaps it is that which is discussed by the author in some subsequent pages of the volume: see Art. 1230. It does not however appear to me that Laplace himself was really wrong.

1626. The memoir then makes no use of Laplace's coefficients, but does use the proposition which reduces the determination of the attraction in any direction to the investigation of a single function, which we now call the *potential*. It may be said to occupy a position intermediate between Laplace's first three memoirs and his subsequent researches. General formulæ are investigated by the aid of expansions according to Taylor's theorem; and they are applied to the complete discussion of a problem which we may enunciate thus: Suppose a nearly spherical mass of fluid in the form of a figure of revolution; let z denote the ordinate of any point measured parallel to the axis of figure from the centre of the spheroid as origin: then the form for equilibrium is determined, when besides the attraction of the mass there is a small force represented by $A+Bz^2+Dz^4+Ez^6+Fz^8$, where $A, B, \ldots$ are constants. The numerical work is laborious, but it is correct.

1627. An application is made to the case of Saturn and his ring. Suppose that Saturn consists of homogeneous rotating fluid; then consider the influence which the ring exerts on the figure of Saturn. We will reproduce some of the investigation, though not with the original notation.

The ring is treated as if it were the perimeter of a circle. Let R denote the radius of this circle, and μ the mass of the ring. Take the centre of the ring as the origin, and the plane of the ring as that of (x, y). Then the potential of the ring at the point (x, y, z) being denoted by V, we have

$$V = \frac{\mu}{2\pi}\int_0^{2\pi} \frac{d\theta}{\sqrt{\{(R\cos\theta - x)^2 + (R\sin\theta - y)^2 + z^2\}}}.$$

Since the planet is supposed to be a figure of revolution, we may without loss of generality, put $y = 0$. We shall also assume that $x^2 + z^2 = r^2$, so that for a first approximation the planet is treated as a sphere of radius r.

Thus $$V = \frac{\mu}{2\pi}\int_0^{2\pi} \frac{d\theta}{\sqrt{\{R^2 + r^2 - 2Rx\cos\theta\}}}.$$

The expression under the integral sign may be expanded in a convergent series; thus putting λ for $\sqrt{(R^2 + r^2)}$ we see that the general term of V is

$$\frac{\mu}{2\pi\lambda}\cdot\frac{1.3.5\ldots(2s-1)}{2^s\,\underline{|s}}\cdot\left(\frac{2Rx}{\lambda^2}\right)^s\int_0^{2\pi}\cos^s\theta\,d\theta.$$

If s is an odd number this vanishes; if s is an even number it is equal to

$$\frac{\mu}{\lambda}\cdot\left(\frac{2Rx}{\lambda^2}\right)^s\cdot\frac{1.3.5\ldots(2s-1)}{2^s\,\underline{|s}}\cdot\frac{(s-1)(s-3)\ldots1}{s(s-2)(s-4)\ldots2}.$$

Let ρ denote the density of Saturn, so that the mass is $\frac{4\pi r^3\rho}{3}$; and suppose that the mass of the ring is $\frac{1}{n}$ of that of Saturn: then $\mu = \frac{4\pi r^3\rho}{3n}$. Thus

$$V = \frac{4\pi r^3\rho}{3n\lambda}\left\{1 + \frac{1^2.3}{2^2.4}\frac{4x^2R^2}{\lambda^4} + \frac{1^2.3^2.5.7}{2^2.4^2.6.8}\frac{16x^4R^4}{\lambda^8} + \ldots\right\}.$$

Let the coefficient of x^{2s} in this expression be denoted by $\frac{4\pi\rho}{3n}\cdot\frac{e_s}{r^{2s-3}}$; then

$$V = \frac{4\pi\rho}{3n}\left\{\frac{r^3}{\lambda} + e_1x^2 + \frac{e_2x^4}{r^2} + \frac{e_3x^6}{r^4} + \ldots\right\}.$$

Here $e_1, e_2, e_3, \ldots$ are abstract numbers independent of the unit of length.

1628. At this stage a step is taken which may be said to be one of the special characteristics of the investigation; we put $r^2 - z^2$ for x^2, and then rearrange in powers of z. Thus if we stop at the term in z^8 we obtain

$$V = V_0 +$$

$$\frac{4\pi\rho}{3n}\left\{-(e_1 + 2e_2 + 3e_3 + 4e_4)z^2 + (e_2 + 3e_3 + 6e_4)\frac{z^4}{r^2} - (e_3 + 4e_4)\frac{z^6}{r^4} + \frac{e_4 z^8}{r^6}\right\}.$$

where V_0 is the value of V when $z = 0$, which is not required for the investigation.

For numerical calculation put $\frac{R}{r} = 2$; then according to the memoir

$$V = V_0 + \frac{2\pi\rho}{n}\left\{-\cdot 09233z^2 + \cdot 04849\frac{z^4}{r^2} - \cdot 01768\frac{z^6}{r^4} + \cdot 00205\frac{z^8}{r^6}\right\}.$$

I am however unable to verify these numerical values. It seems to me that we have the following *exact* results:

$$e_1 = \cdot 024\sqrt{5},$$

$$e_2 = \frac{7}{20}e_1 = \cdot 0084\sqrt{5},$$

$$e_3 = \frac{44}{100}e_2 = \cdot 003696\sqrt{5},$$

$$e_4 = \frac{39}{80}e_3 = \cdot 0018018\sqrt{5};$$

and thus instead of the preceding expression we get

$$V = V_0 + \frac{2\pi\rho}{n}\left\{-\cdot 08809z^2 + \cdot 04517\frac{z^4}{r^2} - \cdot 01625\frac{z^6}{r^4} + \cdot 00269\frac{z^8}{r^6}\right\}.$$

1629. But besides this difference there is another point of importance, namely, that in order to have the coefficient of any power of z^2 correct to five places of decimals it is not sufficient to stop at e_4; we must take in some of the following terms e_5, e_6, ... We shall find for instance that

$$e_5 = \frac{5168}{10000}e_4.$$

Therefore the coefficient of $\frac{a^6}{r^7}$ instead of being only e_4 ought to be $e_4 + 5e_6 + 15e_8 + \ldots$; and the term $5e_6$ alone is greater than $\frac{5}{2}e_4$.

Thus if the coefficients are calculated accurately to five places of decimals, the values will differ decidedly from those which are given in the memoir. I have in consequence not carried my numerical verification of the memoir beyond this point.

I may remark that on the last page ·183 seems to be given as the *ellipticity* of an ellipse, of which the semiaxes are respectively 1 and $\sqrt{(1{\cdot}415)}$: I am then not certain as to the exact meaning of the word *ellipticity* here.

1630. The conclusion obtained in the memoir is that the action of the ring tends to give such a form to Saturn, that the generating curve would fall *within* an ellipse having the same axes. Then it is stated:

...It is remarkable, that this deviation from the elliptic form, is exactly the opposite to that given by the observations of Dr Herschel. This accurate observer, in the Philosophical Transactions for 1805 and 1806, has given a great number of his observations, which shew that Saturn is protuberant between the poles and the equator, and that his longest diameter makes an angle of 43° with the plane of his equator. Here then is a complete discordance between theory and observation; nor is it easy, with our present knowledge of the planet, to suggest anything by which they can be reconciled.

1631. Some extensive trigonometrical operations were carried on in Piedmont and Savoy, which are detailed in a work published at Milan in 1825, entitled *Opérations géodésiques et astronomiques pour la mesure d'un arc du parallèle moyen*, ...

I have not seen this work, which appears to consist of two quarto volumes, with a folio volume of plates. The operations involved a new determination of Beccaria's arc, to which allusion was made in Art. 717. See the article *Figure of the Earth* in the *Encyclopædia Metropolitana*, pages 208 and 212.

1632. I may notice an interesting article entitled *Modern Astronomy*, published in the *North American Review* for April, 1825; this was written by Bowditch. It is a brief sketch of the history of Astronomy during the century preceding the date; and it is very valuable on account of the great learning and ability of the author.

A sentence which relates to our subject may be quoted. After saying that the geodetical measures indicated an ellipticity between $\frac{1}{300}$ and $\frac{1}{310}$ it is added:

It may also be observed, that this oblateness being less than $\frac{1}{230}$, proves by Clairaut's theorem, beforementioned, that the earth increases in density from the surface towards the centre, confirming the proof deduced before from other sources.

This seems to me to ascribe more to Clairaut's theorem than it really contains; from the fact that the ellipticity is less than $\frac{1}{230}$, it follows by the theorem that Clairaut's fraction is greater than $\frac{1}{230}$: but then it does not follow necessarily that the density increases from the surface to the centre. Moreover by the "proof deduced before from other sources," it seems we must understand such results as those of the Schehallien experiments, which shew that the mean density is much greater than the superficial density: but this is not quite the same as we usually understand by the statement that the density increases from the surface to the centre. See Arts. 485 and 1319.

INDEX.

The figures refer to the Articles of the work.

THE END.

www.ingramcontent.com/pod-product-compliance
Lightning Source LLC
Chambersburg PA
CBHW020856210326
41598CB00018B/1691
* 9 7 8 3 7 4 1 1 7 0 3 1 7 *